TRELLISES AND TRELLIS-BASED DECODING ALGORITHMS FOR LINEAR BLOCK CODES

THE KLUWER INTERNATIONAL SERIES IN ENGINEERING AND COMPUTER SCIENCE

TRELLISES AND TRELLIS-BASED DECODING ALGORITHMS FOR LINEAR BLOCK CODES

by

Shu Lin
University of Hawaii at Manoa
Hawaii, USA

Tadao Kasami
Nara Institute of Science and Technology
Nara, Japan

Toru Fujiwara
Osaka University
Osaka, Japan

Marc Fossorier
University of Hawaii at Manoa
Hawaii, USA

KLUWER ACADEMIC PUBLISHERS
Boston / Dordrecht / London

Distributors for North, Central and South America:
Kluwer Academic Publishers
101 Philip Drive
Assinippi Park
Norwell, Massachusetts 02061 USA

Distributors for all other countries:
Kluwer Academic Publishers Group
Distribution Centre
Post Office Box 322
3300 AH Dordrecht, THE NETHERLANDS

Library of Congress Cataloging-in-Publication Data

A C.I.P. Catalogue record for this book is available
from the Library of Congress.

Printed on acid-free paper.

Printed in the United States of America

Contents

Preface

As the demand for data reliability increases, coding for error control becomes increasingly important in data transmission systems. It has become an integral part of almost all data communication system designs. Today error control schemes are being used in a broad range of data communication systems to achieve reliable data transmission.

Soft-decision decoding of block codes has been investigated by many coding theorists for several decades. However, finding computationally efficient and practically implementable soft-decision maximum likelihood decoding algorithms for block codes is still an open and challenging problem. Trellis representation of convolutional codes by Forney (1973) makes it possible to implement maximum likelihood decoding by the use of Viterbi algorithm (1967). As a result, convolutional codes with trellis-based Viterbi decoding have been widely used for error control in digital communications over the last two decades. Trellis characterization of linear block codes by Bahl, Cocke, Jelinek and Raviv (1974) came only one year after Forney's discovery of trellis structure of convolutional codes. However, it took another 4 years before Wolf and Massey (1978) gave a closer look of the trellis structure of linear block codes. Unfortunately, their works did not arouse any enthusiasm. For the next ten years, there was basically no research in this significant area. It was really Forney's landmark paper in 1988 that aroused the enthusiasm for research in trellis structure of block codes. Since then, there have been rapid and exciting developments in the trellis structure of block codes. These developments have stimulated coding theorists to apply the powerful Viterbi algorithm to block codes and to develop new efficient trellis-based decoding algorithms for maximum likelihood decoding or suboptimum decoding. In a short period of time, various trellis-based soft-decoding algorithms, optimum or suboptimum, for linear block codes have been devised. Some are more efficient than the Viterbi algorithm. In fact, some ideas developed in the study of trellis structure of block codes can be used for improving decoding and analyzing the trellis complexity of convolutional codes. At least one new decoding algorithm for convolutional codes which is more efficient than the Viterbi algorithm has been devised. All these new developments provide the practicing communication engineers with more choices in designing error control systems.

The purpose of this book is to put trellises and trellis-based decoding algorithms for linear codes together in a simple and unified form. The approach is to explain the material in an easily understood manner, with minimum of mathematical rigor. No high level of mathematics is included in presenting the material. The book is intended for practicing communication engineers who want to have a fast grasp and understanding of the subject. For this purpose, only material which we consider essential and useful for practical applications are included. This book can also be used as a supplement for a course on error correcting codes or as a text for a course on special topics in coding.

The authors would like to express their special appreciation to Dr. Ian F. Blake, Professor Daniel J. Costello, Jr., and Ms. Carrie Matsuzaki who read the first draft very carefully and made numerous corrections and suggestions for improvements. The authors are deeply in debted to Mr. Hitoshi Tokushige and Dr. Frank Burkert who spent numerous hours to type the entire manuscript of this book. Their generous assistance is deeply appreciated. The authors would also like to thank Dr. Hari T. Moorthy for his contributions in various parts of this book.

The first author would like to thank Professor Joachim Hagenauer for inviting him to give a course on special topics in coding at the Technical University of Munich during his sabbatical leave in Germany, 1996. This book is actually based on a subset of class notes written for this course. The first author particularly wants to thank Ms. Yoshiko Kobayashi and Dr. Frank Burkert for typing these class notes.

The authors are also very grateful to the National Science Foundation, National Aeronautics and Space Administration, and Ministry of Education, Japan for their continuing support of authors' research in the coding field. Without this assistance, the authors' interest in coding could never have developed to the point of writing this book. The first author would also like to thank Alexander von Humboldt Foundation for the research award prize which supported him to conduct research in trellises and trellis-based decoding algorithms for codes during his sabbatical leave at the Technical University of Munich, Germany during the academic year of 1996. This made it possible for him to develop the very first draft of this book.

Finally, the authors would like to give special thanks to their wives for their continuing love and affection throughout this project.

1 INTRODUCTION

1.1 TRELLIS REPRESENTATION OF CODES

A code trellis is a graphical representation of a code, block or convolutional, in which every path represents a codeword (or a code sequence for a convolutional code). This representation makes it possible to implement maximum likelihood decoding (MLD) of a code with reduced decoding complexity. The most well known trellis-based MLD algorithm is the Viterbi algorithm [23, 79, 105]. The trellis representation was first introduced and used for convolutional codes [23]. This representation, together with the Viterbi decoding algorithm, has resulted in a wide range of applications of convolutional codes for error control in digital communications over the last two decades.

The recent search for efficient MLD schemes for linear block codes has motivated some coding theorists to study the trellis structure of these codes so that trellis-based decoding algorithms can be devised to reduce decoding complexity. Trellis representation of linear block codes was first presented in [1] and then in [67, 109]. The first serious study of trellis structure and trellis construction for linear block codes was due to Wolf. In his 1978 paper [109], Wolf presented the first method for constructing trellises for linear block codes and proved that an N-section trellis diagram for a q-ary (N, K) linear block code has at most $q^{\min(K,N-K)}$ states. He also presented a method for labeling the states based on the parity-check matrix of a code. Right after Wolf's work, Massey presented a simple but elegant paper [67] in which he gave a precise definition of a code trellis, derived some fundamental properties, and provided implications of the trellis structure for encoding and decoding of codes. However, these early works in trellis representation of linear block codes did not arouse much enthusiasm, and for the next 10 years, there was basically no research in this area.

There are two major reasons for this inactive period of research in this area. First, most coding theorists at that time believed that block codes did not have simple trellis structure like convolutional codes and maximum likelihood decoding of linear block codes using the Viterbi algorithm was practically impossible, except for very short block codes. Second, since almost all of the linear block codes are constructed algebraically or based on finite geometries, it was the belief of many coding theorists that algebraic decoding was the only way to decode these codes. These two reasons seriously hindered the development of efficient soft-decision decoding methods for linear block codes and their applications to error control in digital communications. This led to a general belief that block codes are inferior to convolutional codes and hence, that they were not useful.

In fact, for more than two decades, most of the practicing communication engineers believed that the rate-1/2 convolutional code of constraint length 7 with Viterbi decoding was the only effective error control coding scheme for digital communications, except for perhaps ARQ schemes. To achieve higher reliability for certain applications such as NASA's satellite and deep space communications, this convolutional code concatenated with a Reed-Solomon outer code was thought the best solution.

It was really Forney's paper in 1988 [24] that aroused enthusiasm for research in the trellis structure of linear block codes. In this paper, Forney showed that some block codes, such as Reed-Muller (RM) codes and some lattice codes, do have relatively simple trellis structures, and he presented a method for constructing sectionalized trellises for linear block codes and asserted that the construction results in minimal trellises with respect to the state complexity (the number of states). Motivated by Forney's work and the desire to achieve maximum likelihood decoding of linear block codes to improve error performance over traditional hard-decision algebraic decoding, there have been significant efforts in studying the trellis structure and devising trellis-based decoding algorithms for linear block codes over the last eight years. Developments have been dramatic and rapid, and the new results are exciting and encouraging. Trellis-based decoding algorithms that are more efficient than the conventional Viterbi decoding algorithm have recently been devised [32, 37] and implementation of trellis-based high-speed decoders for NASA's high-speed satellite communications is now underway [52, 63]. All of these new developments make block codes more competitive with convolutional codes.

1.2 ORGANIZATION OF THE BOOK

Chapter 2 gives a brief review of linear block codes. The goal is to provide the essential background material for the development of trellis structure and trellis-based decoding algorithms for linear block codes in the later chapters. Chapters 3 through 6 present the fundamental concepts, finite-state machine model, state space formulation, basic structural properties, state labeling, construction procedures, complexity, minimality, and sectionalization of trellises. Chapter 7 discusses trellis decomposition and subtrellises for low-weight codewords. Chapter 8 first presents well known methods for constructing long powerful codes from short component codes or component codes of smaller dimensions, and then provides methods for constructing their trellises which include Shannon and Cartesian product techniques. Chapter 9 deals with convolutional codes, puncturing, zero-tail termination and tail-biting. It shows that trellis construction procedures for both block and convolutional codes are essentially the same, except that the trellises for convolutional codes or terminated convolutional codes are time-invariant and the trellises for block codes are in general time-varying. For both types of codes, trellis states are defined based on a certain set of information bits, called the state-defining information set.

Chapters 10 through 13 present various trellis-based decoding algorithms, old and new. Chapter 10 first discusses the application of the well known Viterbi decoding algorithm to linear block codes, optimum sectionalization of a code trellis to minimize computation complexity, and design issues for IC (integrated circuit) implementation of a Viterbi decoder. Then it presents a new decoding algorithm for convolutional codes, named differential trellis decoding (DTD) algorithm. DTD algorithm is devised based on the principle of **compare-select-add** (CSA) which is simply the opposite of the principle of **add-compare-select** (ACS) used in the Viterbi algorithm. This new algorithm is more efficient than the Viterbi decoding algorithm. For rate-1/2 antipodal convolutional codes and their higher rate punctured codes, it requires about 1/3 less real number operations than the Viterbi decoding algorithm. This DTD algorithm can also be applied to trellis decoding of block codes. Chapter 11 presents a trellis-based recursive MLD for linear block codes, the RMLD algorithm. This decoding algorithm is devised based on the **divide and conquer principle**. The implementation of this algorithm does not require the construction of the entire code trellis; only some special one-section trellises of much smaller state and branch complexities for constructing path metric tables recursively are needed. This reduces the decoding complexity significantly and it is more efficient than the Viterbi decoding algorithm. Furthermore, it allows **parallel/pipeline** processing of received sequences to speed up decoding. Chapter 12 presents a suboptimum reliability-based iterative decoding algorithm with a low-weight trellis search for the most likely codeword. This decoding algorithm provides a good trade-off between error performance and decoding complexity. All the decoding algorithms presented in Chapters 10 through 12 are devised to minimize word error probability. Chapter 13 presents decoding algorithms that minimize bit error probability and provide the corresponding soft (reliability) information at the output of the decoder. Decoding algorithms presented are the MAP (**maximum a posteriori probability**) decoding algorithm and the SOVA (**soft-output Viterbi algorithm**) algorithm. Finally, the minimization of bit error probability in trellis-based MLD is discussed.

2 LINEAR BLOCK CODES

Chapter 2 gives a brief review of linear block codes. The goal is to provide the essential background material for the development of trellis structure and trellis-based decoding algorithms for linear block codes in the later chapters. We mainly present the basic concepts of encoding and decoding of linear block codes and state some facts without derivations or proofs. Since in most present digital data communication systems, information is coded in binary digits, '0' or '1', we discuss only linear block codes with symbols from the binary field GF(2). First, linear block codes are defined and described in terms of generator and parity-check matrices. Second, coset partition of a linear block code is discussed, which is needed in analyzing the code trellis structure and construction. Third, the concepts of minimum distance, weight distribution and distance profile are presented, which are needed in the later chapters for presenting decoding algorithms and their error performances. Finally, the concepts of hard-decision, soft-decision, and maximum likelihood decoding are presented.

References 3, 9, 14, 59, 63, 78 and 79 contain excellent treatments of linear block codes.

2.1 GENERATION OF LINEAR BLOCK CODES

In block coding, an information sequence of binary digits (called bits) is divided into message blocks of fixed length; each message block consists of K information bits. There are a total of 2^K distinct messages. Each message is encoded into a codeword (or code sequence) of N bits according to certain rules, where $N \geq K$. Therefore, corresponding to the 2^K possible messages, there are 2^K codewords. This set of 2^K codewords forms a block code of length N. For a block code to be useful, the 2^K codewords must be distinct. Hence, there should be a one-to-one correspondence between a message and a codeword.

Definition 2.1 A binary block code of length N and 2^K codewords is called an (N, K) linear block code if and only if its 2^K codewords form a K-dimensional subspace of the vector space of all the N-tuples over the binary field GF(2). The parameter K is called the **dimension** of the code space.

An (N, K) linear block code C is generated by a $K \times N$ generator matrix over GF(2),

$$G = \begin{bmatrix} \boldsymbol{g}_1 \\ \boldsymbol{g}_2 \\ \vdots \\ \boldsymbol{g}_K \end{bmatrix} = \begin{bmatrix} g_{11} & g_{12} & \cdots & g_{1i} & \cdots & g_{1N} \\ g_{21} & g_{22} & \cdots & g_{2i} & \cdots & g_{2N} \\ \vdots & \vdots & & \vdots & & \vdots \\ g_{K1} & g_{K2} & \cdots & g_{Ki} & \cdots & g_{KN} \end{bmatrix} \tag{2.1}$$

where the rows, $\boldsymbol{g}_1$, $\boldsymbol{g}_2$, ..., $\boldsymbol{g}_K$, are linearly independent over GF(2). The 2^K linear combinations of the K rows of G form the codewords of C. We say that the rows of G span the code C, or C is the **row space** of G. Let

$$\boldsymbol{a} = (a_1, a_2, \ldots, a_K)$$

be a message to be encoded. A natural encoding mapping is that the codeword

$$\boldsymbol{u} = (u_1, u_2, \ldots, u_N)$$

for the message $\boldsymbol{a} = (a_1, a_2, \ldots, a_K)$ is given by

$$\boldsymbol{u} \;=\; \boldsymbol{a} \cdot G$$

$$
\begin{aligned}
&= (a_1, a_2, \ldots, a_K) \begin{bmatrix} \boldsymbol{g}_1 \\ \boldsymbol{g}_2 \\ \vdots \\ \boldsymbol{g}_K \end{bmatrix} \\
&= a_1 \cdot \boldsymbol{g}_1 + a_2 \cdot \boldsymbol{g}_2 + \cdots + a_K \cdot \boldsymbol{g}_K. \qquad (2.2)
\end{aligned}
$$

From (2.1) and (2.2), we find that for $1 \leq i \leq N$, the i-th component of $\boldsymbol{u}$ is given by

$$u_i = a_1 \cdot g_{1i} + a_2 \cdot g_{2i} + \cdots + a_K \cdot g_{Ki}. \qquad (2.3)$$

During an encoding interval, K information bits are encoded into N code bits. These N code bits are shifted onto the channel, one at a time, in N units of time. An encoding interval, denoted Γ, is represented by a set of $N+1$ time instants,

$$\Gamma = \{0, 1, 2, \ldots, N\}. \qquad (2.4)$$

For $1 \leq i \leq N$, the i-th unit of time is the interval from time-$(i-1)$ to time-i. During this interval, the i-th code bit u_i is formed and transmitted. By time-i, the transmission is completed. This interval is called a **bit interval.**

Example 2.1 Consider a binary $(8, 4)$ linear block code which is generated by the following generator matrix:

$$G = \begin{bmatrix} \boldsymbol{g}_1 \\ \boldsymbol{g}_2 \\ \boldsymbol{g}_3 \\ \boldsymbol{g}_4 \end{bmatrix} = \begin{bmatrix} 1 & 1 & 1 & 1 & 1 & 1 & 1 & 1 \\ 0 & 0 & 0 & 0 & 1 & 1 & 1 & 1 \\ 0 & 0 & 1 & 1 & 0 & 0 & 1 & 1 \\ 0 & 1 & 0 & 1 & 0 & 1 & 0 & 1 \end{bmatrix}. \qquad (2.5)$$

If $\boldsymbol{a} = (1101)$ is the message to be encoded, its corresponding codeword, according to (2.2), is given by

$$
\begin{aligned}
\boldsymbol{u} &= 1 \cdot \boldsymbol{g}_1 + 1 \cdot \boldsymbol{g}_2 + 0 \cdot \boldsymbol{g}_3 + 1 \cdot \boldsymbol{g}_4 \\
&= (11111111) + (00001111) + (01010101) \\
&= (10100101).
\end{aligned}
$$

The 16 codewords of this code are listed in Table 2.1.

△△

Table 2.1. The codewords of the code generated by (2.5).

Messages	Codewords	Messages	Codewords
(0000)	(00000000)	(0001)	(01010101)
(1000)	(11111111)	(1001)	(10101010)
(0100)	(00001111)	(0101)	(01011010)
(1100)	(11110000)	(1101)	(10100101)
(0010)	(00110011)	(0011)	(01100110)
(1010)	(11001100)	(1011)	(10011001)
(0110)	(00111100)	(0111)	(01101001)
(1110)	(11000011)	(1111)	(10010110)

A binary (N, K) linear block code C is also uniquely specified by an $(N - K) \times N$ matrix over GF(2), called a **parity-check matrix**,

$$H = \begin{bmatrix} h_{11} & h_{12} & \cdots & h_{1N} \\ h_{21} & h_{22} & \cdots & h_{2N} \\ \vdots & \vdots & \ddots & \vdots \\ h_{N-K,1} & h_{N-K,2} & \cdots & h_{N-K,N} \end{bmatrix}, \qquad (2.6)$$

where the rows are linearly independent. A binary N-tuple $\boldsymbol{u} = (u_1, u_2, \ldots, u_N)$ is a codeword in C if and only if the following condition holds:

$$\boldsymbol{u} \cdot H^T = \mathbf{0}, \qquad (2.7)$$

where $\mathbf{0}$ denotes the all-zero $(N - K)$-tuple, $(0, 0, \ldots, 0)$. Code C is called the **dual** (or **null**) space of H. H itself generates an $(N, N - K)$ linear code, denote $C^\perp$. For any codeword $\boldsymbol{u} = (u_1, u_2, \ldots, u_N) \in C$ and any codeword $\boldsymbol{v} = (v_1, v_2, \ldots, v_N) \in C^\perp$, the **inner product**

$$\begin{aligned} \boldsymbol{u} \cdot \boldsymbol{v} &\triangleq u_1 \cdot v_1 + u_2 \cdot v_2 + \cdots + u_N \cdot v_N \\ &= 0. \end{aligned}$$

$C^\perp$ is called the **dual code** of C and vise versa.

In general, the generator matrix G of a linear block code C is used for encoding, while the parity-check matrix H is used for decoding, particularly for error detection.

Example 2.2 Consider the (8,4) linear block code given in Example 2.1. A parity-check matrix for this code is the generator matrix itself given by (2.5), i.e.,

$$H = G = \begin{bmatrix} 1 & 1 & 1 & 1 & 1 & 1 & 1 & 1 \\ 0 & 0 & 0 & 0 & 1 & 1 & 1 & 1 \\ 0 & 0 & 1 & 1 & 0 & 0 & 1 & 1 \\ 0 & 1 & 0 & 1 & 0 & 1 & 0 & 1 \end{bmatrix}.$$

In this case, $C = C^{\perp}$ and C is said to be **self-dual.**

$\triangle\triangle$

For an (N, K) linear block code C, the ratio $R = K/N$ is called the **code rate** which represents the average number of information bits carried by a code symbol (or the average number of information bits transmitted per channel usage).

2.2 COSET PARTITION OF A LINEAR BLOCK CODE

Consider a binary (N, K) linear block code C with a generator matrix G. Let K_1 be a nonnegative integer such that $0 \leq K_1 \leq K$. A subset of 2^{K_1} codewords in C is said to be a **linear subcode** of C if this subset itself is a K_1-dimensional subspace of the vector space of all the N-tuples over GF(2). Any K_1 rows of the generator matrix G span an (N, K_1) linear subcode of C, and they form a generator matrix for the subcode. If $K_1 = 0$, the subcode consists of only the all-zero codeword $\mathbf{0}$ of C. For $K_1 = K$, the subcode is just the code itself.

Let C_1 be an (N, K_1) linear subcode of C. Then C can be partitioned into 2^{K-K_1} disjoint **cosets** of C_1; each coset is of the following form:

$$\boldsymbol{v}_l \oplus C_1 \triangleq \{\boldsymbol{v}_l + \boldsymbol{u} : \boldsymbol{u} \in C_1\} \tag{2.8}$$

with $1 \leq l \leq 2^{K-K_1}$, where for $\boldsymbol{v}_l \neq \mathbf{0}$, $\boldsymbol{v}_l$ is in C but not in C_1 and for $\boldsymbol{v}_l = \mathbf{0}$, the coset $\mathbf{0} \oplus C_1$ is just the subcode C_1 itself. This partition of C with respect to C_1 is denoted with C/C_1, and the codewords $\boldsymbol{v}_l$ for $1 \leq l \leq 2^{K-K_1}$ are called the **coset representatives**. Any codeword in a coset can be used as the coset representative without changing the composition (the codewords) of the coset. Important properties of cosets are:

(1) The sum of two codewords in a coset is a codeword in the subcode C_1.

(2) Let $\boldsymbol{x}$ and $\boldsymbol{y}$ be two codewords in cosets $\boldsymbol{v}_i \oplus C_1$ and $\boldsymbol{v}_j \oplus C_1$, respectively, where $i \neq j$. Then the sum $\boldsymbol{x}+\boldsymbol{y}$ is a codeword in the coset $(\boldsymbol{v}_i+\boldsymbol{v}_j)\oplus C_1$ with $\boldsymbol{v}_i + \boldsymbol{v}_j$ as the coset representative.

The set of representatives for the cosets in the partition C/C_1 is denoted $[C/C_1]$ which is called the **coset representative space** for the partition C/C_1. Code C can be expressed as the **direct-sum** of C_1 and $[C/C_1]$ as follows:

$$C = [C/C_1] \oplus C_1 \triangleq \{\boldsymbol{v}+\boldsymbol{u} : \boldsymbol{v} \in [C/C_1] \text{ and } \boldsymbol{u} \in C_1\}. \tag{2.9}$$

Let G_1 be the subset of K_1 rows of the generator matrix G which generates the subcode C_1. Then the 2^{K-K_1} codewords generated by the $K - K_1$ rows in the set $G \setminus G_1$ can be used as the representatives for the cosets in the partition C/C_1. These 2^{K-K_1} codewords form an $(N, K - K_1)$ linear subcode of C.

Let C_2 be an (N, K_2) linear subcode of C_1 with $0 \leq K_2 \leq K_1$. We can further partition each coset $\boldsymbol{v}_l \oplus C_1$ in the partition C/C_1 based on C_2 into $2^{K_1-K_2}$ cosets of C_2; each coset consists of the following codewords in C:

$$\boldsymbol{v}_l \oplus (\boldsymbol{w}_k \oplus C_2) \triangleq \{\boldsymbol{v}_l + \boldsymbol{w}_k + \boldsymbol{u} : \boldsymbol{u} \in C_2\} \tag{2.10}$$

with $1 \leq l \leq 2^{K-K_1}$ and $1 \leq k \leq 2^{K_1-K_2}$ where for $\boldsymbol{w}_k \neq \boldsymbol{0}$, $\boldsymbol{w}_k$ is a codeword in C_1 but not in C_2. We denote this partition with $C/C_1/C_2$. This partition consists of 2^{K-K_2} cosets of C_2. Now C can be expressed as the following direct-sum:

$$C = [C/C_1] \oplus [C_1/C_2] \oplus C_2. \tag{2.11}$$

Let $C_1, C_2, \ldots, C_m$ be a sequence of linear subcodes of C with dimensions $K_1, K_2, \ldots, K_m$, respectively, such that

$$C \supseteq C_1 \supseteq C_2 \supseteq \cdots \supseteq C_m \tag{2.12}$$

and

$$K \geq K_1 \geq K_2 \geq \cdots \geq K_m \geq 0. \tag{2.13}$$

Then we can form a chain of partitions,

$$C/C_1, C/C_1/C_2, \ldots, C/C_1/C_2/\cdots/C_m, \tag{2.14}$$

and C can be expressed as the following direct-sum:

$$C = [C/C_1] \oplus [C_1/C_2] \oplus \cdots \oplus [C_{m-1}/C_m] \oplus C_m. \tag{2.15}$$

2.3 THE MINIMUM DISTANCE AND WEIGHT DISTRIBUTION OF A LINEAR BLOCK CODE

Let $\boldsymbol{u}$ and $\boldsymbol{v}$ be two N-tuples over GF(2). The **Hamming distance** between $\boldsymbol{u}$ and $\boldsymbol{v}$, denoted $d(\boldsymbol{u}, \boldsymbol{v})$, is defined as the number of places where they differ. The **minimum (Hamming) distance** of a block code C, denoted $d_{\min}(C)$, is defined as the minimum Hamming distance between all distinct pairs of codewords in C, i.e.,

$$d_{\min}(C) \triangleq \min\{d(\boldsymbol{u}, \boldsymbol{v}) : \boldsymbol{u}, \boldsymbol{v} \in C, \boldsymbol{u} \neq \boldsymbol{v}\}. \tag{2.16}$$

The (Hamming) **weight** of an N-tuple $\boldsymbol{v}$, denoted $w(\boldsymbol{v})$, is defined as the number of nonzero components of $\boldsymbol{v}$. It follows from the definition of Hamming distance and the fact that the sum of two N-tuples over GF(2) is an another N-tuple over GF(2) that the Hamming distance between two N-tuples, $\boldsymbol{u}$ and $\boldsymbol{v}$, is equal to the Hamming weight of the sum of $\boldsymbol{u}$ and $\boldsymbol{v}$, i.e.,

$$d(\boldsymbol{u}, \boldsymbol{v}) = w(\boldsymbol{u} + \boldsymbol{v}). \tag{2.17}$$

For a linear block code C, it follows from (2.16) and (2.17) that

$$\begin{aligned} d_{\min}(C) &= \min\{w(\boldsymbol{u} + \boldsymbol{v}) : \boldsymbol{u}, \boldsymbol{v} \in C, \boldsymbol{u} \neq \boldsymbol{v}\} \\ &= \min\{w(\boldsymbol{x}) : \boldsymbol{x} \in C, \boldsymbol{x} \neq \boldsymbol{0}\} \\ &\triangleq w_{\min}(C). \end{aligned} \tag{2.18}$$

The parameter $w_{\min}(C) \triangleq \min\{w(\boldsymbol{x}) : \boldsymbol{x} \in C, \boldsymbol{x} \neq \boldsymbol{0}\}$ is called the **minimum weight** of C. Eq.(2.18) simply says that the minimum distance of a linear block code is equal to the minimum weight of its nonzero codewords. The minimum weight of the $(8,4)$ linear block code given in Table 2.1 is 4; therefore, its minimum distance is 4.

Let C be an (N,K) linear block code. For $0 \leq i \leq N$, let A_i be the number of codewords with weight i. The numbers $A_0, A_1, A_2, \ldots, A_N$ are called the **weight distribution** of C. It is clear that $A_0 = 1$. The weight distribution of the $(8,4)$ linear block code given in Table 2.1 is

$$A_0 = 1,\ A_1 = A_2 = A_3 = 0,\ A_4 = 14,\ A_5 = A_6 = A_7 = 0,\ A_8 = 1.$$

The weight distribution is often expressed as a polynomial,

$$A(X) = A_0 + A_1 X + \cdots + A_N X^N,$$

which is called the **weight enumerator** of C. Let $W = \{0, w_1, w_2, \ldots, w_m\}$ denote the set of all weights of codewords in C such that:

(i) $0 < w_1 < w_2 < \cdots < w_m \leq N$; and

(ii) For $1 \leq i \leq m$, the number of codewords in C with weight w_i is not equal to zero.

This set is called the **weight profile** of C. The weight profile of the $(8,4)$ linear block code given by Table 2.1 is $\{0,4,8\}$. Let $\boldsymbol{u}$ be any codeword in C. The weight distribution of C actually gives the distribution of distances of codewords in C from the codeword $\boldsymbol{u}$. The weight profile W of C gives the profile of distances of codewords in C from the codeword $\boldsymbol{u}$.

The error performance of a linear block code is determined by its minimum distance and weight distribution. For an (N,K) linear block code with minimum distance $d_{\min}$, we often use the notation $(N,K,d_{\min})$ to represent the code. Therefore, the code given by Table 2.1 is an $(8,4,4)$ linear block code.

2.4 DECODING

Suppose an (N,K) linear block code C is used for error control over an additive white Gaussian noise (AWGN) channel. Let $\boldsymbol{u} = (u_1, u_2, \ldots, u_N)$ be the codeword to be transmitted. Before the transmission, a modulator maps each code bit into an elementary signal waveform. Binary PSK or FSK are commonly used signal waveforms for transmitting the bits in a codeword. The resultant signal sequence is then transmitted over the channel and corrupted by noise. At the receiving end, the received signal sequence is processed by a demodulator and sampled at the end of each signal (bit) interval. This results in sequence of real numbers,

$$\boldsymbol{r} = (r_1, r_2, \ldots, r_N),$$

which is called the **received sequence**. For $1 \leq i \leq N$, the i-th received component r_i is the sum of a fixed real number c_i and a Gaussian random variable n_i of zero-mean and variance $N_0/2$ where c_i corresponds to the transmitted code bit u_i at time-i. These received components may or may not be quantized. At one extreme, the demodulator can be used to make **firm decisions** on whether each transmitted code bit is a '0' or a '1'. Thus the output is quantized to two levels, denoted as 0 and 1. We say that the demodulator has made a "**hard decision**" on each transmitted code bit. This hard decision results in a binary received sequence,

$$\boldsymbol{z} = (z_1, z_2, \ldots, z_N),$$

which may contain **transmission errors**, i.e., for some i, $z_i \neq u_i$. This binary hard-decision sequence is fed into a decoder which attempts to correct the transmission errors (if any) and recover the transmitted codeword $\boldsymbol{u}$. Since the decoder operates on the hard decisions made by the demodulator, the decoding process is called **hard-decision decoding**. At the other extreme, the unquantized outputs from the demodulator can be fed directly into the decoder for processing. We refer to the resulting decoding as **soft-decision decoding**. Since the decoder makes use of the additional information contained in the unquantized received samples to recover the transmitted codeword, soft-decision decoding provides better error performance than hard-decision decoding. Decoding based on the quantized outputs from the demodulator, where the number of quantization levels exceeds two, is also referred to as soft-decision decoding. Soft-decision decoding provides better error performance than hard-decision decoding; however, hard-decision decoding is much simpler to implement. Various hard-decision decoding algorithms based on the algebraic structures of linear block codes have been devised. These hard-decision decoding algorithms are also termed algebraic decoding algorithms. Recently, effective soft-decision decoding algorithms have been devised, and they achieve either optimum error performance or suboptimum error performance with reduced decoding complexity.

Let $\boldsymbol{u}^*$ be the estimate of the transmitted codeword at the output of the decoder. If the codeword $\boldsymbol{u}$ was transmitted, a decoding error occurs if and only if $\boldsymbol{u}^* \neq \boldsymbol{u}$. Given that $\boldsymbol{r}$ is received, the conditional error probability of the decoder is defined as

$$P(E \mid \boldsymbol{r}) \triangleq P(\boldsymbol{u}^* \neq \boldsymbol{u} \mid \boldsymbol{r}). \tag{2.19}$$

The error probability of the decoder is then given by

$$P(E) = \sum_{\boldsymbol{r}} P(E \mid \boldsymbol{r})P(\boldsymbol{r}). \tag{2.20}$$

$P(\boldsymbol{r})$ is independent of the decoding rule used since $\boldsymbol{r}$ is produced prior to decoding. Hence, an optimum decoding rule (i.e., one that minimizes $P(E)$) must minimize $P(E \mid \boldsymbol{r}) = P(\boldsymbol{u}^* \neq \boldsymbol{u} \mid \boldsymbol{r})$ for all $\boldsymbol{r}$. Since minimizing $P(\boldsymbol{u}^* \neq \boldsymbol{u} \mid \boldsymbol{r})$ is equivalent to maximizing $P(\boldsymbol{u}^* = \boldsymbol{u} \mid \boldsymbol{r})$, $P(E \mid \boldsymbol{r})$ is minimized for a given $\boldsymbol{r}$ by choosing $\boldsymbol{u}^*$ as the codeword that maximizes

$$P(\boldsymbol{u} \mid \boldsymbol{r}) = \frac{P(\boldsymbol{r} \mid \boldsymbol{u})P(\boldsymbol{u})}{P(\boldsymbol{r})}, \tag{2.21}$$

that is, $\boldsymbol{u}^*$ is chosen as the **most likely** codeword given $\boldsymbol{r}$ is received. If all the codewords are equally likely, maximizing (2.21) is equivalent to maximizing $P(\boldsymbol{r} \mid \boldsymbol{u})$. For an AWGN channel,

$$P(\boldsymbol{r} \mid \boldsymbol{u}) = \prod_{i=1}^{N} P(r_i \mid u_i), \tag{2.22}$$

since each received symbol depends on the corresponding transmitted symbol. A decoder that chooses its estimate to maximize (2.22) is called a **maximum likelihood decoder** and the decoding process is called the **maximum likelihood decoding** (**MLD**). Maximizing (2.22) is equivalent to maximizing

$$\log P(\boldsymbol{r} \mid \boldsymbol{u}) = \sum_{i=1}^{N} \log P(r_i \mid u_i) \tag{2.23}$$

which is called the **log-likelihood function**.

Suppose BPSK signaling is used. Assume that each signal has unit energy. Let $\boldsymbol{u} = (u_1, u_2, \ldots, u_N)$ be the codeword to be transmitted. The modulator maps this codeword into a bipolar sequence represented by

$$\boldsymbol{c} = (c_1, c_2, \ldots, c_N)$$

where for $1 \leq i \leq N$,

$$c_i = 2u_i - 1. \tag{2.24}$$

From (2.24), we see that

$$c_i = \begin{cases} -1, & \text{if } u_i = 0, \\ +1, & \text{if } u_i = 1. \end{cases} \tag{2.25}$$

The **squared Euclidean distance** between the received sequence $\boldsymbol{r} = (r_1, r_2, \ldots, r_N)$ and $\boldsymbol{c}$ is given by

$$|\boldsymbol{r} - \boldsymbol{c}|^2 \triangleq \sum_{i=1}^{N} (r_i - c_i)^2. \tag{2.26}$$

For an AWGN channel, maximizing the log-likelihood function is equivalent to minimizing the squared Euclidean distance between $\boldsymbol{r}$ and $\boldsymbol{c}$. If we expand the right-hand side of (2.26), we have

$$|\boldsymbol{r} - \boldsymbol{c}|^2 = \sum_{i=1}^{N} r_i^2 - 2\sum_{i=1}^{N} r_i \cdot c_i + \sum_{i=1}^{N} c_i^2. \tag{2.27}$$

In computing $|\boldsymbol{r} - \boldsymbol{c}|^2$ for all codewords in C, $\sum_{i=1}^{N} r_i^2$ is a common term and $\sum_{i=1}^{N} c_i^2 = N$. Therefore, minimizing $|\boldsymbol{r} - \boldsymbol{c}|^2$ of (2.26) is equivalent to maximizing

$$\begin{aligned} \boldsymbol{r} \cdot \boldsymbol{c} &\triangleq \sum_{i=1}^{N} r_i \cdot c_i \\ &= \sum_{i=1}^{N} r_i \cdot (2u_i - 1). \end{aligned} \tag{2.28}$$

The inner product given by (2.28) is called the **correlation** between the received sequence $\boldsymbol{r}$ and the codeword $\boldsymbol{u}$.

Furthermore, (2.28) can be expanded as follows:

$$\boldsymbol{r} \cdot \boldsymbol{c} = 2\sum_{i=1}^{N} r_i \cdot u_i - \sum_{i=1}^{N} r_i. \tag{2.29}$$

Since the second term $\sum_{i=1}^{N} r_i$ in (2.29) is a common term in computing $\boldsymbol{r} \cdot \boldsymbol{c}$ for all codewords in C, maximizing $\boldsymbol{r} \cdot \boldsymbol{c}$ is equivalent to maximizing

$$\boldsymbol{r} \cdot \boldsymbol{u} \triangleq \sum_{i=1}^{N} r_i \cdot u_i. \tag{2.30}$$

The inner product given by (2.30) is called the **binary correlation** between the received sequence $\boldsymbol{r}$ and the codeword $\boldsymbol{u}$.

Summarizing the above, MLD can be stated in four equivalent ways:

(1) **Log-likelihood function**

Decode the received sequence $\boldsymbol{r}$ into a codeword $\boldsymbol{u}$ for which the log-likelihood function

$$\log P(\boldsymbol{r} \mid \boldsymbol{u}) = \sum_{i=1}^{N} \log P(r_i \mid u_i)$$

is maximized.

(2) **Squared Euclidean distance**

Decode the received sequence $\boldsymbol{r}$ into a codeword $\boldsymbol{u}$ for which the squared Euclidean distance

$$|\boldsymbol{r} - \boldsymbol{u}|^2 \triangleq \sum_{i=1}^{N} (r_i - (2u_i - 1))^2$$

is minimized.

(3) **Correlation function**

Decode the received sequence $\boldsymbol{r}$ into a codeword $\boldsymbol{u}$ for which the correlation function

$$m(\boldsymbol{r}, \boldsymbol{u}) \triangleq \sum_{i=1}^{N} r_i \cdot (2u_i - 1)$$

is maximized.

(4) **Binary correlation function**
Decode the received sequence $\boldsymbol{r}$ into a codeword $\boldsymbol{u}$ for which the binary correlation function

$$b(\boldsymbol{r}, \boldsymbol{u}) = \sum_{i=1}^{N} r_i \cdot u_i$$

is maximized.

2.5 REED-MULLER CODES

Reed-Muller (RM) codes form a class of multiple error-correction codes. These codes were discovered by Muller in 1954 [78], but the first decoding algorithm for these codes was devised by Reed, also in 1954 [83]. They are finite geometry codes and rich in algebraic and geometric structures. The purpose of including these codes in this reviewing chapter is that they have very simple and regular trellis structures and their trellises can be easily constructed. These codes can be decoded effectively with trellis-based decoding algorithms. Furthermore, they provide good example codes. Throughout this book, many example codes are RM codes.

For any nonnegative integers m and r with $0 \leq r \leq m$, there exists a binary r-th order RM code, denoted $\text{RM}_{r,m}$, with the following parameters:

Length	$N_{r,m} = 2^m$
Dimension	$K_{r,m} = 1 + \binom{m}{1} + \cdots + \binom{m}{r}$
Minimum distance	$d_{r,m} = 2^{m-r}$.

In the following, we first present the original construction of RM codes and then we describe an alternate construction for these codes. For $1 \leq i \leq m$, let $\boldsymbol{v}_i$ be a 2^m-tuple over GF(2) of the following form:

$$\boldsymbol{v}_i = (\underbrace{0\cdots0}_{2^{i-1}}, \underbrace{1\cdots1}_{2^{i-1}}, \underbrace{0\cdots0}_{2^{i-1}}, \ldots, \underbrace{1\cdots1}_{2^{i-1}}) \tag{2.31}$$

which consists of 2^{m-i+1} alternate all-zero and all-one 2^{i-1}-tuples. For $m = 3$, we have the following three 8-tuples:

$$\begin{aligned} \boldsymbol{v}_3 &= (00001111), \\ \boldsymbol{v}_2 &= (00110011), \\ \boldsymbol{v}_1 &= (01010101). \end{aligned}$$

Let $\boldsymbol{a} = (a_1, a_2, \ldots, a_N)$ and $\boldsymbol{b} = (b_1, b_2, \ldots, b_N)$ be two binary N-tuples. Define the following logic (**boolean**) product of $\boldsymbol{a}$ and $\boldsymbol{b}$,

$$\boldsymbol{a} \cdot \boldsymbol{b} \triangleq (a_1 \cdot b_1, a_2 \cdot b_2, \ldots, a_N \cdot b_N),$$

where '$\cdot$' denotes the logic product (or **AND** operation), i.e. $a_i \cdot b_i = 1$ if and only if both a_i and b_i are '1'. For $m = 3$,

$$\boldsymbol{v}_3 \cdot \boldsymbol{v}_1 = (00000101).$$

For simplicity, we use $\boldsymbol{ab}$ for $\boldsymbol{a} \cdot \boldsymbol{b}$.

Let $\mathbf{1}$ denote the all-one 2^m-tuple, $\mathbf{1} = (1, 1, \ldots, 1)$. For $1 \leq i_1 < i_2 < \cdots < i_l \leq m$, the product

$$\boldsymbol{v}_{i_1} \boldsymbol{v}_{i_2} \cdots \boldsymbol{v}_{i_l}$$

is said to have degree l. Since the weights of $\boldsymbol{v}_1, \boldsymbol{v}_2, \ldots, \boldsymbol{v}_m$ are even and powers of 2, it can be shown that the weight of the product $\boldsymbol{v}_{i_1} \boldsymbol{v}_{i_2} \cdots \boldsymbol{v}_{i_l}$ is also even and a power of 2, in fact 2^{m-l}.

The r-th order RM code, $\text{RM}_{r,m}$, of length 2^m is generated by the following set of vectors:

$$\begin{aligned} G_{\text{RM}}(r, m) &= \{\mathbf{1}, \boldsymbol{v}_1, \boldsymbol{v}_2, \ldots, \boldsymbol{v}_m, \boldsymbol{v}_1\boldsymbol{v}_2, \boldsymbol{v}_1\boldsymbol{v}_3, \ldots, \boldsymbol{v}_{m-1}\boldsymbol{v}_m, \\ &\qquad \ldots \text{up to products of degree } r\}. \end{aligned} \tag{2.32}$$

There are

$$K_{r,m} = 1 + \binom{m}{1} + \binom{m}{2} + \cdots + \binom{m}{r}$$

vectors in $G_{\text{RM}}(r, m)$ and they are linearly independent. If the vectors in $G_{\text{RM}}(r, m)$ are arranged as rows of a matrix, then the matrix is a generator matrix of the RM code, $\text{RM}_{r,m}$. For $0 \leq l \leq r$, there are exactly $\binom{m}{l}$ rows in $G_{\text{RM}}(r, m)$ of weight 2^{m-l}. All the codewords of the RM code, $\text{RM}_{r,m}$ with $0 \leq r < m$, have even weights. It is also clear that the $(r-1)$-th order RM code, $\text{RM}_{r-1,m}$, is a proper subcode of the r-th order RM code, $\text{RM}_{r,m}$.

Example 2.3 Let $m = 4$. The 2nd order RM code of length 16 is generated by the following 11 vectors:

$$
\begin{array}{ll}
\boldsymbol{v}_0 = \mathbf{1} & 1111111111111111 \\
\boldsymbol{v}_4 & 0000000011111111 \\
\boldsymbol{v}_3 & 0000111100001111 \\
\boldsymbol{v}_2 & 0011001100110011 \\
\boldsymbol{v}_1 & 0101010101010101 \\
\boldsymbol{v}_3\boldsymbol{v}_4 & 0000000000001111 \\
\boldsymbol{v}_2\boldsymbol{v}_4 & 0000000000110011 \\
\boldsymbol{v}_1\boldsymbol{v}_4 & 0000000001010101 \\
\boldsymbol{v}_2\boldsymbol{v}_3 & 0000001100000011 \\
\boldsymbol{v}_1\boldsymbol{v}_3 & 0000010100000101 \\
\boldsymbol{v}_1\boldsymbol{v}_2 & 0001000100010001
\end{array}
$$

This is a $(16, 11)$ code with minimum distance 4.

△△

The code given in Example 2.1 is the 1st order RM code, $\mathrm{RM}_{1,3}$, of length 8. Let

$$G_{(2,2)} \triangleq \begin{bmatrix} 1 & 1 \\ 0 & 1 \end{bmatrix} \tag{2.33}$$

be a 2×2 matrix over GF(2). The **two-fold Kronecker product** of $G_{(2,2)}$ is defined as

$$
\begin{aligned}
G_{(2^2,2^2)} &\triangleq \begin{bmatrix} 1 & 1 \\ 0 & 1 \end{bmatrix} \otimes \begin{bmatrix} 1 & 1 \\ 0 & 1 \end{bmatrix} \\
&= \begin{bmatrix} 1 & 1 & 1 & 1 \\ 0 & 1 & 0 & 1 \\ 0 & 0 & 1 & 1 \\ 0 & 0 & 0 & 1 \end{bmatrix}
\end{aligned} \tag{2.34}
$$

where $\otimes$ denotes the **Kronecker product**. The 3-fold Kronecker product of $G_{(2,2)}$ is defined as

$$G_{(2^3,2^3)} \triangleq \begin{bmatrix} 1 & 1 \\ 0 & 1 \end{bmatrix} \otimes \begin{bmatrix} 1 & 1 \\ 0 & 1 \end{bmatrix} \otimes \begin{bmatrix} 1 & 1 \\ 0 & 1 \end{bmatrix}$$

represents the logic product of $\boldsymbol{v}_i$ and $\boldsymbol{v}_j$, represented by $g(x_1, x_2, \ldots, x_m) = x_i$ and $h(x_1, x_2, \ldots, x_m) = x_j$, respectively. For $1 < i_1 < i_2 < \cdots < i_r \leq m$, the boolean function

$$f(x_1, x_2, \ldots, x_m) = x_{i_1} x_{i_2} \cdots x_{i_r} \tag{2.45}$$

represents the logic product of $\boldsymbol{v}_{i_1}, \boldsymbol{v}_{i_2}, \ldots,$ and $\boldsymbol{v}_{i_r}$. Therefore, the generator vectors of the r-th order RM code of length $N = 2^m$ are represented by the boolean functions in the following set

$$\begin{aligned} B(r, m) = \{&1, x_1, x_2, \ldots, x_m, x_1x_2, x_1x_3, \ldots, x_{m-1}x_m, \\ &\ldots \text{ up to all products of } r \text{ variables}\}. \end{aligned} \tag{2.46}$$

Let $P(r, m)$ denote the set of all boolean functions (or polynomials) of degree r or less with m variables. Then

$$\mathrm{RM}_{r,m} = \{\boldsymbol{b}(f) : f \in P(r, m)\}. \tag{2.47}$$

Finally, we want to point out that the dual code of the r-th order RM code, $\mathrm{RM}_{r,m}$, is the $(m - r - 1)$-th order RM code, $\mathrm{RM}_{m-r-1,m}$.

3 TRELLIS REPRESENTATION OF LINEAR BLOCK CODES

A code trellis is a graphical representation of a code, block or convolutional, in which every path represents a codeword (or code sequence). This representation makes it possible to implement maximum likelihood decoding (MLD) of a code with a significant reduction in decoding complexity. Chapter 3 presents the fundamental concepts and basic structural properties of trellises for linear block codes. An encoder with finite memory for a linear code is modeled as a finite-state machine. With this model, representation of the dynamic behavior of the encoder by a trellis diagram is easy to conceive. During an encoding interval, the state of the encoder at a specific time instant is simply defined by the information bits stored in the memory which affect both the past and future outputs of the encoder. To facilitate the construction of a code trellis, the generator matrix of a code is put in trellis oriented form. From this trellis oriented generator matrix, some basic structural properties can be derived.

3.1 TRELLIS REPRESENTATION OF CODES

An encoder for a linear code C with a finite memory, for which the output code bits at any time instant during an encoding interval $\Gamma = \{0, 1, 2, \ldots\}$ are uniquely determined by the current input information bits and the **state** of the encoder at the time can be modeled as a finite-state machine. The dynamic behavior of such an encoder can be **graphically** represented by a state diagram expanded in time, called a **trellis diagram** (or simply trellis) as shown in Figure 3.1.

The encoder starts from some initial state, denoted σ_0. At any time instant i during its encoding interval Γ, the encoder resides in one and only one allowable state in a finite set. In the trellis diagram, the set of allowable states at time-i is represented by a set of vertices (or nodes) at the i-th level, one for each allowable state. The encoder moves from one allowable state at one time instant to another allowable state at the next time instant in one unit of time. This is called a **state transition** which, in the trellis diagram, is represented by a directed edge (or branch) connecting the starting state to the destination state. Each edge is labeled with the code bits that are generated during the state transition. The set of allowable states at a given time instant i is called the **state space** of the encoder at time-i, denoted $\Sigma_i(C)$. A state $\sigma_i \in \Sigma_i(C)$ is said to be reachable if there exists an information sequence that takes the encoder from the initial state σ_0 to state σ_i at time-i. Every state of the encoder is reachable from the initial state σ_0. In the trellis, every vertex at level-i for $i \in \Gamma$ is connected by a path from the initial state σ_0. The label sequence of this path is a code sequence (or a prefix of a code sequence). Every vertex in the trellis has at least one incoming edge except for the initial state and at least one outgoing edge except for a state called the final state. Encoding of an information sequence is equivalent to tracing a path in the trellis starting from the initial vertex σ_0. If the encoding interval Γ is semi infinite, the trellis continues indefinitely; otherwise it terminates at a **final state**, denoted σ_f. Convolutional codes have semi infinite trellises, while the trellises for linear block codes terminate at the end of each encoding interval.

For $i \in \Gamma$, let I_i and O_i denote the input information block and its corresponding output code block, respectively, during the interval from time-i to

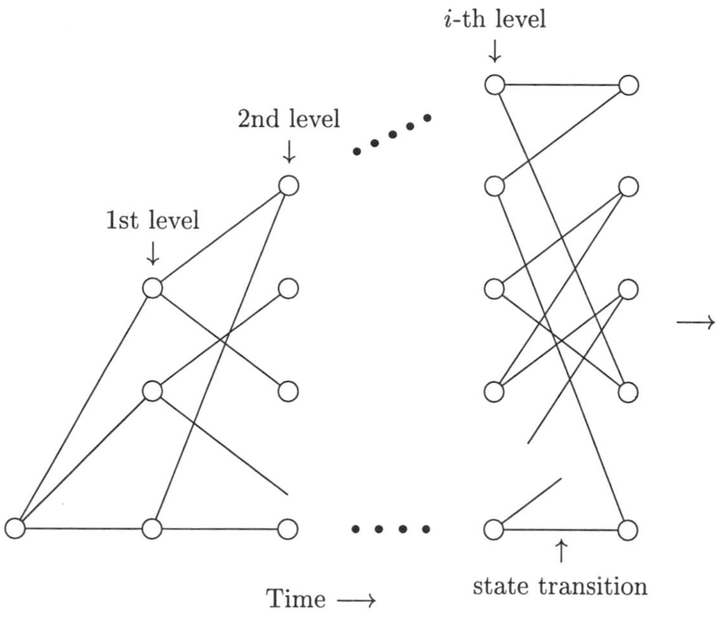

Figure 3.1. Trellis representation of a finite state encoder.

time-$(i+1)$. Then the dynamic behavior of the encoder for a linear code is governed by two functions:

(i) **Output function,**

$$O_i = f_i(\sigma_i, I_i),$$

where $f_i(\sigma_i, I_i) \neq f_i(\sigma_i, I_i')$ for $I_i \neq I_i'$.

(ii) **State transition function,**

$$\sigma_{i+1} = g_i(\sigma_i, I_i),$$

where $\sigma_i \in \Sigma_i(C)$ and $\sigma_{i+1} \in \Sigma_{i+1}(C)$ are called the **current** and **next states**, respectively. In the trellis diagram for C, the current and next states are connected by an edge (σ_i, σ_{i+1}) labeled with O_i.

A code trellis is said to be **time-invariant** if there exists a finite period $\{0, 1, \ldots, \nu\} \subset \Gamma$ and a state space $\Sigma(C)$ such that

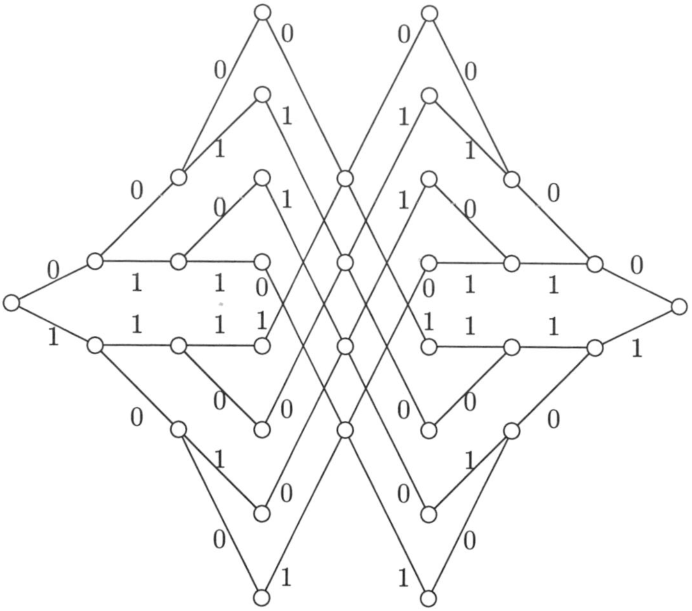

Figure 3.2. A time-varying trellis diagram for a block code.

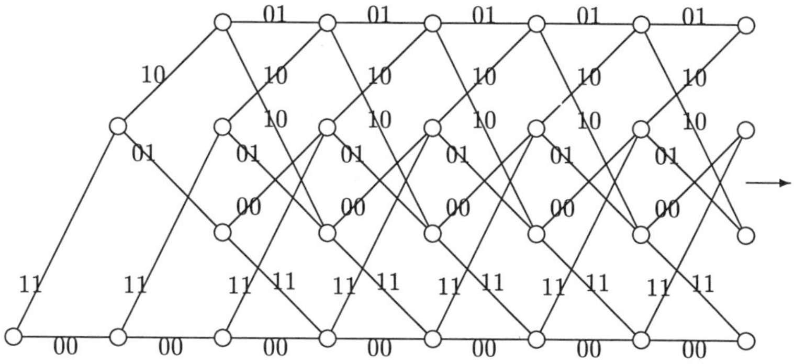

Figure 3.3. A time-invariant trellis diagram.

(1) $\Sigma_i(C) \subset \Sigma(C)$ for $0 \leq i < \nu$ and $\Sigma_i(C) = \Sigma(C)$ for $i \geq \nu$ and

(2) $f_i = f$ and $g_i = g$ for all $i \in \Gamma$.

A code trellis that is not time-invariant is said to be **time-varying**. A trellis diagram for a linear block code is, in general, time-varying. However, a trellis diagram for a convolutional code is usually time-invariant. Figure 3.2 and 3.3 depict a time-varying trellis diagram for a block code and a time-invariant trellis diagram for a convolutional code, respectively.

3.2 BIT-LEVEL TRELLISES FOR BINARY LINEAR BLOCK CODES

Consider a binary (N, K) linear block code C with generator and parity-check matrices, G and H, respectively. During each encoding interval, a message of K information bits is shifted into the encoder memory and encoded into a codeword of N code bits. The N code bits are formed and shifted onto the channel in N bit times. Therefore, the encoding span Γ is finite and consists of $N + 1$ time instants,

$$\Gamma = \{0, 1, 2, \ldots, N\}.$$

C can be represented by an N-section trellis diagram over the time span Γ. Let $\mathcal{E}(C)$ denote the encoder for C.

Definition 3.1 An N-section trellis diagram for a binary linear block code C of length N, denoted T, is a directed graph consisting of $N+1$ levels of vertices (called states) and edges (called branches) such that:

(1) For $0 \leq i \leq N$, the vertices at the i-th level represent the states in the state space $\Sigma_i(C)$ of the encoder $\mathcal{E}(C)$ at time-i. At time-0 (or the 0-th level) there is only one vertex, denoted σ_0, called the **initial vertex** (or **state**). At time-N (or the N-th level), there is only one vertex, denoted σ_f, called the **final vertex** (or **state**).

(2) For $0 < i \leq N$, a branch in the i-th section of the trellis T connects a state $\sigma_{i-1} \in \Sigma_{i-1}(C)$ to a state $\sigma_i \in \Sigma_i(C)$ and is labeled with a code bit u_i that represents the encoder output in the bit interval from time-$(i-1)$ to time-i. A branch represents a state transition.

(3) Except for the initial state, every state has at least one, but no more than two, incoming branches. Except for the final state, every state has at least one, but no more than two, outgoing branches. The initial state has no incoming branches. The final state has no outgoing branches. Two branches diverging from the same state have different labels.

(4) There is a **directed** path from the initial state σ_0 to the final state σ_f with a label sequence $(u_1, u_2, \ldots, u_N)$ if and only if $(u_1, u_2, \ldots, u_N)$ is a codeword in C.

△△

Two states in the code trellis are said to be **adjacent** if they are connected by a branch. During one encoding interval Γ, the encoder starts from the initial state σ_0, transverses a sequence of states

$$(\sigma_0, \sigma_1, \ldots, \sigma_i, \ldots, \sigma_f),$$

generates a code sequence

$$(u_1, u_2, \ldots, u_i, \ldots, u_N),$$

and then reaches the final state σ_f. The bit-level 8-section trellis diagram for the $(8,4)$ linear block code given in Example 2.1 (Table 2.1) is shown in Figure 3.2.

For $0 \leq i \leq N$, let $|\Sigma_i(C)|$ denote the cardinality of the state space $\Sigma_i(C)$. Then, the sequence,

$$(|\Sigma_0(C)|, |\Sigma_1(C)|, \ldots, |\Sigma_{N-1}(C)|, |\Sigma_N(C)|),$$

is called the **state space complexity profile**, which is a measure of the **state complexity** of the N-section code trellis T. We will show later that for $0 \leq i \leq N$, $|\Sigma_i(C)|$ is a power of 2. Define

$$\rho_i(C) \triangleq \log_2 |\Sigma_i(C)|,$$

which is called the **state space dimension** at time-i. When there is no confusion, we simply use ρ_i for $\rho_i(C)$ for simplicity. The sequence,

$$(\rho_0, \rho_1, \ldots, \rho_N),$$

is called **state space dimension profile**. From Figure 3.2, we see that the state space complexity and dimension profiles for the $(8, 4)$ code given in Example 2.1 are $(1, 2, 4, 8, 4, 8, 4, 2, 1)$ and $(0, 1, 2, 3, 2, 3, 2, 1, 0)$, respectively.

3.3 TRELLIS ORIENTED GENERATOR MATRIX

To facilitate the code trellis construction, we put the generator matrix G in a special form. Let $\boldsymbol{u} = (u_1, u_2, \ldots, u_N)$ be a nonzero binary N-tuple. The first nonzero component of $\boldsymbol{u}$ is called the **leading** '1' of $\boldsymbol{u}$ and the last nonzero component of $\boldsymbol{u}$ is called the **trailing** '1' of $\boldsymbol{u}$.

A generator matrix G for C is said to be in **trellis oriented form** (TOF) if the following two conditions hold:

(1) The leading '1' of each row appears in a column before the leading '1' of any row below it.

(2) No two rows have their trailing "ones" in the same column.

Any generator matrix for C can be put in TOF by two steps of **Gaussian elimination**.

Example 3.1 Consider the $(8, 4)$ RM code given in Example 2.1 with following generator matrix,

$$\begin{bmatrix} 1 & 1 & 1 & 1 & 1 & 1 & 1 & 1 \\ 0 & 0 & 0 & 0 & 1 & 1 & 1 & 1 \\ 0 & 0 & 1 & 1 & 0 & 0 & 1 & 1 \\ 0 & 1 & 0 & 1 & 0 & 1 & 0 & 1 \end{bmatrix}.$$

It is not in TOF. By interchanging the second and the fourth rows, we have

$$\begin{bmatrix} 1 & 1 & 1 & 1 & 1 & 1 & 1 & 1 \\ 0 & 1 & 0 & 1 & 0 & 1 & 0 & 1 \\ 0 & 0 & 1 & 1 & 0 & 0 & 1 & 1 \\ 0 & 0 & 0 & 0 & 1 & 1 & 1 & 1 \end{bmatrix}.$$

Add the fourth row of the above matrix to the first, second and third rows. These additions result in the following matrix in TOF:

$$G = \begin{bmatrix} \boldsymbol{g}_1 \\ \boldsymbol{g}_2 \\ \boldsymbol{g}_3 \\ \boldsymbol{g}_4 \end{bmatrix} = \begin{bmatrix} 1 & 1 & 1 & 1 & 0 & 0 & 0 & 0 \\ 0 & 1 & 0 & 1 & 1 & 0 & 1 & 0 \\ 0 & 0 & 1 & 1 & 1 & 1 & 0 & 0 \\ 0 & 0 & 0 & 0 & 1 & 1 & 1 & 1 \end{bmatrix}.$$

△△

The span of a row $\boldsymbol{g} = (g_1, g_2, \ldots, g_N)$ in a trellis oriented generator matrix (TOGM) G is defined as the **smallest interval** $\{i, i+1, \ldots, j\}$ which contains all the nonzero bits of $\boldsymbol{g}$. This is denoted as $\text{span}(\boldsymbol{g}) \triangleq [i, j]$. For a row $\boldsymbol{g}$ in a TOGM G whose span is $[i, j]$, the **active span** of $\boldsymbol{g}$, denoted $\text{aspan}(\boldsymbol{g})$, is defined as $\text{aspan}(\boldsymbol{g}) \triangleq [i, j-1]$ for $i < j$ and $\text{aspan}(\boldsymbol{g}) \triangleq \emptyset$ (empty set) for $i = j$.

Let $\boldsymbol{g}_1, \boldsymbol{g}_2, \ldots, \boldsymbol{g}_K$ be the K rows of a TOGM G with

$$\boldsymbol{g}_l = (g_{l1}, g_{l2}, \ldots, g_{lN})$$

for $1 \leq l \leq K$. Then

$$G = \begin{bmatrix} \boldsymbol{g}_1 \\ \boldsymbol{g}_2 \\ \vdots \\ \boldsymbol{g}_K \end{bmatrix} = \begin{bmatrix} g_{11} & g_{12} & \cdots & g_{1N} \\ g_{21} & g_{22} & \cdots & g_{2N} \\ \vdots & \vdots & \ddots & \vdots \\ g_{K1} & g_{K2} & \cdots & g_{KN} \end{bmatrix}.$$

Let $(a_1, a_2, \ldots, a_K)$ be the block of K information bits (called a message) to be encoded. The corresponding codeword is given by

$$\begin{aligned} \boldsymbol{u} &= (u_1, u_2, \ldots, u_N) \\ &= a_1 \cdot \boldsymbol{g}_1 + a_2 \cdot \boldsymbol{g}_2 + \cdots + a_K \cdot \boldsymbol{g}_K. \end{aligned}$$

We see that the l-th information bit a_l affects the output $\boldsymbol{u}$ of the encoder $\mathcal{E}(C)$ over the span of the l-th row $\boldsymbol{g}_l$ of the TOGM G. This $\text{span}(\boldsymbol{g}_l)$ may be regarded as the **constraint length** of the code associated with the l-th input information bit a_l.

At time-i with $1 \leq i \leq N$, the number of information bits that affect the next output code bit u_{i+1} is equal to the number of rows in G whose active spans contain i. These information bits define the state of the encoder at time-i.

3.4 STATE SPACE FORMULATION

In the following, we give a mathematical formulation of the state space of the N-section trellis for an (N, K) linear block code C over GF(2) with a TOGM G.

At time-i, $0 \leq i \leq N$, the rows of G are divided into three disjoint subsets:

(1) G_i^p consists of those rows of G whose spans are contained in the interval $[1, i]$.

(2) G_i^f consists of those rows of G whose spans are contained in the interval $[i+1, N]$.

(3) G_i^s consists of those rows of G whose active spans contain i.

Let A_i^p, A_i^f and A_i^s denote the subsets of information bits that correspond to the rows of G_i^p, G_i^f and G_i^s, respectively. The information bits in A_i^p do not affect the encoder outputs after time-i, and hence they become the **past** with respect to time-i. The information bits in A_i^f only affect the encoder outputs after time-i. Since the active spans of the rows in G_i^s contain the time instant i, the information bits in A_i^s affect not only the past encoder outputs up to time-i but also the future encoder outputs beyond time-i. We say that the information bits in A_i^s define a state of encoder $\mathcal{E}(C)$ for the code C at time-i. Let $\rho_i \triangleq |A_i^s| = |G_i^s|$. Then there are 2^{ρ_i} distinct states that the encoder $\mathcal{E}(C)$ can occupy at time-i; each state is defined by a specific combination of the ρ_i information bits in A_i^s. These states form the state space $\Sigma_i(C)$ of the encoder $\mathcal{E}(C)$ (or simply of the code C). The parameter ρ_i is the dimension of the state space $\Sigma_i(C)$. In the trellis representation of C, the states in $\Sigma_i(C)$ are represented by 2^{ρ_i} vertices at the i-th level of the trellis.

Example 3.2 Consider the TOGM G for the $(8, 4)$ RM code given in Example 3.1. The spans of the four rows are: $\text{span}(\boldsymbol{g}_1) = [1, 4]$, $\text{span}(\boldsymbol{g}_2) = [2, 7]$, $\text{span}(\boldsymbol{g}_3) = [3, 6]$, and $\text{span}(\boldsymbol{g}_4) = [5, 8]$. Their active spans are therefore: $\text{aspan}(\boldsymbol{g}_1) = [1, 3]$, $\text{aspan}(\boldsymbol{g}_2) = [2, 6]$, $\text{aspan}(\boldsymbol{g}_3) = [3, 5]$ and $\text{aspan}(\boldsymbol{g}_4) = [5, 7]$. For each i with $0 \leq i \leq 8$, counting the number of rows which are active at time-i yields the state space dimension profile $(0, 1, 2, 3, 2, 3, 2, 1, 0)$.

△△

The above formulation of a state space actually provides a sequential machine model for the encoder $\mathcal{E}(C)$.

3.5 STATE TRANSITION AND OUTPUT

For $0 \leq i \leq N$, suppose the encoder $\mathcal{E}(C)$ is in state $\sigma_i \in \Sigma_i(C)$. From time-i to time-$(i+1)$, $\mathcal{E}(C)$ generates a code bit u_{i+1} and moves from state σ_i to a state $\sigma_{i+1} \in \Sigma_{i+1}(C)$. Let

$$G_i^s = \{\boldsymbol{g}_1^{(i)}, \boldsymbol{g}_2^{(i)}, \ldots, \boldsymbol{g}_{\rho_i}^{(i)}\} \tag{3.1}$$

and

$$A_i^s = \{a_1^{(i)}, a_2^{(i)}, \ldots, a_{\rho_i}^{(i)}\} \tag{3.2}$$

where $\rho_i = |G_i^s|$. The current state σ_i of the encoder is defined by a specific combination of the information bits in A_i^s.

Let $\boldsymbol{g}^*$ be the row in G_i^f whose leading '1' is at position-$(i+1)$. The uniqueness of this row $\boldsymbol{g}^*$ (if it exists) is guaranteed by the first condition in the definition of a generator matrix in TOF given in section 3.3. Let g_{i+1}^* denote the $(i+1)$-th component of $\boldsymbol{g}^*$. Then $g_{i+1}^* = 1$. Let a^* denote the information bit that corresponds to row $\boldsymbol{g}^*$. It follows from (2.3) and the structure of the TOGM G that the output code bit u_{i+1} generated during the bit interval between time-i and time-$(i+1)$ is given by

$$u_{i+1} = a^* + \sum_{l=1}^{\rho_i} a_l^{(i)} g_{l,i+1}^{(i)}, \tag{3.3}$$

where $g_{l,i+1}^{(i)}$ is the $(i+1)$-th component of $\boldsymbol{g}_l^{(i)}$ in G_i^s. Note that a^* begins to affect the output of the encoder $\mathcal{E}(C)$ at time-$(i+1)$. For this reason, the bit a^* is regarded as the **current input information bit**. The second term in (3.3) is the contribution from the state σ_i defined by the information bits in $A_i^s = \{a_1^{(i)}, a_2^{(i)}, \ldots, a_{\rho_i}^{(i)}\}$ which are stored in memory. From (3.3), we see that the current output u_{i+1} is uniquely determined by the current state σ_i of the encoder $\mathcal{E}(C)$ and the current input a^*. The output bit u_{i+1} can have two possible values depending on the current input information bit a^*; each value takes the encoder $\mathcal{E}(C)$ to a different state at time-$(i+1)$. That is, there are two possible transitions from the current state σ_i to two states in $\Sigma_{i+1}(C)$ at time-$(i+1)$. In the code trellis, there are two edges (or branches) diverging from the vertex σ_i labeled with '0' and '1', respectively.

Suppose there is no such row $\boldsymbol{g}^*$ in G_i^f. Then the output code bit is given by

$$u_{i+1} = \sum_{l=1}^{\rho_i} a_l^{(i)} \cdot g_{l,i+1}^{(i)}. \tag{3.4}$$

In this case, we may regard that the current input information bit a^* is being set to "0", i.e. $a^* = 0$ (this is called a **dummy** information bit). The output code bit u_{i+1} can take only one value given by (3.4) and there is **only one** possible transition from the current state σ_i to a state in $\Sigma_{i+1}(C)$. In the trellis T, there is only one branch diverging from the vertex σ_i.

Example 3.3 Again we consider the $(8,4)$ code with its TOGM G given in Example 3.1. Consider time-2. Then we find that $G_2^p = \emptyset, G_2^f = \{\boldsymbol{g}_3, \boldsymbol{g}_4\}$ and $G_2^s = \{\boldsymbol{g}_1, \boldsymbol{g}_2\}$. Therefore, the information bits a_1 and a_2 define the state of the encoder at time-2 and there are 4 distinct states defined by four combinations of values of a_1 and a_2, $\{00, 01, 10, 11\}$. We also see that $\boldsymbol{g}^* = \boldsymbol{g}_3$. Therefore, the current input information bit is $a^* = a_3$. The current output code bit u_3 is given by

$$\begin{aligned} u_3 &= a_3 + a_1 \cdot g_{13} + a_2 \cdot g_{23} \\ &= a_3 + a_1. \end{aligned}$$

For every state defined by a_1 and a_2, u_3 has two possible values depending on a_3. In the trellis, there are two branches diverging from each state at time-2, as shown in Figure 3.2.

Now consider time-3. At $i = 3$, we find that $G_3^p = \emptyset$, $G_3^f = \{\boldsymbol{g}_4\}$ and $G_3^s = \{\boldsymbol{g}_1, \boldsymbol{g}_2, \boldsymbol{g}_3\}$. Therefore, the information bits a_1, a_2 and a_3 define 8 states at time-3, as shown in Figure 3.2. There is no row $\boldsymbol{g}^*$ in G_3^f with leading '1' at position (or time) $i = 4$. Hence we set the current input information bit $a^* = 0$. The output code bit u_4 is given by

$$\begin{aligned} u_4 &= a_1 \cdot g_{14} + a_2 \cdot g_{24} + a_3 \cdot g_{34} \\ &= a_1 + a_2 + a_3. \end{aligned}$$

In the trellis, there is only one branch diverging from each of the 8 states, as shown in Figure 3.2.

△△

Let $\boldsymbol{g}^0$ be the row in G_i^s whose trailing '1' is at the position-$(i+1)$. (Note that this row $\boldsymbol{g}^0$ may not exist.) The uniqueness of the row $\boldsymbol{g}^0$ (if it exists) is guaranteed by the second condition of a generator matrix in TOF given in Section 3.3. Let a^0 be the information bit in A_i^s that corresponds to row $\boldsymbol{g}^0$. Then at time-$(i+1)$,

$$G_{i+1}^s = (G_i^s \backslash \{\boldsymbol{g}^0\}) \cup \{\boldsymbol{g}^*\} \tag{3.5}$$

and

$$A_{i+1}^s = (A_i^s \backslash \{a^0\}) \cup \{a^*\}. \tag{3.6}$$

The information bits in A_{i+1}^s define the state space $\Sigma_{i+1}(C)$ at time-$(i+1)$. The change from A_i^s to A_{i+1}^s defines a state transition from the current state σ_i defined by A_i^s to the next state σ_{i+1} defined by A_{i+1}^s. Therefore from A_i^s, A_{i+1}^s, (3.3) and (3.4), we can construct the N-section code trellis T for C.

The construction of the N-section trellis T is carried out serially, section by section. Suppose the trellis has been constructed up to section-i. Now we want to construct the $(i+1)$-th section from time-i to time-$(i+1)$. The state space $\Sigma_i(C)$ is known. The $(i+1)$-th section is constructed by taking the following steps:

(1) Determine G_{i+1}^s and A_{i+1}^s from (3.5) and (3.6). Form the state space $\Sigma_{i+1}(C)$ at time-$(i+1)$.

(2) For each state $\sigma_i \in \Sigma_i(C)$, determine its state transition(s) following the state transition rules given above. Connect σ_i to its adjacent state(s) in $\Sigma_{i+1}(C)$ by edge(s).

(3) For each state transition, determine the output code bit u_{i+1} from the output function of (3.3) or (3.4), and label the corresponding edge in the trellis with u_{i+1}.

3.6 TIME-VARYING STRUCTURE

During the encoding interval $\Gamma = \{0, 1, \ldots, N\}$, the output function of the encoder $\mathcal{E}(C)$ changes between (3.3) and (3.4). Also, the set $\{g_{1,i+1}^{(i)}, g_{2,i+1}^{(i)}, \ldots, g_{\rho_i,i+1}^{(i)}\}$ in the summations of (3.3) and (3.4) may change from one time instant to another. This is because each column in the TOGM is, in general, not a downward shift of the column before it. Therefore, the output function of $\mathcal{E}(C)$ is time-varying. As the encoder $\mathcal{E}(C)$ moves from time-i to time-$(i+1)$, its state space may also change, i.e., $\Sigma_{i+1}(C) \neq \Sigma_i(C)$. Consequently, the trellis for $\mathcal{E}(C)$ is time-varying.

To describe the time-varying state space of $\mathcal{E}(C)$, there are four cases to consider.

Case I: There is no such row $\boldsymbol{g}^0$ in G_i^s, but there is a row $\boldsymbol{g}^*$ in G_i^f. As the encoder moves from time-i to time-$(i+1)$, the active span of $\boldsymbol{g}^*$ contains the time instant $i+1$. Therefore, $\boldsymbol{g}^*$ is added to the set G_i^s to form G_{i+1}^s. The information bit a^* that corresponds to $\boldsymbol{g}^*$ is now in the encoder memory and is included in determining the next and future states of the encoder. The next state σ_{i+1} is determined by the information bits

$$a_1^{(i)}, a_2^{(i)}, \ldots, a_{\rho_i}^{(i)}, a^*.$$

Since $|G_{i+1}^s| = |G_i^s| + 1$, $\rho_{i+1} = \rho_i + 1$. This results in state space expansion.

Case II: There is a row $\boldsymbol{g}^0 \in G_i^s$ and a row $\boldsymbol{g}^* \in G_i^f$. When the encoder moves from time-i to time-$(i+1)$, the span of $\boldsymbol{g}^0$ moves into the interval $[1, i+1]$ and $\boldsymbol{g}^0$ is replaced by $\boldsymbol{g}^*$ in G_{i+1}^s. In this case, the information bit a^0 that corresponds to $\boldsymbol{g}^0$ becomes part of the past with respect to time-$(i+1)$ and will not affect the encoder outputs further; however, the information bit a^* is now in the memory and is included in determining the next and future states of the encoder. Assuming that $a^0 = a_1^{(i)}$, the next state σ_{i+1} of the encoder is then determined by the information bits

$$a_2^{(i)}, a_3^{(i)}, \ldots, a_{\rho_i}^{(i)}, a^*.$$

Therefore, from time-i to time-$(i+1)$, the state space of the encoder and its dimension remain the same, i.e., $\rho_{i+1} = \rho_i$.

Case III: There is no such row $\boldsymbol{g}^0$ in G_i^s and no such row $\boldsymbol{g}^*$ in G_i^f. In this case, $G_{i+1}^s = G_i^s$ and there in no change in the state space dimension as the encoder moves from time-i to time-$(i+1)$. The next state is determined by the same set of information bits as at time-i, i.e.,

$$a_1^{(i)}, a_2^{(i)}, \ldots, a_{\rho_i}^{(i)}.$$

Case IV: There exists a row $\boldsymbol{g}^0 \in G_i^s$ but there is no such row $\boldsymbol{g}^* \in G_i^f$. In this case, $\boldsymbol{g}^0$ is excluded from G_i^s to form G_{i+1}^s and its corresponding information bit a^0 becomes part of the past as the encoder moves from time-i to time-$(i+1)$. Assuming that $a^0 = a_1^{(i)}$, the state σ_{i+1} of the encoder is determined by the information bits

$$a_2^{(i)}, a_3^{(i)}, \ldots, a_{\rho_i}^{(i)}.$$

Consequently, $\left|G_{i+1}^s\right| = |G_i^s| - 1$ and $\rho_{i+1} = \rho_i - 1$. This results in state space reduction.

Example 3.4 Consider the $(8,4)$ code given in Example 3.1. From its TOGM G, we see that for $i = 0, 1$ and 2, there is no such row $\boldsymbol{g}^0$ in G_i^s, but there is a row $\boldsymbol{g}^*$ in G_i^f. Hence there is state space expansion from time-0 to time-3 as shown Figure 3.2. We note that there is such a row $\boldsymbol{g}^0$ in G_3^s and there is no such row $\boldsymbol{g}^*$ in G_3^f. Therefore, there is state space reduction from time-3 to time-4, as shown in Figure 3.2.

$\triangle\triangle$

From the above analysis of the N-section trellis for an (N, K) linear block code C, we have the following observations. At time-i, with $0 \leq i \leq N$,

(1) The information bits in A_i^p become the past and do not affect the future outputs of the encoder beyond time-i.

(2) The information bits in A_i^f affect the encoder outputs only beyond time-i, i.e., they are the future input information bits.

(3) The information bits in A_i^s are the bits stored in the encoder memory that define the encoder state at time-i.

The above observations make the formulation of a trellis diagram for a block code the same as for a convolutional code [62].

3.7 STRUCTURAL PROPERTIES

For $0 \leq i < j \leq N$, let $C_{i,j}$ denote the subcode of C consisting of those codewords in C whose nonzero components are confined to the span of $j-i$ consecutive positions in the set $\{i+1, i+2, \ldots, j\}$. Clearly, every codeword in $C_{i,j}$ is of the form,

$$(\underbrace{0, 0, \ldots, 0}_{i}, u_{i+1}, u_{i+2}, \ldots, u_j, \underbrace{0, 0, \ldots, 0}_{N-j}).$$

It follows from the definition of $C_{i,j}$ and the structure of the TOGM G for C that $C_{i,j}$ is spanned by those rows in G whose spans are contained in the interval $[i+1, j]$. The two subcodes, $C_{0,i}$ and $C_{i,N}$, are spanned by the rows in G_i^p and G_i^f, respectively, and they are called the **past** and **future** subcodes of C with respect to time-i.

For a linear code D, let $k(D)$ denote its dimension. Then, $k(C_{0,i}) = |G_i^p|$ and $k(C_{i,N}) = |G_i^f|$. Recall that the dimension of the state space $\Sigma_i(C)$ at time-i is

$$\begin{aligned} \rho_i(C) &= |G_i^s| = K - |G_i^p| - |G_i^f| \\ &= K - k(C_{0,i}) - k(C_{i,N}). \end{aligned} \tag{3.7}$$

This gives a relationship between the state space dimension $\rho_i(C)$ at time-i and the dimensions of the past and future subcodes, $C_{0,i}$ and $C_{i,N}$, of C with respect to time-i.

Note that $C_{0,i}$ and $C_{i,N}$ have only the all-zero codeword $\mathbf{0}$ in common. The direct-sum of $C_{0,i}$ and $C_{i,N}$, denoted $C_{0,i} \oplus C_{i,N}$, is a subcode of C with dimension

$$k(C_{0,i}) + k(C_{i,N}).$$

Let $C/(C_{0,i} \oplus C_{i,N})$ denote the **partition** of C with respect to $C_{0,i} \oplus C_{i,N}$. Then this partition consists of

$$\begin{aligned} |C/(C_{0,i} \oplus C_{i,N})| &= 2^{K-k(C_{0,i})-k(C_{i,N})} \\ &= 2^{\rho_i} \end{aligned} \tag{3.8}$$

cosets of $C_{0,i} \oplus C_{i,N}$. Eq.(3.8) says that the number of states in the state space $\Sigma_i(C)$ at time-i is equal to the number of cosets in the partition $C/(C_{0,i} \oplus C_{i,N})$.

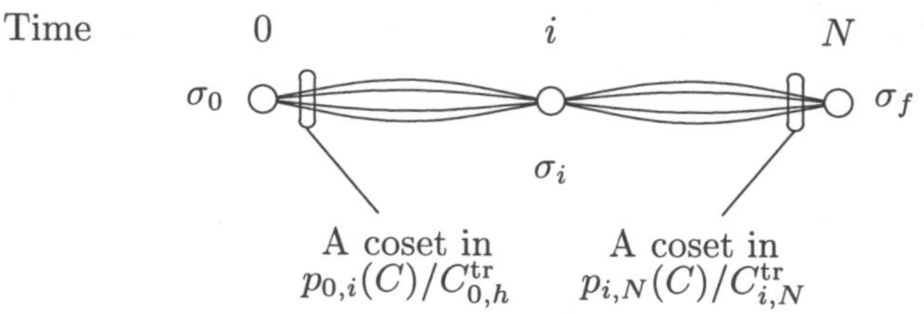

Figure 3.4. The paths in the code trellis that represent the $2^{K-\rho_i}$ codewords in $\boldsymbol{v} \oplus (C_{0,i} \oplus C_{i,N})$.

Let S_i denote the subspace of C that is spanned by the rows in G_i^s. Then each codeword in S_i is given by

$$\begin{aligned} \boldsymbol{v} &= (a_1^{(i)}, a_2^{(i)}, \ldots, a_{\rho_i}^{(i)}) \cdot G_i^s \\ &= a_1^{(i)} \cdot \boldsymbol{g}_1^{(i)} + a_2^{(i)} \cdot \boldsymbol{g}_2^{(i)} + \cdots + a_{\rho_i}^{(i)} \cdot \boldsymbol{g}_{\rho_i}^{(i)} \end{aligned} \tag{3.9}$$

where $a_l^{(i)} \in A_i^s$ for $1 \leq l \leq \rho_i$. The 2^{ρ_i} codewords in S_i can be used as the representatives for the cosets in the partition $C/(C_{0,i} \oplus C_{i,N})$. Therefore, S_i is the coset representative space for the partition $C/(C_{0,i} \oplus C_{i,N})$. From (3.9), we see that there is **one-to-one correspondence** between $\boldsymbol{v}$ and the state $\sigma_i \in \Sigma_i(C)$ defined by $(a_1^{(i)}, a_2^{(i)}, \ldots, a_{\rho_i}^{(i)})$. Since there is a **one-to-one correspondence** between $\boldsymbol{v}$ and a coset in $C/(C_{0,i} \oplus C_{i,N})$, therefore, there is a **one-to-one correspondence** between a state in the state space $\Sigma_i(C)$ and a coset in the partition $C/(C_{0,i} \oplus C_{i,N})$.

With the above one-to-one correspondence in the trellis T, the codeword $\boldsymbol{v}$ given by (3.9) is represented by a path that passes through the state σ_i defined by the information bits, $a_1^{(i)}$, $a_2^{(i)}$, ..., $a_{\rho_i}^{(i)}$ (i.e., a path that connects the initial state σ_0 to the final state σ_f through the state σ_i). If we fix the information bits, $a_1^{(i)}$, $a_2^{(i)}$, ..., $a_{\rho_i}^{(i)}$, and allow the other $K - \rho_i$ information bits to vary, we obtain $2^{K-\rho_i}$codewords of C in the coset

$$\boldsymbol{v} \oplus (C_{0,i} \oplus C_{i,N}) \triangleq \{\boldsymbol{v} + \boldsymbol{u} : \boldsymbol{u} \in C_{0,i} \oplus C_{i,N}\} \tag{3.10}$$

with $\boldsymbol{v}$ as the coset representative. In the trellis, these $2^{K-\rho_i}$ codewords are represented by paths that connect the initial state σ_0 to the final state σ_f

through the state σ_i at time-i defined by the information bits, $a_1^{(i)}$, $a_2^{(i)}$, ..., $a_{\rho_i}^{(i)}$, as shown in Figure 3.4. Note that

$$K - \rho_i = k(C_{0,i}) + k(C_{i,N}) \tag{3.11}$$

which is simply the dimension of $C_{0,i} \oplus C_{i,N}$.

For $0 \le i < j \le N$, let $p_{i,j}(C)$ denote the linear code of length $j-i$ obtained from C by removing the first i and last $N-j$ components of each codeword in C. This code is called a **punctured** (or **truncated**) code of C. Let $C_{i,j}^{\text{tr}}$ denote the punctured code of the subcode $C_{i,j}$, i.e.,

$$C_{i,j}^{\text{tr}} \triangleq p_{i,j}(C_{i,j}). \tag{3.12}$$

It follows from the structure of the TOGM G that

$$k(p_{i,j}(C)) = K - k(C_{0,i}) - k(C_{j,N}) \tag{3.13}$$

and

$$k(C_{i,j}^{\text{tr}}) = k(C_{i,j}). \tag{3.14}$$

Consider the punctured code $p_{0,i}(C)$. Partition $p_{0,i}(C)$ based on $C_{0,i}^{\text{tr}}$. It follows from (3.13) and (3.14) that the partition $p_{0,i}(C)/C_{0,i}^{\text{tr}}$ consists of

$$2^{K-k(C_{0,i})-k(C_{i,N})} = 2^{\rho_i}$$

cosets of $C_{0,i}^{\text{tr}}$. We can readily see that there is a **one-to-one correspondence** between the cosets in $p_{0,i}(C)/C_{0,i}^{\text{tr}}$ and the cosets in $C/(C_{0,i} \oplus C_{i,N})$, and hence a **one-to-one correspondence** between the cosets in $p_{0,i}(C)/C_{0,i}^{\text{tr}}$ and the states in the state space $\Sigma_i(C)$. The codewords in a coset in $p_{0,i}(C)/C_{0,i}^{\text{tr}}$ are simply the prefixes of the codewords in its corresponding coset in $C/(C_{0,i} \oplus C_{i,N})$. Hence the codewords in a coset of $p_{0,i}(C)/C_{0,i}^{\text{tr}}$ will take the encoder $\mathcal{E}(C)$ to a unique state $\sigma_i \in \Sigma_i(C)$. In the trellis T, they are the paths connecting the initial state σ_0 to the state σ_i as shown in Figure 3.4. Let $L(\sigma_0, \sigma_i)$ denote the paths in the trellis T that connect the initial state σ_0 to the state σ_i in $\Sigma_i(C)$. Then $L(\sigma_0, \sigma_i)$ is a coset in the partition $p_{0,i}(C)/C_{0,i}^{\text{tr}}$.

Now we consider the punctured code $p_{i,N}(C)$. Partition $p_{i,N}(C)$ based on $C_{i,N}^{\text{tr}} = p_{i,N}(C_{i,N})$. Then it follow from (3.13) and (3.14) that the partition

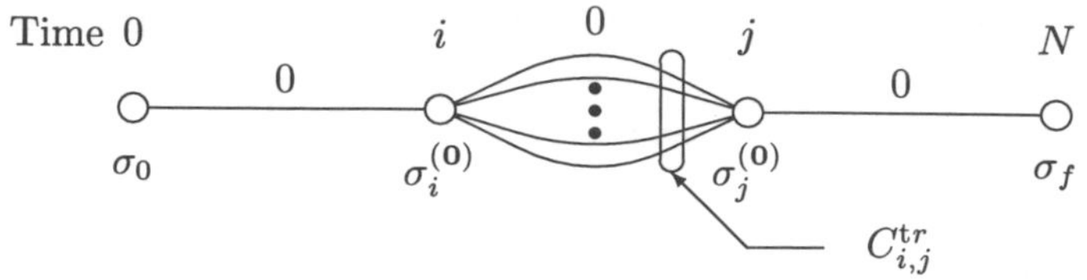

Figure 3.5. Paths in the code trellis that represent the codewords in $C_{i,j}$.

$p_{i,N}(C)/C_{i,N}^{\rm tr}$ consists of

$$2^{K-k(C_{0,i})-k(C_{i,N})} = 2^{\rho_i}$$

cosets of $C_{i,N}^{\rm tr}$. Again we see that there is a **one-to-one correspondence** between the cosets in $p_{i,N}(C)/C_{i,N}^{\rm tr}$ and the cosets in $C/(C_{0,i} \oplus C_{i,N})$, and hence a **one-to-one correspondence** between the cosets in $p_{i,N}(C)/C_{i,N}^{\rm tr}$ and the states in the state space $\Sigma_i(C)$. In the trellis, the codewords in a coset $p_{i,N}(C)/C_{i,N}^{\rm tr}$ form the paths that connect a state $\sigma_i \in \Sigma_i(C)$ to the final state σ_f as shown in Figure 3.4. Let $L(\sigma_i, \sigma_f)$ denote the paths in the trellis T that connect the state $\sigma_i \in \Sigma_i(C)$ to the final state σ_f. Then $L(\sigma_i, \sigma_f)$ is a coset in the partition $p_{i,N}(C)/C_{i,N}^{\rm tr}$.

For $0 \le i < j \le N$, let $\sigma_i^{(0)}$ and $\sigma_j^{(0)}$ denote two states on the all-zero path $\mathbf{0}$ in the trellis T at time-i and time-j, respectively. Let $L(\sigma_i^{(0)}, \sigma_j^{(0)})$ denote the paths of length $j-i$ in T that connect $\sigma_i^{(0)}$ to $\sigma_j^{(0)}$. Consider the paths in T that start from the initial state σ_0, follow the all-zero path $\mathbf{0}$ to the state $\sigma_i^{(0)}$, transverse through the paths in $L(\sigma_i^{(0)}, \sigma_j^{(0)})$ to the state $\sigma_j^{(0)}$, then follow the all-zero path $\mathbf{0}$ until they reach the final state σ_f as shown in Figure 3.5. These paths represent the codewords in the subcode $C_{i,j}$ of C. This implies that

$$L(\sigma_i^{(0)}, \sigma_j^{(0)}) = C_{i,j}^{\rm tr}. \tag{3.15}$$

Let $\boldsymbol{v} = (v_1, v_2, \ldots, v_N)$ be a path in the code trellis T. For $0 \le i < j \le N$, let $\sigma_i^{(\boldsymbol{v})}$ and $\sigma_j^{(\boldsymbol{v})}$ be two states on the path $\boldsymbol{v}$ at time-i and time-j, respectively. Let $L(\sigma_i^{(\boldsymbol{v})}, \sigma_j^{(\boldsymbol{v})})$ denote the paths of length $j-i$ that connect $\sigma_i^{(\boldsymbol{v})}$ to $\sigma_j^{(\boldsymbol{v})}$. Consider the paths in T that start form the initial state σ_0, follow the path $\boldsymbol{v}$ to the state $\sigma_i^{(\boldsymbol{v})}$, transverse through the paths in $L(\sigma_i^{(\boldsymbol{v})}, \sigma_j^{(\boldsymbol{v})})$, then follow the

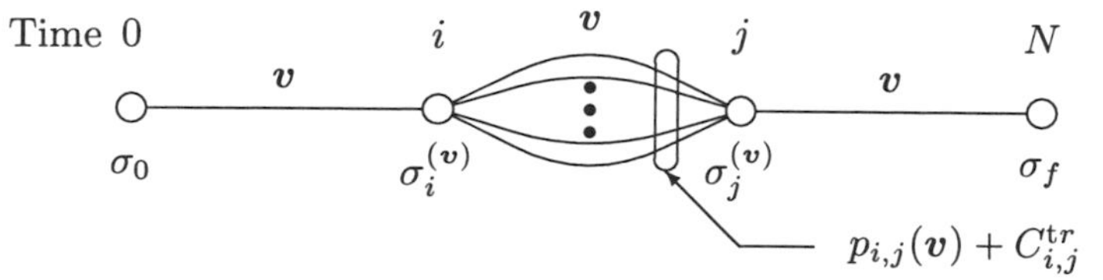

Figure 3.6. Paths in the code trellis that represent the codewords in the coset $\boldsymbol{u} \oplus C_{i,j}$.

path $\boldsymbol{v}$ until they reach the final state σ_f as shown in Figure 3.6. These paths represent the codewords in the following coset of $C_{i,j}$:

$$\boldsymbol{v} \oplus C_{i,j} \triangleq \{\boldsymbol{v} + \boldsymbol{u} : \boldsymbol{u} \in C_{i,j}\}. \tag{3.16}$$

This is a coset in the partition $C/C_{i,j}$. This implies that $L(\sigma_i^{(\boldsymbol{v})}, \sigma_j^{(\boldsymbol{v})})$ is a coset of $C_{i,j}^{\mathrm{tr}}$ in $p_{i,j}(C)$, i.e.,

$$L(\sigma_i^{(\boldsymbol{v})}, \sigma_j^{(\boldsymbol{v})}) = p_{i,j}(\boldsymbol{v}) + C_{i,j}^{\mathrm{tr}} \in p_{i,j}(C)/C_{i,j}^{\mathrm{tr}}, \tag{3.17}$$

where $p_{i,j}(\boldsymbol{v})$ denotes the vector of length $j-i$ obtained from $\boldsymbol{v}$ by removing the first i and last $N-j$ components of $\boldsymbol{v}$. For any two connected states $\sigma_i \in \Sigma_i(C)$ and $\sigma_j \in \Sigma_j(C)$ with $0 \leq i < j \leq N$, they must be on a path in the trellis T. It follows from (3.17) that

$$L(\sigma_i, \sigma_j) \in p_{i,j}(C)/C_{i,j}^{\mathrm{tr}}. \tag{3.18}$$

Therefore, the number of paths that connect a state $\sigma_i \in \Sigma_i(C)$ to a state $\sigma_j \in \Sigma_j(C)$ is given by

$$|L(\sigma_i, \sigma_j)| = \begin{cases} 2^{k(C_{i,j}^{\mathrm{tr}})}, & \text{if } \sigma_i \text{ and } \sigma_j \text{ are connected,} \\ 0, & \text{if } \sigma_i \text{ and } \sigma_j \text{ are not connected.} \end{cases} \tag{3.19}$$

For $0 \leq i < j < k \leq N$, let $\Sigma_j(\sigma_i, \sigma_k)$ denote the set of states in $\Sigma_j(C)$ through which the path in $L(\sigma_i, \sigma_k)$ connect the state σ_i to the state σ_k as shown in Figure 3.7. Let

$$L(\sigma_i, \sigma_j) \circ L(\sigma_j, \sigma_k) \triangleq \{\boldsymbol{u} \circ \boldsymbol{v} : \boldsymbol{u} \in L(\sigma_i, \sigma_j), \boldsymbol{v} \in L(\sigma_j, \sigma_k)\} \tag{3.20}$$

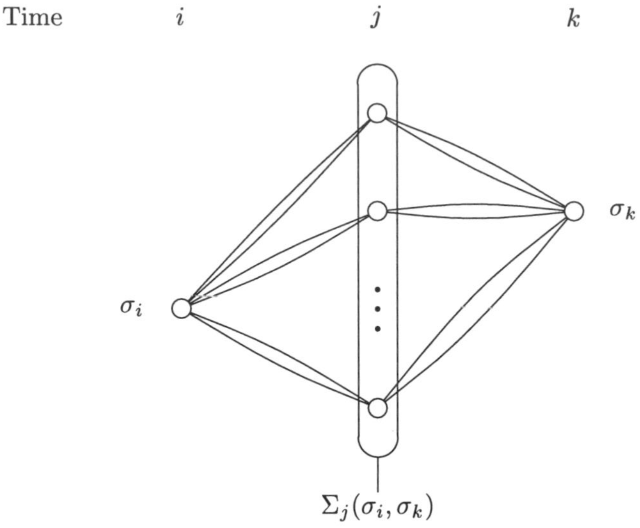

Figure 3.7. The state set $\Sigma_j(\sigma_i, \sigma_k)$.

where $\boldsymbol{u} \circ \boldsymbol{v}$ denotes the concatenation of two sequences $\boldsymbol{u}$ and $\boldsymbol{v}$. In the trellis, $L(\sigma_i, \sigma_j) \circ L(\sigma_j, \sigma_k)$ consists of those paths in $L(\sigma_i, \sigma_k)$ that connect the state σ_i to the state σ_k through the state σ_j. Then,

$$L(\sigma_i, \sigma_k) = \bigcup_{\sigma_j \in \Sigma_j(\sigma_i, \sigma_k)} L(\sigma_i, \sigma_j) \circ L(\sigma_j, \sigma_k). \tag{3.21}$$

The above developments give some fundamental structural properties of an N-section trellis T for a linear block code C.

4 STATE LABELING, TRELLIS CONSTRUCTION PROCEDURES AND TRELLIS SYMMETRY

The construction of a code trellis can be facilitated by labeling the states at each level of the trellis. State labeling is also necessary in the implementation of a trellis-based decoding algorithm. This chapter presents three methods for labeling the states of the N-section trellis for an (N, K) linear block code. The first two methods are based on the information set that defines the state space at a particular encoding time instant and the third method is based on the parity-check matrix of the code. The first two methods are more efficient than the third one for codes with $K < N - K$; however, the third method is more efficient for codes with $N - K < K$. Based on these labeling methods, construction procedures for the N-section trellis for an (N, K) linear block code are presented. Also presented in this chapter is the mirror symmetry structure of a code trellis. This symmetry structure is useful in decoding.

4.1 STATE LABELING BY THE STATE DEFINING INFORMATION SET

In a code trellis, each state is labeled by a fixed sequence (or given a name). This can be accomplished by using a K-tuple $\boldsymbol{\lambda}$ with components corresponding to the K information bits, $a_1, a_2, \ldots, a_K$, in a message. At time-i, all the components of $\boldsymbol{\lambda}$ are set to zero except for the components at the positions corresponding to the information bits in $A_i^s = \{a_1^{(i)}, a_2^{(i)}, \ldots, a_{\rho_i}^{(i)}\}$. Every combination of the ρ_i bits at the positions corresponding to the information bits in A_i^s gives the label $l(\sigma_i)$ for the state σ_i defined by the information bits, $a_1^{(i)}, a_2^{(i)}, \ldots, a_{\rho_i}^{(i)}$.

Example 4.1 Consider the $(8,4)$ code given in Example 3.1. At time-4, we find that $A_4^s = \{a_2, a_3\}$. There are 4 states corresponding to 4 combinations of a_2 and a_3. Therefore, the label for each of these 4 states is given by $(0, a_2, a_3, 0)$. △△

The construction of the N-section trellis for an (N, K) linear block code C can be carried out as follows. Suppose the trellis T has been constructed up to section-i. At this point, G_i^s, A_i^s and $\Sigma_i(C)$ are known. Each state $\sigma_i \in \Sigma_i(C)$ is labeled by a K-tuple. The $(i+1)$-th section is constructed by taking the following steps:

(1) Determine G_{i+1}^s and A_{i+1}^s from (3.5) and (3.6).

(2) Form the state space $\Sigma_{i+1}(C)$ at time-$(i+1)$ and label each state in $\Sigma_{i+1}(C)$ based on A_{i+1}^s. The state in $\Sigma_{i+1}(C)$ form the vertices of the code trellis T at the $(i+1)$-th level.

(3) For each state $\sigma_i \in \Sigma_i(C)$ at time-i, determine its transition(s) to the state(s) in $\Sigma_{i+1}(C)$ based on the information bits of a^* and a^0. For each transition from a state $\sigma_i \in \Sigma_i(C)$ to a state $\sigma_{i+1} \in \Sigma_{i+1}(C)$, connect the state σ_i to the state σ_{i+1} by an edge (σ_i, σ_{i+1}).

(4) For each state transition (σ_i, σ_{i+1}), determine the output code bit u_{i+1} and label the edge (σ_i, σ_{i+1}) with u_{i+1}.

Recall that at time-i, there are two branches diverging from a state in $\Sigma_i(C)$ if there exists a current information bit a^*. One branch corresponds to $a^* =$

Table 4.1. State defining sets and state labels for the 8-section trellis for the $(8,4)$ linear block code.

i	G_i^s	a^*	a^0	A_i^s	State Label
0	$\emptyset$	a_1	$-$	$\emptyset$	(0000)
1	$\{g_1\}$	a_2	$-$	$\{a_1\}$	(a_1000)
2	$\{g_1, g_2\}$	a_3	$-$	$\{a_1, a_2\}$	(a_1a_200)
3	$\{g_1, g_2, g_3\}$	$-$	a_1	$\{a_1, a_2, a_3\}$	$(a_1a_2a_30)$
4	$\{g_2, g_3\}$	a_4	$-$	$\{a_2, a_3\}$	$(0a_2a_30)$
5	$\{g_2, g_3, g_4\}$	$-$	a_3	$\{a_2, a_3, a_4\}$	$(0a_2a_3a_4)$
6	$\{g_2, g_4\}$	$-$	a_2	$\{a_2, a_4\}$	$(0a_20a_4)$
7	$\{g_4\}$	$-$	a_4	$\{a_4\}$	$(000a_4)$
8	$\emptyset$	$-$	$-$	$\emptyset$	(0000)

0 and the other corresponds to $a^* = 1$. For the convenience of graphical representation, in the code trellis T, we use the upper branch to represent $a^* = 0$ and the lower branch to represent $a^* = 1$. If a^* is a dummy information bit, then there is only one branch diverging from each state in $\Sigma_i(C)$. This single branch represents a dummy information bit. Using the above representation, we can easily extract the information bits from each path in the trellis (the dummy information bits are deleted).

Example 4.2 Consider the state labeling and trellis construction for the $(8,4)$ RM code given in Example 3.1 whose TOGM G is repeated below,

$$G = \begin{bmatrix} g_1 \\ g_2 \\ g_3 \\ g_4 \end{bmatrix} = \begin{bmatrix} 1 & 1 & 1 & 1 & 0 & 0 & 0 & 0 \\ 0 & 1 & 0 & 1 & 1 & 0 & 1 & 0 \\ 0 & 0 & 1 & 1 & 1 & 1 & 0 & 0 \\ 0 & 0 & 0 & 0 & 1 & 1 & 1 & 1 \end{bmatrix}.$$

For $0 \leq i \leq 8$, we determine the submatrix G_i^s and the state defining information set A_i^s as listed in Table 4.1. From A_i^s, we form the label for each state in $\Sigma_i(C)$ as shown in Table 4.1. The state transitions from time-i to time-$(i+1)$ are determined by the change from A_i^s to A_{i+1}^s. Following the trellis

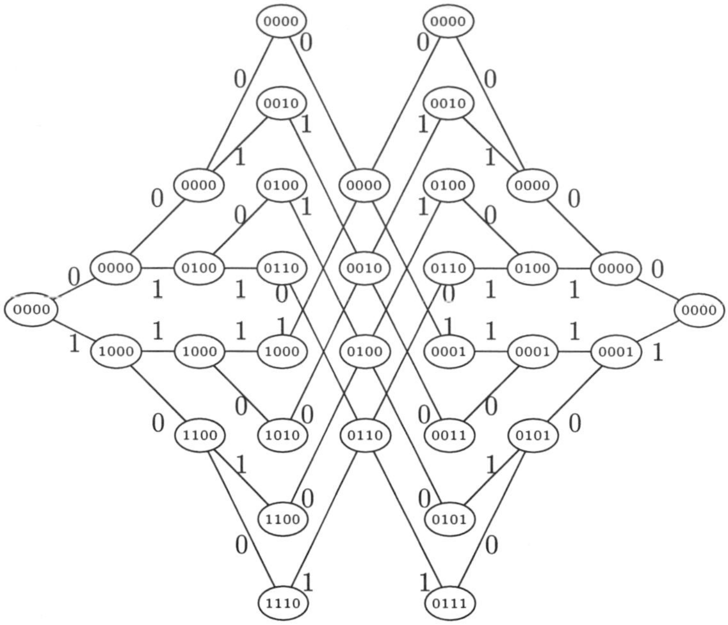

Figure 4.1. The 8-section trellis diagram for the $(8,4)$ RM code with state labeling by the state defining information set.

construction procedure given above, we obtain the 8-section trellis diagram for the $(8,4)$ RM code as shown in Figure 4.1. Each state in the trellis is labeled by a 4-tuple.

△△

In many cases, we do not need K bits for labeling the states of the N-section trellis for a binary (N,K) linear block code C. Let $(\rho_0, \rho_1, \ldots, \rho_N)$ be the state space dimension profile of the trellis. Define

$$\rho_{\max}(C) \triangleq \max_{0 \le i \le N} \rho_i \tag{4.1}$$

which is simply the maximum state space dimension of the trellis. From (3.7), we find that $\rho_{\max}(C) \le K$. In general, $\rho_{\max}$ is smaller than K. Since the number of states at any level of the trellis is less than or at most equal to

Table 4.2. State labeling for the $(8,4)$ RM code using $\rho_{\max}(C) = 3$ bits.

i	A_i^s	State Label
0	$\emptyset$	(000)
1	$\{a_1\}$	$(a_1 00)$
2	$\{a_1, a_2\}$	$(a_1 a_2 0)$
3	$\{a_1, a_2, a_3\}$	$(a_1 a_2 a_3)$
4	$\{a_2, a_3\}$	$(a_2 a_3 0)$
5	$\{a_2, a_3, a_4\}$	$(a_2 a_3 a_4)$
6	$\{a_2, a_4\}$	$(a_2 a_4 0)$
7	$\{a_4\}$	$(a_4 00)$
8	$\emptyset$	(000)

$2^{\rho_{\max}(C)}$, $\rho_{\max}(C)$ bits are sufficient for labeling the states in the trellis. Consider the state space $\Sigma_i(C)$ at time-i with $0 \leq i \leq N$ which is defined by the set $\{a_1^{(i)}, a_2^{(i)}, \ldots, a_{\rho_i}^{(i)}\}$ of ρ_i information bits. For each state $\sigma_i \in \Sigma_i(C)$, we form a $\rho_{\max}(C)$-tuple, denoted $l(\sigma_i)$, in which the first ρ_i components are simply $a_1^{(i)}, a_2^{(i)}, \ldots, a_{\rho_i}^{(i)}$ and the remaining $\rho_{\max}(C) - \rho_i$ components are set to 0, i.e.,

$$l(\sigma_i) \triangleq (a_1^{(i)}, a_2^{(i)}, \ldots, a_{\rho_i}^{(i)}, 0, 0, \ldots, 0). \tag{4.2}$$

Then $l(\sigma_i)$ is the label for the state σ_i.

Example 4.3 Again we consider the $(8,4)$ RM code given in Example 4.2. From the TOGM G of the code, we find the state space dimension profile of the 8-section trellis for the code to be $(0,1,2,3,2,3,2,1,0)$. Hence $\rho_{\max}(C) = 3$. Using 3 bits for labeling the states as described above, the state labels are given in Table 4.2. Compared to the state labeling given in Example 4.2, one bit is saved.

△△

4.2 STATE LABELING BY PARITY-CHECK MATRIX

Consider a binary (N, K) linear block code C with a parity-check matrix

$$H = [\boldsymbol{h}_1, \boldsymbol{h}_2, \ldots, \boldsymbol{h}_j, \ldots, \boldsymbol{h}_N], \tag{4.3}$$

where, for $1 \leq j \leq N$, $\boldsymbol{h}_j$ denotes the j-th column of H and is a binary $(N-K)$-tuple. A binary N-tuple $\boldsymbol{c}$ is a codeword in C if and only if

$$\boldsymbol{c} \cdot H^T = (\underbrace{0, 0, \ldots, 0}_{N-K}), \tag{4.4}$$

where H^T denotes the transpose of H. C is called the **null space** of H.

Let $\mathbf{0}_{N-K}$ denote the all-zero $(N-K)$-tuple $(0, 0, \ldots, 0)$. For $1 \leq i \leq N$, let H_i denote the submatrix that consists of the first i columns of H, i.e.,

$$H_i = [\boldsymbol{h}_1, \boldsymbol{h}_2, \ldots, \boldsymbol{h}_i]. \tag{4.5}$$

It is clear that the rank of H_i is at most $N-K$, i.e.,

$$\text{Rank}(H_i) \leq N - K. \tag{4.6}$$

Then for each codeword $\boldsymbol{c} \in C_{0,i}^{\text{tr}}$,

$$\boldsymbol{c} \cdot H_i^T = \mathbf{0}_{N-K}. \tag{4.7}$$

$C_{0,i}^{\text{tr}}$ is the **null space** of H_i.

Now we consider the partition

$$p_{0,i}(C)/C_{0,i}^{\text{tr}}.$$

Let D be a coset in $p_{0,i}/C_{0,i}^{\text{tr}}$ and $D \neq C_{0,i}^{\text{tr}}$. For every vector $\boldsymbol{a} \in D$,

$$\boldsymbol{a} \cdot H_i^T = (s_1, s_2, \ldots, s_{N-K}) \neq \mathbf{0}_{N-K} \tag{4.8}$$

and is the same for all vectors in D, i.e., for $\boldsymbol{a}_1, \boldsymbol{a}_2 \in D$ and $\boldsymbol{a}_1 \neq \boldsymbol{a}_2$,

$$\boldsymbol{a}_1 \cdot H_i^T = \boldsymbol{a}_2 \cdot H_i^T = (s_1, s_2, \ldots, s_{N-K}). \tag{4.9}$$

The $(N-K)$-tuple $(s_1, s_2, \ldots, s_{N-K})$ is called the **label** for the coset D. Let D_1 and D_2 be two different cosets in $p_{0,i}(C)/C_{0,i}^{\text{tr}}$. Let $\boldsymbol{a}_1 \in D_1$ and $\boldsymbol{a}_2 \in D_2$. It follows from the theory of linear block codes that $\boldsymbol{a}_1 \neq \boldsymbol{a}_2$ and

$$\boldsymbol{a}_1 \cdot H_i^T \neq \boldsymbol{a}_2 \cdot H_i^T.$$

This says that different cosets in $p_{0,i}(C)/C_{0,i}^{\text{tr}}$ have different labels.

Recall the mathematical formulation of the state spaces of a code trellis. There is a one-to-one correspondence between a state σ in the state space $\Sigma_i(C)$ at time-i and a coset $D \in p_{0,i}(C)/C_{0,i}^{\text{tr}}$, and the codewords of $p_{0,i}(C)$ in D form the paths that connect the initial state σ_0 to state σ. This one-to-one correspondence leads to the definition of a state label.

Let $L(\sigma_0, \sigma)$ denote the set of paths in the code trellis for C that connect the initial state σ_0 to a state σ in the state space $\Sigma_i(C)$ at time-i.

Definition 4.1 For $0 \leq i \leq N$, the label of a state $\sigma \in \Sigma_i(C)$ based on a parity-check matrix H of C, denoted $l(\sigma)$, is defined as the binary $(N-K)$-tuple

$$l(\sigma) \triangleq \boldsymbol{a} \cdot H_i^T = (s_1, s_2, \ldots, s_{N-K}), \tag{4.10}$$

for any $\boldsymbol{a} \in L(\sigma_0, \sigma)$. For $i = 0, H_i = \emptyset$ and the initial state σ_0 is labeled with the all-zero $(N-K)$-tuple, $\mathbf{0}_{N-K}$. For $i = N, L(\sigma_0, \sigma_f) = C$ and the final state σ_f is also labeled with $\mathbf{0}_{N-K}$.

△△

It follows from the above definition of a state label, the one-to-one correspondence between the states in $\Sigma_i(C)$ and the cosets in $p_{0,i}(C)/C_{0,i}^{\text{tr}}$ for $0 \leq i \leq N$, and (4.10) that every state $\sigma \in \Sigma_i(C)$ has a **unique label** and different states have different labels.

For $0 \leq i < N$, let σ_i and σ_{i+1} be two adjacent states with $\sigma_i \in \Sigma_i(C)$ and $\sigma_{i+1} \in \Sigma_{i+1}(C)$. Let u_{i+1} be the label of the branch in the code trellis that connects state σ_i to state σ_{i+1}. The label u_{i+1} is simply the encoder output bit in the interval from time-i to time-$(i+1)$ and is given by (3.3) or (3.4). For every path $(u_1, u_2, \ldots, u_i) \in L(\sigma_0, \sigma_i)$, the path $(u_1, u_2, \ldots, u_i, u_{i+1})$ obtained by concatenating $(u_1, u_2, \ldots, u_i)$ with the branch u_{i+1} is a path that connects the initial state σ_0 to the state σ_{i+1} through the state σ_i. Hence, $(u_1, u_2, \ldots, u_i, u_{i+1}) \in L(\sigma_0, \sigma_{i+1})$. Then it follows from the definition of a state label that

$$\begin{aligned} l(\sigma_{i+1}) &= (u_1, u_2, \ldots, u_i, u_{i+1}) \cdot H_{i+1}^T \\ &= (u_1, u_2, \ldots, u_i) \cdot H_i^T + u_{i+1} \cdot \boldsymbol{h}_{i+1}^T \\ &= l(\sigma_i) + u_{i+1} \cdot \boldsymbol{h}_{i+1}^T. \end{aligned} \tag{4.11}$$

Eq.(4.11) simply says that given the starting state labeled $l(\sigma_i)$ at time-i and the output code bit u_{i+1} during the interval between time-i and time-$(i+1)$, the destination state labeled $l(\sigma_{i+1})$ at time-$(i+1)$ is uniquely determined.

Now we present a procedure for constructing the N-section trellis diagram for a binary (N, K) linear block code C by state labeling using the parity-check matrix of the code. Let $\boldsymbol{u} = (u_1, u_2, \ldots, u_N)$ be a binary N-tuple. For $0 < i \leq N$, let $p_{0,i}(\boldsymbol{u})$ denote the **prefix** of $\boldsymbol{u}$ that consists of the first i components, i.e.,

$$p_{0,i}(\boldsymbol{u}) = (u_1, u_2, \ldots, u_i). \tag{4.12}$$

Suppose that trellis has been completed up to the i-th section (or time-i). At this point, the rows of the TOGM G in the set $G_i^s = \{\boldsymbol{g}_1^{(i)}, \boldsymbol{g}_2^{(i)}, \ldots, \boldsymbol{g}_{\rho_i}^{(i)}\}$ and their corresponding information bits $a_1^{(i)}, a_2^{(i)}, \ldots, a_{\rho_i}^{(i)}$ uniquely define a state $\sigma_i \in \Sigma_i(C)$. Let

$$\boldsymbol{u} = a_1^{(i)} \cdot \boldsymbol{g}_1^{(i)} + a_2^{(i)} \cdot \boldsymbol{g}_2^{(i)} + \cdots + a_{\rho_i}^{(i)} \cdot \boldsymbol{g}_{\rho_i}^{(i)}.$$

Then $p_{0,i}(\boldsymbol{u})$ is a path connecting the initial state σ_0 to the state σ_i defined by $a_1^{(i)}, a_2^{(i)}, \ldots, a_{\rho_i}^{(i)}$. The label of state σ_i is given by

$$l(\sigma_i) = p_{0,i}(\boldsymbol{u}) \cdot H_i^T.$$

The construction of the $(i+1)$-section of the code trellis is accomplished by taking the following four steps:

(1) Identify the special row $\boldsymbol{g}^*$ (if any) in the submatrix G_i^f and its corresponding information bit a^*. Identify the special row $\boldsymbol{g}^0$ (if any) in the submatrix G_i^s. Form the submatrix G_{i+1}^s by including $\boldsymbol{g}^*$ in G_i^s and excluding $\boldsymbol{g}^0$ from G_i^s.

(2) Determine the set of information bits, $A_{i+1}^s = \{a_1^{(i+1)}, a_2^{(i+1)}, \ldots, a_{\rho_{i+1}}^{(i+1)}\}$, that correspond to the rows in G_{i+1}^s. Define and label the states in $\Sigma_{i+1}(C)$.

(3) For each state $\sigma_i \in \Sigma_i(C)$, form the next output code bit u_{i+1} from either (3.3) (if there is such a row $\boldsymbol{g}^*$ in G_i^f at time-i) or (3.4) (if there is no such row $\boldsymbol{g}^*$ in G_i^f at time-i).

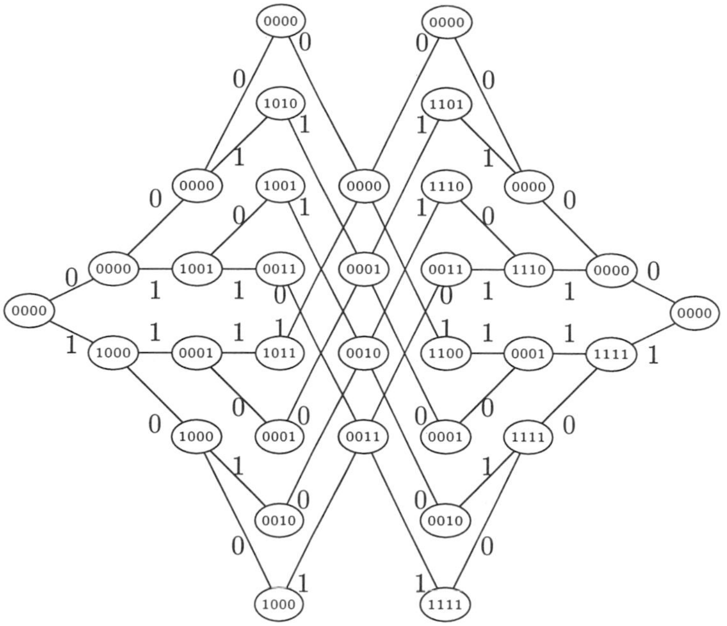

Figure 4.2. 8-section trellis for $(8, 4)$ RM code with state labeling by parity-check matrix.

(4) For each possible value of u_{i+1} (two if computed from (3.3) and one if computed from (3.4)), connect the state σ_i to the state $\sigma_{i+1} \in \Sigma_{i+1}(C)$ with label

$$l(\sigma_{i+1}) = l(\sigma_i) + u_{i+1} \cdot \boldsymbol{h}_{i+1}^T.$$

The connecting branch, denoted $L(\sigma_i, \sigma_{i+1})$, is labeled with u_{i+1}. This completes the construction of the $(i + 1)$-th section of the trellis.

Repeat the above steps until the entire code trellis is constructed.

Example 4.4 Consider the $(8, 4)$ RM code given in Example 3.1. This code is **self dual**. Therefore, a generator matrix is also a parity-check matrix. Suppose we choose the parity-check matrix as follows:

$$H = \begin{bmatrix} 1 & 1 & 1 & 1 & 1 & 1 & 1 & 1 \\ 0 & 0 & 0 & 0 & 1 & 1 & 1 & 1 \\ 0 & 0 & 1 & 1 & 0 & 0 & 1 & 1 \\ 0 & 1 & 0 & 1 & 0 & 1 & 0 & 1 \end{bmatrix}.$$

Using this parity-check matrix for labeling and following the above trellis construction steps, we obtain the 8-section trellis with state labels shown in Figure 4.2. To illustrate the construction process, we assume that the trellis has been completed up to time-3. At this time instant, $G_3^s = \{\boldsymbol{g}_1, \boldsymbol{g}_2, \boldsymbol{g}_3\}$ and $A_3^s = \{a_1, a_2, a_3\}$ are known. The eight states in $\Sigma_3(C)$ are defined by the eight combinations of a_1, a_2 and a_3. These 8 states and their labels are given below:

	states defined by (a_1, a_2, a_3)	state labels
$\sigma_3^{(0)}$	(000)	(0000)
$\sigma_3^{(1)}$	(001)	(1010)
$\sigma_3^{(2)}$	(010)	(1001)
$\sigma_3^{(3)}$	(011)	(0011)
$\sigma_3^{(4)}$	(100)	(1011)
$\sigma_3^{(5)}$	(101)	(0001)
$\sigma_3^{(6)}$	(110)	(0010)
$\sigma_3^{(7)}$	(111)	(1000)

Now we want to construct the 4-th section of the trellis up to time-4. At time-3, from the TOGM G, we find that $\boldsymbol{g}^0 = \boldsymbol{g}_1$ and there is no such row $\boldsymbol{g}^*$ with leading '1' at time-4. Therefore, $G_4^s = \{\boldsymbol{g}_2, \boldsymbol{g}_3\}$ and $A_4^s = \{a_2, a_3\}$. The four states in $\Sigma_4(C)$ at time-4 are defined by the four combinations of a_2 and a_3. The four codewords generated by the rows in G_4^s are:

(a_2, a_3)	paths
$(0, 0)$	$\boldsymbol{u}_0 = (00000000)$
$(0, 1)$	$\boldsymbol{u}_1 = (00111100)$
$(1, 0)$	$\boldsymbol{u}_2 = (01011010)$
$(1, 1)$	$\boldsymbol{u}_3 = (01100110)$

l state σ_0 to the four states, denot

$$(0000),$$
$$) = (0011),$$
$$_{,4}(\boldsymbol{u}_2) = (0101),$$
$$p_{0,4}(\boldsymbol{u}_3) = (0110).$$

is

$$H = \begin{bmatrix} 1 & 1 & 1 & 1 \\ 0 & 0 & 0 & 0 \\ 0 & 0 & 1 & 1 \\ 0 & 1 & 0 & 1 \end{bmatrix}.$$

From $p_{0,4}(\boldsymbol{u}_j)$, with $0 \leq j \leq 3$ and H_4, we can determine the labels for the four states, $\sigma_4^{(0)}, \sigma_4^{(1)}, \sigma_4^{(2)}$ and $\sigma_4^{(3)}$, in $\Sigma_4(C)$ which are given below:

	states defined by (a_2, a_3)	state labels
$\sigma_4^{(0)}$	(00)	(0000)
$\sigma_4^{(1)}$	(01)	(0001)
$\sigma_4^{(2)}$	(10)	(0010)
$\sigma_4^{(3)}$	(11)	(0011)

The four states and their labels are shown in Figure 4.3 at time-4. Now suppose the encoder is in the state $\sigma_3^{(5)}$ with label $l(\sigma_3^{(5)}) = (0001)$ at time-3. Since no such row $\boldsymbol{g}^*$ exists at $i = 3$, the output code bit u_4 is computed from (3.4) as follows:

$$\begin{aligned} u_4 &= 1 \cdot g_{14} + 0 \cdot g_{24} + 1 \cdot g_{34} \\ &= 1 \cdot 1 + 0 \cdot 1 + 1 \cdot 1 \\ &= 0. \end{aligned}$$

Then the state $\sigma_3^{(5)}$ is connected to the state in $\Sigma_4(C)$ with label

$$\begin{aligned} l(\sigma_3^{(5)}) + u_4 \cdot \boldsymbol{h}_4^T &= (0001) + 0 \cdot (1011) \\ &= (0001), \end{aligned}$$

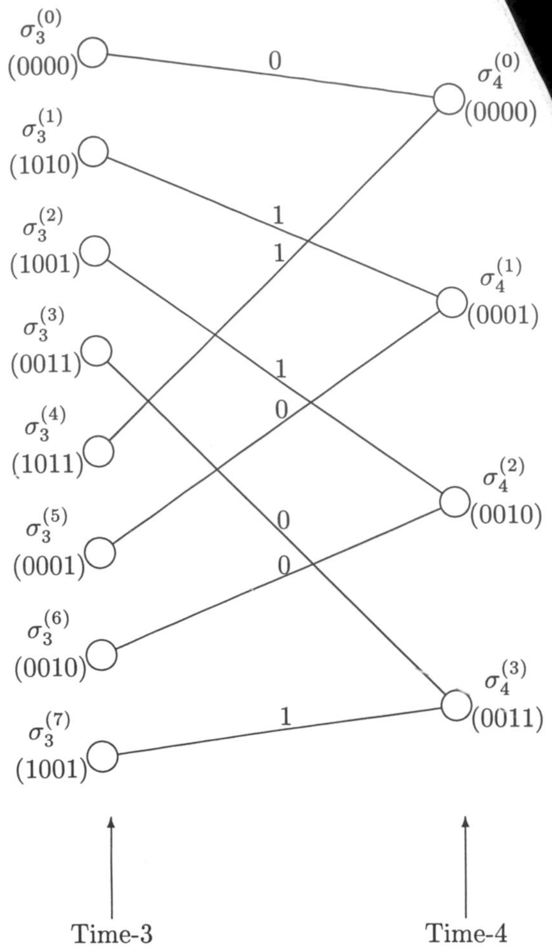

Figure 4.3. State labels at the two ends of the 4-th section of the trellis for $(8, 4)$ RM code.

which is state $\sigma_4^{(1)}$. The connecting branch is labeled with $u_4 = 0$. The connections from the other states in $\Sigma_3(C)$ to the states in $\Sigma_4(C)$ are accomplished in the same manner.

△△

State labeling based on the state defining information sets requires K (or $\rho_{\max}(C)$) bits to label each state of the trellis; however, state labeling based on the parity-check matrix requires $N - K$ bits to label each state of the trellis. Therefore, labeling method-1 is more efficient for codes with $K < N - K$ while labeling method-2 is more efficient for codes with $K > N - K$.

4.3 STRUCTURAL SYMMETRY

Consider a binary (N, K) linear block code C with even length N and TOGM

$$G = \begin{bmatrix} \boldsymbol{g}_1 \\ \boldsymbol{g}_2 \\ \vdots \\ \boldsymbol{g}_K \end{bmatrix} = \begin{bmatrix} g_{11} & g_{12} & \cdots & g_{1N} \\ g_{21} & g_{22} & \cdots & g_{2N} \\ \vdots & \vdots & \ddots & \vdots \\ g_{K1} & g_{K2} & \cdots & g_{KN} \end{bmatrix}.$$

Let T denote the N-section trellis diagram for C. Suppose the TOGM G has the following symmetry property: For each row $\boldsymbol{g}$ in G with $\text{span}(\boldsymbol{g}) = [a, b]$, there exists a row $\boldsymbol{g}'$ in G with $\text{span}(\boldsymbol{g}') = [N + 1 - b, N + 1 - a]$. With this symmetry property in G, we can readily see that for $0 \le i < N/2$, the number of rows in G that are active at time-$(N - i)$ is equal to the number of rows in G that are active at time-i. This implies that

$$|\Sigma_{N-i}(C)| = |\Sigma_i(C)|$$

for $0 \le i < N/2$. We can permute the rows of G such that the resultant matrix, denoted G', is in a **reverse trellis oriented form**:

(1) The trailing '1' of each row appears in a column before the trailing '1' of any row below it.

(2) No two rows have their leading "ones" in the same column.

If we rotate the matrix G' by 180° counter clockwisely, we obtain a matrix G'' in which the i-th row $\boldsymbol{g}''_i$ is simply the $(K + 1 - i)$-th row $\boldsymbol{g}'_{K+1-i}$ of G' in reverse order (the trailing '1' of $\boldsymbol{g}'_{K+1-i}$ becomes the leading '1' of $\boldsymbol{g}''_i$ and the leading '1' of $\boldsymbol{g}'_{K+1-i}$ becomes the trailing '1' of $\boldsymbol{g}''_i$). From the above, we see that G'' and G are structurally identical in the sense that

$$\text{span}(\boldsymbol{g}''_i) = \text{span}(\boldsymbol{g}_i)$$

for $1 \leq i \leq K$. Consequently, the N-section trellis T for C has the following **mirror symmetry** [101]: The last $N/2$ sections of T form the mirror image of the first $N/2$ sections of T (not including the path labels).

Example 4.5 Consider the (8,4) RM code given in Example 4.2 with TOGM

$$G = \begin{bmatrix} \boldsymbol{g}_1 \\ \boldsymbol{g}_2 \\ \boldsymbol{g}_3 \\ \boldsymbol{g}_4 \end{bmatrix} = \begin{bmatrix} 1 & 1 & 1 & 1 & 0 & 0 & 0 & 0 \\ 0 & 1 & 0 & 1 & 1 & 0 & 1 & 0 \\ 0 & 0 & 1 & 1 & 1 & 1 & 0 & 0 \\ 0 & 0 & 0 & 0 & 1 & 1 & 1 & 1 \end{bmatrix}.$$

We find that $\text{span}(\boldsymbol{g}_1) = [1,4]$, $\text{span}(\boldsymbol{g}_4) = [5,8]$, and $\boldsymbol{g}_1$ and $\boldsymbol{g}_4$ are symmetrical with each other. Row $\boldsymbol{g}_2$ has span $[2,7]$ and is symmetrical with itself. Row $\boldsymbol{g}_3$ has span $[3,6]$ and is also symmetrical with itself. Suppose we permute the second and third rows of G. We obtain the following matrix in reverse trellis oriented form:

$$G' = \begin{bmatrix} \boldsymbol{g}'_1 \\ \boldsymbol{g}'_2 \\ \boldsymbol{g}'_3 \\ \boldsymbol{g}'_4 \end{bmatrix} = \begin{bmatrix} 1 & 1 & 1 & 1 & 0 & 0 & 0 & 0 \\ 0 & 0 & 1 & 1 & 1 & 1 & 0 & 0 \\ 0 & 1 & 0 & 1 & 1 & 0 & 1 & 0 \\ 0 & 0 & 0 & 0 & 1 & 1 & 1 & 1 \end{bmatrix}.$$

Rotating G' 180° counter clockwisely, we obtain the following matrix:

$$\begin{bmatrix} \boldsymbol{g}''_1 \\ \boldsymbol{g}''_2 \\ \boldsymbol{g}''_3 \\ \boldsymbol{g}''_4 \end{bmatrix} = \begin{bmatrix} 1 & 1 & 1 & 1 & 0 & 0 & 0 & 0 \\ 0 & 1 & 0 & 1 & 1 & 0 & 1 & 0 \\ 0 & 0 & 1 & 1 & 1 & 1 & 0 & 0 \\ 0 & 0 & 0 & 0 & 1 & 1 & 1 & 1 \end{bmatrix}.$$

We find that G'' and G are in fact identical, not just structurally identical. Therefore, the 8-section trellis T for the (8,4) RM code has mirror symmetry with respect to the boundary location 4, the last four sections form the mirror image of the first four sections as shown in Figures 3.2 and 4.1.

△△

For the case that N is odd, if the TOGM G of a binary (N, K) code C has the mirror symmetry property, then the last $(N-1)/2$ sections of the N-section trellis T for C form the mirror image of the first $(N-1)/2$ sections of T.

The four paths that connect the initial state σ_0 to the four states, denoted $\sigma_4^{(0)}$, $\sigma_4^{(1)}$, $\sigma_4^{(2)}$ and $\sigma_4^{(3)}$, in $\Sigma_4(C)$ are:

$$\begin{aligned} p_{0,4}(\boldsymbol{u}_0) &= (0000), \\ p_{0,4}(\boldsymbol{u}_1) &= (0011), \\ p_{0,4}(\boldsymbol{u}_2) &= (0101), \\ p_{0,4}(\boldsymbol{u}_3) &= (0110). \end{aligned}$$

The submatrix H_4 is

$$H = \begin{bmatrix} 1 & 1 & 1 & 1 \\ 0 & 0 & 0 & 0 \\ 0 & 0 & 1 & 1 \\ 0 & 1 & 0 & 1 \end{bmatrix}.$$

From $p_{0,4}(\boldsymbol{u}_j)$, with $0 \leq j \leq 3$ and H_4, we can determine the labels for the four states, $\sigma_4^{(0)}, \sigma_4^{(1)}, \sigma_4^{(2)}$ and $\sigma_4^{(3)}$, in $\Sigma_4(C)$ which are given below:

	states defined by (a_2, a_3)	state labels
$\sigma_4^{(0)}$	(00)	(0000)
$\sigma_4^{(1)}$	(01)	(0001)
$\sigma_4^{(2)}$	(10)	(0010)
$\sigma_4^{(3)}$	(11)	(0011)

The four states and their labels are shown in Figure 4.3 at time-4. Now suppose the encoder is in the state $\sigma_3^{(5)}$ with label $l(\sigma_3^{(5)}) = (0001)$ at time-3. Since no such row $\boldsymbol{g}^*$ exists at $i = 3$, the output code bit u_4 is computed from (3.4) as follows:

$$\begin{aligned} u_4 &= 1 \cdot g_{14} + 0 \cdot g_{24} + 1 \cdot g_{34} \\ &= 1 \cdot 1 + 0 \cdot 1 + 1 \cdot 1 \\ &= 0. \end{aligned}$$

Then the state $\sigma_3^{(5)}$ is connected to the state in $\Sigma_4(C)$ with label

$$\begin{aligned} l(\sigma_3^{(5)}) + u_4 \cdot \boldsymbol{h}_4^T &= (0001) + 0 \cdot (1011) \\ &= (0001), \end{aligned}$$

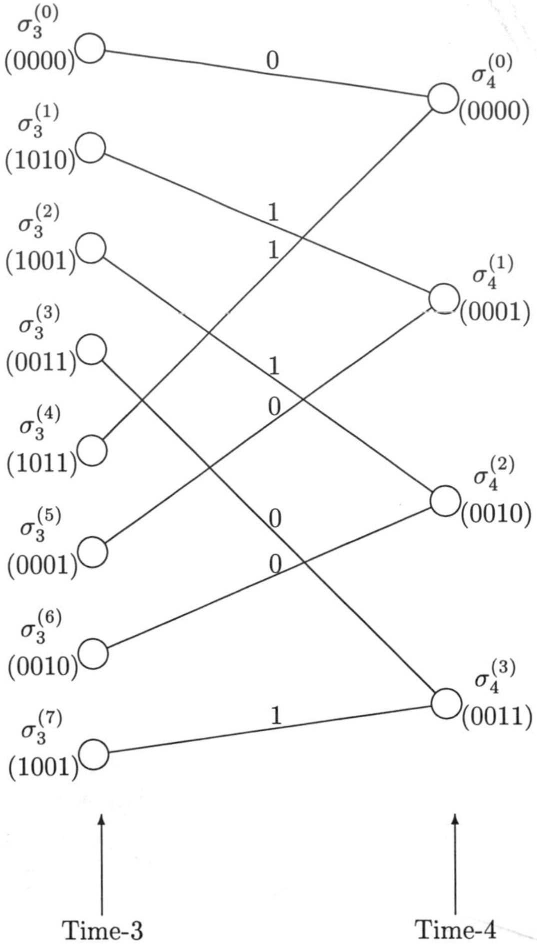

Figure 4.3. State labels at the two ends of the 4-th section of the trellis for $(8, 4)$ RM code.

which is state $\sigma_4^{(1)}$. The connecting branch is labeled with $u_4 = 0$. The connections from the other states in $\Sigma_3(C)$ to the states in $\Sigma_4(C)$ are accomplished in the same manner.

△△

State labeling based on the state defining information sets requires K (or $\rho_{\max}(C)$) bits to label each state of the trellis; however, state labeling based on the parity-check matrix requires $N - K$ bits to label each state of the trellis. Therefore, labeling method-1 is more efficient for codes with $K < N - K$ while labeling method-2 is more efficient for codes with $K > N - K$.

4.3 STRUCTURAL SYMMETRY

Consider a binary (N, K) linear block code C with even length N and TOGM

$$G = \begin{bmatrix} g_1 \\ g_2 \\ \vdots \\ g_K \end{bmatrix} = \begin{bmatrix} g_{11} & g_{12} & \cdots & g_{1N} \\ g_{21} & g_{22} & \cdots & g_{2N} \\ \vdots & \vdots & \ddots & \vdots \\ g_{K1} & g_{K2} & \cdots & g_{KN} \end{bmatrix}.$$

Let T denote the N-section trellis diagram for C. Suppose the TOGM G has the following symmetry property: For each row $\boldsymbol{g}$ in G with $\text{span}(\boldsymbol{g}) = [a, b]$, there exists a row $\boldsymbol{g}'$ in G with $\text{span}(\boldsymbol{g}') = [N+1-b, N+1-a]$. With this symmetry property in G, we can readily see that for $0 \leq i < N/2$, the number of rows in G that are active at time-$(N - i)$ is equal to the number of rows in G that are active at time-i. This implies that

$$|\Sigma_{N-i}(C)| = |\Sigma_i(C)|$$

for $0 \leq i < N/2$. We can permute the rows of G such that the resultant matrix, denoted G', is in a **reverse trellis oriented form**:

(1) The trailing '1' of each row appears in a column before the trailing '1' of any row below it.

(2) No two rows have their leading "ones" in the same column.

If we rotate the matrix G' by 180° counter clockwisely, we obtain a matrix G'' in which the i-th row $\boldsymbol{g}''_i$ is simply the $(K+1-i)$-th row $\boldsymbol{g}'_{K+1-i}$ of G' in reverse order (the trailing '1' of $\boldsymbol{g}'_{K+1-i}$ becomes the leading '1' of $\boldsymbol{g}''_i$ and the leading '1' of $\boldsymbol{g}'_{K+1-i}$ becomes the trailing '1' of $\boldsymbol{g}''_i$). From the above, we see that G'' and G are structurally identical in the sense that

$$\text{span}(\boldsymbol{g}''_i) = \text{span}(\boldsymbol{g}_i)$$

for $1 \leq i \leq K$. Consequently, the N-section trellis T for C has the following **mirror symmetry** [101]: The last $N/2$ sections of T form the mirror image of the first $N/2$ sections of T (not including the path labels).

Example 4.5 Consider the $(8,4)$ RM code given in Example 4.2 with TOGM

$$G = \begin{bmatrix} \boldsymbol{g}_1 \\ \boldsymbol{g}_2 \\ \boldsymbol{g}_3 \\ \boldsymbol{g}_4 \end{bmatrix} = \begin{bmatrix} 1 & 1 & 1 & 1 & 0 & 0 & 0 & 0 \\ 0 & 1 & 0 & 1 & 1 & 0 & 1 & 0 \\ 0 & 0 & 1 & 1 & 1 & 1 & 0 & 0 \\ 0 & 0 & 0 & 0 & 1 & 1 & 1 & 1 \end{bmatrix}.$$

We find that span$(\boldsymbol{g}_1) = [1,4]$, span$(\boldsymbol{g}_4) = [5,8]$, and $\boldsymbol{g}_1$ and $\boldsymbol{g}_4$ are symmetrical with each other. Row $\boldsymbol{g}_2$ has span $[2,7]$ and is symmetrical with itself. Row $\boldsymbol{g}_3$ has span $[3,6]$ and is also symmetrical with itself. Suppose we permute the second and third rows of G. We obtain the following matrix in reverse trellis oriented form:

$$G' = \begin{bmatrix} \boldsymbol{g}'_1 \\ \boldsymbol{g}'_2 \\ \boldsymbol{g}'_3 \\ \boldsymbol{g}'_4 \end{bmatrix} = \begin{bmatrix} 1 & 1 & 1 & 1 & 0 & 0 & 0 & 0 \\ 0 & 0 & 1 & 1 & 1 & 1 & 0 & 0 \\ 0 & 1 & 0 & 1 & 1 & 0 & 1 & 0 \\ 0 & 0 & 0 & 0 & 1 & 1 & 1 & 1 \end{bmatrix}.$$

Rotating G' 180° counter clockwisely, we obtain the following matrix:

$$\begin{bmatrix} \boldsymbol{g}''_1 \\ \boldsymbol{g}''_2 \\ \boldsymbol{g}''_3 \\ \boldsymbol{g}''_4 \end{bmatrix} = \begin{bmatrix} 1 & 1 & 1 & 1 & 0 & 0 & 0 & 0 \\ 0 & 1 & 0 & 1 & 1 & 0 & 1 & 0 \\ 0 & 0 & 1 & 1 & 1 & 1 & 0 & 0 \\ 0 & 0 & 0 & 0 & 1 & 1 & 1 & 1 \end{bmatrix}.$$

We find that G'' and G are in fact identical, not just structurally identical. Therefore, the 8-section trellis T for the $(8,4)$ RM code has mirror symmetry with respect to the boundary location 4, the last four sections form the mirror image of the first four sections as shown in Figures 3.2 and 4.1.

$\triangle\triangle$

For the case that N is odd, if the TOGM G of a binary (N, K) code C has the mirror symmetry property, then the last $(N-1)/2$ sections of the N-section trellis T for C form the mirror image of the first $(N-1)/2$ sections of T.

For the case that $G'' = G$, the N-section trellis T of C has **full mirror symmetry** structure [101]. For N even, the last $N/2$ sections of T in **reverse direction** (the final state σ_f is being regarded as the initial state) is completely identical to the first $N/2$ sections of T (including the path labels). The 8-section trellis of the $(8, 4)$ RM code has full mirror symmetry as shown in Figure 4.1. For N odd, the last $(N-1)/2$ sections of T in reverse direction are completely identical to the first $(N-1)/2$ sections of T (including the path labels).

5 TRELLIS COMPLEXITY

This chapter is devoted to analyzing the complexity of an N-section trellis diagram for an (N, K) linear block code. Trellis complexity is, in general, measured in terms of the state and branch complexities. These two complexities determine the storage and computation requirements of a trellis-based decoding algorithm, such as the Viterbi decoding algorithm. The state complexity of a trellis is measured by its state space dimension profile and the branch complexity is measured by the total number of branches (or edges) in the trellis. In Section 5.1, a simple upper bound on the maximum state space dimension is derived. It is proved that the state complexity of a linear block code is the same as that of its dual code. In Section 5.2, the concepts of a minimal trellis diagram and optimum bit permutation in terms of state complexity are introduced. It is proved that the trellis construction based on a TOGM results in a minimal trellis. In Section 5.3, the branch complexity of an N-section trellis diagram is analyzed. Finally, in Section 5.4, the general structure of N-section trellis diagrams for cyclic codes is given. It it shown that the maximum state space dimension meets the upper bound derived in Section 5.1.

5.1 STATE COMPLEXITY

For a binary (N, K) linear block code C, the state complexity of an N-section bit-level code trellis is measured by its state space dimension profile

$$(\rho_0, \rho_1, \rho_2, \ldots, \rho_N),$$

where for $0 \leq i \leq N$,

$$\rho_i = \log_2 |\Sigma_i(C)|.$$

Let $\rho_{\max}(C)$ denote the maximum among the state space dimensions, i.e.,

$$\rho_{\max}(C) = \max_{0 \leq i \leq N} \rho_i.$$

Using the construction method described in Chapter 3, the state space dimension at time-i is given by (3.7),

$$\rho_i = K - k(C_{0,i}) - k(C_{i,N}),$$

for $0 \leq i \leq N$. Since $k(C_{0,i})$ and $k(C_{i,N})$ are nonnegative, we have

$$\rho_{\max}(C) \leq K. \tag{5.1}$$

However, it follows from (4.6) and the definition and uniqueness of a state label at any time-i (see (4.10)) that

$$|\Sigma_i(C)| \leq 2^{N-K}$$

and

$$\rho_i \leq N - K \tag{5.2}$$

for $0 \leq i \leq N$. Eq.(5.2) implies that

$$\rho_{\max}(C) \leq N - K. \tag{5.3}$$

Combining (5.1) and (5.3), we have the following upper bound on the maximum state complexity:

$$\rho_{\max}(C) \leq \min\{K, N - K\}. \tag{5.4}$$

This bound was first proved by Wolf [109]. In general, this bound is quite loose. However, for cyclic (or shortened cyclic) codes, this bound gives the exact state complexity. For noncyclic codes, tighter upper bounds on $\rho_{\max}(C)$ have been obtained.

If the Viterbi algorithm is applied to the N-section trellis of a code, then the maximum numbers of survivors and path metrics needed to be stored are both $2^{\rho_{\max}(C)}$. Therefore, the parameter $\rho_{\max}(C)$ is a key measure of the decoding complexity (or trellis complexity).

For $0 \leq i \leq \min\{K, N-K\}$, it follows from the structure of a TOGM G that the number of rows in G whose active spans contain the time index i is no greater than i. For $i \geq \max\{K, N-K\}$, since there is one-to-one correspondence between the states in $\Sigma_i(C)$ and cosets in the partition $p_{i,N}(C)/C^{\text{tr}}_{i,N}$,

$$\begin{aligned} \rho_i &= k(p_{i,N}(C)) - k(C^{\text{tr}}_{i,N}) \\ &\leq k(p_{i,N}(C)) \\ &\leq N - i \\ &\leq \min\{K, N-K\}. \end{aligned} \tag{5.5}$$

Therefore, for $0 \leq i \leq N$, we have the following upper bound on ρ_i:

$$\rho_i \leq \min\{i, K, N-K, N-i\}. \tag{5.6}$$

Let $C^\perp$ denote the dual code of C. Then $C^\perp$ is an $(N, N-K)$ linear block code. Consider the N-section trellis diagram for $C^\perp$. For $0 \leq i \leq N$, let $\Sigma_i(C^\perp)$ denote the state space of $C^\perp$ at time-i. Then there is a one-to-one correspondence between the states in $\Sigma_i(C^\perp)$ and the cosets in the partition $p_{0,i}(C^\perp)/C^{\perp,\text{tr}}_{0,i}$ where $C^{\perp,\text{tr}}_{0,i}$ denotes the truncation of $C^\perp_{0,i}$ in the interval $[1, i]$. Therefore, the dimension of $\Sigma_i(C^\perp)$ is given by

$$\rho_i(C^\perp) = k(p_{0,i}(C^\perp)) - k(C^{\perp,\text{tr}}_{0,i}). \tag{5.7}$$

Note that $p_{0,i}(C^\perp)$ is the dual code of $C^{\text{tr}}_{0,i}$ and $C^{\perp,\text{tr}}_{0,i}$ is the dual code of $p_{0,i}(C)$. Therefore,

$$\begin{aligned} k(p_{0,i}(C^\perp)) &= i - k(C^{\text{tr}}_{0,i}) \\ &= i - k(C_{0,i}) \end{aligned} \tag{5.8}$$

and

$$k(C_{0,i}^{\perp,\mathrm{tr}}) = i - k(p_{0,i}(C)). \tag{5.9}$$

It follows from (5.7), (5.8) and (5.9) that

$$\rho_i(C^{\perp}) = K - k(C_{0,i}) - k(C_{i,N}). \tag{5.10}$$

From (3.7) and (5.10), we find that for $0 \leq i \leq N$,

$$\rho_i(C^{\perp}) = \rho_i(C). \tag{5.11}$$

This says that C and its dual code $C^{\perp}$ have the same state complexity.

5.2 MINIMAL TRELLISES

An N-section trellis is said to be **minimal** if the total number of states in the trellis is minimum. A minimal trellis is **unique** within isomorphism [77], i.e., two minimal trellises for the same code are isomorphic (**structurally identical**). The above definition of minimality is commonly used in the literature. However, a more meaningful and useful definition of minimality of a trellis is in terms of its state space dimension profile. An N-section trellis is said to be a **minimum state space dimension trellis** if the state space dimension at each time of the trellis is minimum. A more precise definition is given as follows. Let T be an N-section trellis for an (N, K) code C with state space dimension profile $(\rho_0, \rho_1, \ldots, \rho_N)$. T is said to be minimal if, for any other N-section trellis T' for C with state space dimension profile $(\rho_0', \rho_1', \ldots, \rho_N')$, the following inequality holds:

$$\rho_i \leq \rho_i',$$

for $0 \leq i \leq N$.

Suppose a minimum state space dimension trellis T exists. Then, it is clear that T is a minimal trellis in total number of states. The formulation of state spaces given in Section 3.4 results in a minimum state space dimension trellis (or minimal trellis) for an (N, K) linear block code. This will be proved in Theorem 5.1. This says that a minimum state space dimension trellis T exists for any linear block code C. From the uniqueness of a minimal trellis in total number of states within graph isomorphism, the minimal trellis is a minimum state space dimension trellis. This gives the equivalence between the two definitions of minimality of a trellis for a linear block code.

Theorem 5.1 Let C be a binary (N, K) linear block code with trellis oriented generator matrix G. The N-section trellis T for C constructed based on G is a minimum state space dimension trellis.

Proof: We only need to prove that for $1 \leq i < N$, the number of states, 2^{ρ_i}, at time-i in the trellis T is minimum over all the trellises for C, where

$$\rho_i = K - k(C_{0,i}) - k(C_{i,N}).$$

Let C_i^s denote the linear subcode of C that is spanned by the rows in the submatrix G_i^s of G. Then $|C_i^s| = 2^{\rho_i}$. For two different codewords $\boldsymbol{u}$ and $\boldsymbol{v}$ in C_i^s, it follows from condition (1) of a TOGM that

$$p_{0,i}(\boldsymbol{u}) \neq p_{0,i}(\boldsymbol{v}).$$

This implies that

$$|\{p_{0,i}(\boldsymbol{u}) : \boldsymbol{u} \in C_i^s\}| = 2^{\rho_i}.$$

Suppose there is a trellis T' for C whose number of states at time-i is less than 2^{ρ_i}. Then, there must be two different codewords $\boldsymbol{u}$ and $\boldsymbol{v}$ in C_i^s such that: (1) there are two paths connecting the initial state to a state σ at time-i in T' whose label sequences are $p_{0,i}(\boldsymbol{u})$ and $p_{0,i}(\boldsymbol{v})$, respectively; and (2) there is a path connecting the state σ to the final state in T' whose label sequence is $p_{i,N}(\boldsymbol{u})$. Without loss of generality, we assume $\boldsymbol{u} \neq \boldsymbol{0}$.

Let $\boldsymbol{u}'$ denote the binary N-tuple such that $p_{0,i}(\boldsymbol{u}') = p_{0,i}(\boldsymbol{v})$ and $p_{i,N}(\boldsymbol{u}') = p_{i,N}(\boldsymbol{u})$. Since $\boldsymbol{u}'$ is a path in T' connecting the initial state to the final state, it follows from condition (4) of Definition 3.1 of an N-section trellis for a linear block code that $\boldsymbol{u}'$ is a codeword in C. Therefore, $\boldsymbol{u} + \boldsymbol{u}' \in C$. Note that $p_{0,i}(\boldsymbol{u}+\boldsymbol{u}') = p_{0,i}(\boldsymbol{u}) + p_{0,i}(\boldsymbol{v}) \neq \boldsymbol{0}$ and $p_{i,N}(\boldsymbol{u}+\boldsymbol{u}') = p_{i,N}(\boldsymbol{u}) + p_{i,N}(\boldsymbol{u}') = \boldsymbol{0}$. This implies that $\boldsymbol{u} + \boldsymbol{u}' \in C_{0,i}$. There are three cases to be considered:

(1) Suppose $\boldsymbol{u}' \in C_{0,i}$. This implies that $\boldsymbol{u} \in C_{0,i}$ which is a contradiction to the hypothesis that $\boldsymbol{u} \in C_i^s$ and $\boldsymbol{u} \neq \boldsymbol{0}$.

(2) Suppose $\boldsymbol{u}' \in C_i^s$. This implies that $\boldsymbol{u} + \boldsymbol{u}' \in C_i^s$. Since $\boldsymbol{u} + \boldsymbol{u}' \neq \boldsymbol{0}$, $\boldsymbol{u} + \boldsymbol{u}'$ can not be in both C_i^s and $C_{0,i}$. This results in a contradiction.

(3) Suppose $\boldsymbol{u}' \in C_{i,N}$. This implies that $\boldsymbol{u} \in C_{0,i} \oplus C_{i,N}$, which is not possible.

Therefore, $\boldsymbol{u} + \boldsymbol{u}'$ can not be a codeword in C. This results in a contradiction to our earlier hypothesis that there exists an N-section trellis T' for C whose number of states at time-i is less than 2^{ρ_i}. Therefore, the hypothesis is invalid and 2^{ρ_i} gives the minimum number of states at time-i for $1 \leq i < N$.

△△

It follows from Theorem 5.1 that Eq.(3.7) gives the minimum state space dimension ρ_i with $0 \leq i \leq N$ for an N-section trellis for an (N, K) linear block code. From (3.7), we see that the state space dimension ρ_i at time-i depends on the dimensions of the past and future codes, $C_{0,i}$ and $C_{i,N}$. For a given code C, $k(C_{0,i})$ and $k(C_{i,N})$ are fixed.

Given an (N, K) linear block code C, a permutation of the orders of the bit (or symbol) positions results in an equivalent code C' with the same weight distribution. Different permutations of the bit positions may result in different dimensions, $k(C_{0,i})$ and $k(C_{i,N})$, of the past and future subcodes, $C_{0,i}$ and $C_{i,N}$, and hence different state space dimensions ρ_i at time-i. A permutation that yields the smallest state space dimension at every time of the code trellis is called an **optimum** permutation (or bit ordering). It is clear that an optimum permutation reduces the state complexity and is often desirable. Optimum permutation is **hard to find**, however optimum permutations for RM codes are known [45] but they are unknown for other classes of codes.

5.3 BRANCH COMPLEXITY

The branch complexity of an N-section trellis diagram for an (N, K) linear block code C is defined as the total number of branches in the trellis. This complexity determines the number of additions required in a trellis-based decoding algorithm to decode a received sequence.

Consider the N-section trellis diagram T for C which is constructed based on the rules and procedures described in Chapters 3 and 4. Recall that at time-i with $0 \leq i < N$, there are two branches diverging from a state in $\Sigma_i(C)$ if there exists a row $\boldsymbol{g}^*$ in G_i^f; and there is only one branch diverging from a state in $\Sigma_i(C)$ if there exists no such row $\boldsymbol{g}^*$ in G_i^f. Define

$$I_i(\boldsymbol{g}^*) \triangleq \begin{cases} 1, & \text{if } \boldsymbol{g}^* \notin G_i^f, \\ 2, & \text{if } \boldsymbol{g}^* \in G_i^f. \end{cases} \tag{5.12}$$

Let E denote the total number of branches in the N-section trellis T. Then

$$\begin{aligned} E &= \sum_{i=0}^{N-1} |\Sigma_i(C)| \cdot I_i(\boldsymbol{g}^*) \\ &= \sum_{i=0}^{N-1} 2^{\rho_i} \cdot I_i(\boldsymbol{g}^*). \end{aligned} \tag{5.13}$$

Example 5.1 Again we consider the $(8,4)$ linear block code given in Example 3.1. From Table 4.1, we find that

$$I_0(\boldsymbol{g}^*) = I_1(\boldsymbol{g}^*) = I_2(\boldsymbol{g}^*) = I_4(\boldsymbol{g}^*) = 2$$

and

$$I_3(\boldsymbol{g}^*) = I_5(\boldsymbol{g}^*) = I_6(\boldsymbol{g}^*) = I_7(\boldsymbol{g}^*) = 1.$$

The state space dimension profile of the 8-section trellis for the code is (0, 1, 2, 3, 2, 3, 2, 1, 0). From (5.13), we have

$$\begin{aligned} E &= 2^0 \cdot 2 + 2^1 \cdot 2 + 2^2 \cdot 2 + 2^3 \cdot 1 + 2^2 \cdot 2 + 2^3 \cdot 1 + 2^2 \cdot 1 + 2^1 \cdot 1 \\ &= 2 + 4 + 8 + 8 + 8 + 8 + 4 + 2 \\ &= 44. \end{aligned}$$

△△

An N-section trellis diagram for an (N, K) linear block code is said to be a **minimal branch (or edge) trellis diagram** if it has the smallest branch complexity. A minimal trellis diagram has the smallest branch complexity [69]. Branch complexity also depends on the bit ordering of a code. Proper permutation of the bit positions of a code may result in a significant reduction in branch complexity. A permutation that results in minimal branch complexity is called an optimum permutation. From (5.13), we can readily see that a permutation which minimizes each product in the summation of (5.13) is a minimal edge trellis diagram. A good permutation in terms of branch complexity should have the following property: for $0 \le i < N$, when ρ_i is large, $I_i(\boldsymbol{g}^*)$ should be equal to 1.

5.4 TRELLIS STRUCTURE OF CYCLIC CODES

Consider an (N, K) cyclic code C over GF(2) generated by the following polynomial [62],

$$\boldsymbol{g}(X) = 1 + g_1 X + g_2 X^2 + \cdots + g_{N-K-1} X^{N-K-1} + X^{N-K},$$

where for $1 \leq i < N - K$, $g_i \in \text{GF}(2)$. A generator matrix for this cyclic code is given by

$$G = \begin{bmatrix} 1 & g_1 & g_2 & \cdots & \cdots & g_{N-K-1} & 1 & 0 & 0 & \cdots & 0 \\ 0 & 1 & g_1 & g_2 & \cdots & \cdots & g_{N-K-1} & 1 & 0 & \cdots & 0 \\ & & & & & \ddots & & & & & \\ 0 & 0 & \cdots & 0 & 1 & g_1 & g_2 & \cdots & \cdots & g_{N-K-1} & 1 \end{bmatrix}. \quad (5.14)$$

The K rows of G are simply the K cyclic shifts of the first row. This generator matrix has the following properties:

(1) It is in trellis oriented form.

(2) For $1 \leq i \leq K$, the span of the i-th row $\boldsymbol{g}_i$ is

$$\text{span}(\boldsymbol{g}_i) = [i, N - K + i] .$$

(3) The active spans of all the rows have the same length, $N - K$.

Now we consider the bit-level trellis structure for this (N, K) cyclic code. There are two cases to be considered: $K > N - K$ and $K \leq N - K$. Consider the case for which $K > N - K$. For $1 \leq i \leq N - K$, the number of rows whose active spans contain the time index i is i. These rows are simply the first i rows. For $N - K < i \leq K$, the number of rows whose active spans contain the time index i is $N - K$. For $K < i \leq N$, the number of rows whose active spans contain the time index-i is $N - i$. Since $i > K$,

$$N - i < N - K.$$

From the above analysis, we see that the maximum state space dimension is $\rho_{\max}(C) = N - K$ and the state space profile is

$$(0, 1, \ldots, N - K - 1, N - K, \ldots, N - K, N - K - 1, \ldots, 1, 0).$$

Now consider the second case for which $K \leq N - K$. For $1 \leq i \leq K$, the number of rows whose active spans contain the time index i is i (the first i rows). For $K \leq i \leq N - K$, the number of rows whose active spans contain the time index i is K. For $N - K < i \leq N$, the number of rows whose active spans contain i is $N - i$. Since $i > N - K$, $N - i < K$. From the above analysis, we find that the maximum state space dimension is

$$\rho_{\max}(C) = K,$$

and the state space dimension profile is

$$(0, 1, \ldots, K-1, K, \ldots, K, K-1, \ldots, 1, 0).$$

Putting the results of the above two cases together, we conclude that for an (N, K) cyclic code, the maximum state space dimension is

$$\rho_{\max}(C) = \min\{K, N - K\}.$$

This is to say that a code in cyclic form has the worst state complexity (i.e., it meets the upper bound on the state complexity).

The generator polynomial $\boldsymbol{g}(X)$ of an (N, K) binary cyclic code C divides $X^N + 1$ [62]. Let

$$X^N + 1 = \boldsymbol{g}(X)\boldsymbol{h}(X).$$

Then $\boldsymbol{h}(X)$ is a polynomial of degree K of the following form:

$$\boldsymbol{h}(X) = 1 + h_1 X + h_2 X^2 + \cdots + h_{K-1} X^{K-1} + X^K$$

with $h_i \in \mathrm{GF}(2)$ for $1 \leq i < K$. This polynomial is called the parity check polynomial. The dual code $C^\perp$ of C is an $(N, N-K)$ cyclic code with generator polynomial

$$X^K \boldsymbol{h}(X^{-1}) = 1 + h_{K-1} X + \cdots + h_1 X^{K-1} + X^K.$$

The generator matrix for the dual code $C^\perp$ is

$$H = \begin{bmatrix} 1 \; h_{K-1} \; h_{K-2} & \cdots & \cdots \; \cdots\cdots \; \cdots & 1 & 0 \cdots\cdots 0 \\ 0 \;\; 1 \;\; h_{K-1} \; h_{K-2} & \cdots & \cdots\cdots \; \cdots & \cdots & 1 \;\; 0 \cdots 0 \\ 0 \;\; 0 \;\; 1 \;\; h_{K-1} \; h_{K-2} & \cdots\cdots & \cdots & \cdots & \cdots \; 1 \cdots 0 \\ \vdots & & & & \vdots \\ 0 \;\; 0 \;\; \cdots & \cdots & \cdots \; 0 \;\; 1 \;\; h_{K-1} \; h_{K-2} & \cdots\cdots & \cdots 1 \end{bmatrix} \quad (5.15)$$

which is in trellis oriented form. The trellis structure for $C^{\perp}$ can be analyzed in the same manner as for C.

The trellis of a cyclic code has mirror symmetry, i.e., the right-half and left-half of the trellis with respect to the center are structurally identical.

To reduce the state complexity of a cyclic code, a permutation of the bit position is needed [45, 46].

The branch complexity of the N-section trellis diagram T constructed based on the TOGM G given by (5.14) can be evaluated easily. First we note that

$$I_i(\boldsymbol{g}^*) = \begin{cases} 2, & \text{for } 0 \leq i < K, \\ 1, & \text{otherwise.} \end{cases}$$

Suppose $K < N - K$. Then branch complexity of the trellis T is

$$\begin{aligned} E &= \sum_{i=0}^{K-1} 2^i \cdot 2 + (N-2K) \cdot 2^K + \sum_{i=0}^{K-1} 2^{K-i} \cdot 1 \\ &= 2 \cdot (2^K - 1) + (N-2K) \cdot 2^K + 2 \cdot (2^K - 1) \\ &= (N - 2K + 4) \cdot 2^K - 4. \end{aligned} \tag{5.16}$$

For $K \geq N - K$, we have

$$\begin{aligned} E &= \sum_{i=0}^{N-K-1} 2^i \cdot 2 + 2 \cdot (2K - N) \cdot 2^{N-K} + \sum_{i=0}^{N-K-1} 2^{N-K-i} \\ &= (4K - 2N + 4) \cdot 2^{N-K} - 4. \end{aligned} \tag{5.17}$$

5.5 TRELLISES FOR NONBINARY LINEAR BLOCK CODES

The methods for constructing trellises for binary linear block codes can be generalized for constructing trellises for nonbinary linear block codes with symbols from GF(q) in a straightforward manner. The symbol-level N-section trellis diagram for an (N, K) linear block code C over GF(q) has the following basic properties: (1) every branch is labeled with a code symbol from GF(q); (2) except for the initial state, every state has at least one, but no more than q, incoming branches; (3) except for the final state, every state has at least one, but no more than q, outgoing branches; and (4) the initial state has no incoming branch and the final state has no outgoing branch. In the definition of a trellis oriented generator matrix, the leading "1" and trailing "1" of a row are replaced by leading and trailing "nonzero components", respectively. The maximum state space dimension $\rho_{\max}(C)$ of the minimal N-section trellis for C is upper bounded by

$$\rho_{\max}(C) \leq \min\{K, N-K\},$$

and the maximum number of states, $|\Sigma(C)|_{\max}$, is upper bounded by

$$|\Sigma(C)|_{\max} \leq q^{\min\{K,N-K\}}.$$

For Reed-Solomon (RS) codes over GF(q), the above equalities hold, i.e.,

$$\rho_{\max}(C) = \min\{K, N-K\},$$

and

$$|\Sigma(C)|_{\max} = q^{\min\{K,N-K\}}.$$

6 TRELLIS SECTIONALIZATION

So far, we have only considered bit-level N-section trellis diagrams for linear block codes of length N. In a bit-level trellis diagram, every time instant in the encoding interval $\Gamma = \{0, 1, 2, \ldots, N\}$ is a section boundary location and every branch represents a code bit. It is possible to sectionalize a bit-level trellis with section boundary locations at selected instants in the encoding interval Γ. This sectionalization results in a trellis in which a branch may represent multiple code bits and two adjacent states may be connected by multiple branches. Proper sectionalization may result in useful trellis structural properties and allow us to devise efficient trellis-based decoding algorithms. This chapter is devoted in analyzing sectionalized trellis diagrams for linear block codes. Section 6.1 presents the concepts and rules for trellis sectionalization. In Section 6.2, the branch complexity and state connectivity are analyzed and expressed in terms of the dimensions of codes related to the code being considered. In Section 6.3, construction of a sectionalized trellis diagram for a linear block code based on the trellis oriented generator matrix is presented. Section 6.4 studies the parallel structure of a sectionalized trellis diagram.

6.1 SECTIONALIZATION OF A CODE TRELLIS

For a positive integer $L \leq N$, let

$$U \triangleq \{h_0, h_1, h_2, \ldots, h_L\} \tag{6.1}$$

be a subset of $L+1$ time instants in the encoding interval $\Gamma = \{0, 1, 2, \ldots, N\}$ for an (N, K) linear block code C with $0 = h_0 < h_1 < h_2 < \cdots < h_L = N$. An L-section trellis diagram for C with **section boundaries** at the locations (time instants) in U, denoted $T(U)$, can be obtained from the N-section trellis T by: (1) deleting every state in $\Sigma_h(C)$ for $h \in \{0, 1, \ldots, N\} \setminus U$ and every branch to or from a deleted state, and (2) for $1 \leq j \leq L$, connecting a state $\sigma \in \Sigma_{h_{j-1}}$ to a state $\sigma' \in \Sigma_{h_j}$ by a branch with label α if and only if there is a path with label α from state σ to state σ' in the N-section trellis T. In an L-section trellis with boundary locations in $U = \{h_0, h_1, \ldots, h_L\}$, a branch from a state in $\Sigma_{h_{j-1}}(C)$ to a state in $\Sigma_{h_j}(C)$ represents $(h_j - h_{j-1})$ code symbols.

A **subgraph** of a trellis diagram is called a **subtrellis**. The subtrellis of $T(U)$ which consists of the state space $\Sigma_{h_{j-1}}(C)$ at time-h_{j-1}, state space $\Sigma_{h_j}(C)$ at time-h_j, and all the branches between the states in $\Sigma_{h_{j-1}}(C)$ and $\Sigma_{h_j}(C)$, is called the j-th **section** of $T(U)$. The length of the j-th section is $h_j - h_{j-1}$. If the lengths of all the sections of an L-section code trellis $T(U)$ are the same, $T(U)$ is said to be **uniformly sectionalized**. In an L-section trellis diagram with $L < N$, two adjacent states may be connected by multiple branches (called **parallel branches**) with different labels.

Let $\rho_{h_j} \triangleq \log_2 |\Sigma_{h_j}(C)|$ be the dimension of the state space $\Sigma_{h_j}(C)$ at time-h_j. Then

$$(\rho_0, \rho_{h_1}, \rho_{h_2}, \ldots, \rho_{h_{L-1}}, \rho_N)$$

is the state space dimension profile of the L-section code trellis $T(U)$ with section boundary set $U = \{0, h_1, h_2, \ldots, h_{L-1}, N\}$. From (3.7), we have

$$\rho_{h_j} = K - k(C_{0,h_j}) - k(C_{h_j,N}) \,. \tag{6.2}$$

If we choose the section boundaries, $U = \{h_0, h_1, \ldots, h_L\}$, at the places where $\rho_{h_1}, \rho_{h_2}, \ldots, \rho_{h_{L-1}}$ are small, then the resultant L-section code trellis $T(U)$ has

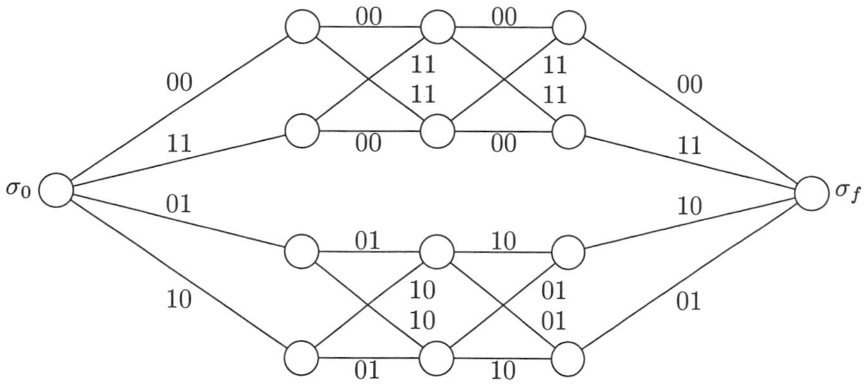

Figure 6.1. A 4-section trellis for the $(8,4)$ RM code.

a small state space dimension profile. The maximum state space dimension is

$$\rho_{L,\max}(C) \triangleq \max_{0 \le j \le L} \rho_{h_j}. \tag{6.3}$$

In implementing a trellis-based decoder, such as a Viterbi decoder, a proper choice of the section boundary locations results in a significant reduction in decoding complexity.

Example 6.1 Again, we consider the $(8,4)$ RM code given in Example 3.1 whose 8-section trellis diagram is shown in Figure 3.2 (or Figure 4.1). Suppose we choose $L = 4$ and the section boundary set $U = \{0,2,4,6,8\}$. Following the above rules of sectionalization of a code trellis, we obtain a uniform 4-section trellis diagram as shown in Figure 6.1, in which every branch represents 2 code bits. The state space dimension profile for this 4-section trellis is $(0,2,2,2,0)$ and the maximum state space dimension is $\rho_{4,\max}(C) = 2$. It is a 4-section, 4-state code trellis. From Figure 6.1, we notice that the right-half of the trellis (the third and fourth sections) is the **mirror image** of the left-half of the trellis (the first and second sections). This **mirror symmetry** allows **bidirectional** decoding. Furthermore, the code trellis consists of two **parallel** and **structurally** identical (**isomorphic**) subtrellises without cross connections between them. This **parallel structure** allows us to devise two identical 2-state (Viterbi) decoders to process the trellis in parallel. The mirror symmetry and parallel structure not only simplify the decoding complexity but also speed up the decoding process. For large code trellises, these structural properties are very important in IC (integrated circuit) implementations.

△△

An L-section trellis diagram obtained from a minimal N-section trellis diagram by deleting states and branches at places other than the section boundary locations is minimal, i.e., it is a minimal L-section trellis diagram for a given section boundary set U.

6.2 BRANCH COMPLEXITY AND STATE CONNECTIVITY

Consider the j-th section of a minimal L-section trellis diagram $T(U)$ with section boundary set $U = \{h_0, h_1, \ldots, h_L\}$ for an (N, K) linear code C. The boundaries of this section are h_{j-1} and h_j. Each branch in this section is labeled with $h_j - h_{j-1}$ bits. Let σ and σ' be two adjacent states in the state spaces $\Sigma_{h_{j-1}}(C)$ and $\Sigma_{h_j}(C)$, respectively. Let $L(\sigma, \sigma')$ denote the set of parallel branches connecting σ and σ'. Let $L(\sigma_0, \sigma)$ denote the set of paths connecting the initial state σ_0 to the state σ. Sometimes, it is convenient to regard the parallel branches, $L(\sigma, \sigma')$, between two states as a single branch. This single branch is called a **composite branch**, and $L(\sigma, \sigma')$ is called a **composite branch label**.

The branch complexity of a trellis section is measured by: (1) the size of a composite branch; (2) the number of distinct composite branches in the trellis section; and (3) the total number of composite branches in the trellis section. The overall branch complexity of the trellis is then the sum of the trellis section branch complexities. These three branch complexity parameters can be expressed in terms of the dimensions of C_{h_{j-1},h_j}, $C_{0,h_{j-1}}$, $C_{h_j,N}$, and $p_{h_{j-1},h_j}(C)$ which can be obtained from the TOGM G of C.

Let $\sigma_{h_{j-1}}$ and σ_{h_j} be two adjacent states with $\sigma_{h_{j-1}} \in \Sigma_{h_{j-1}}(C)$ and $\sigma_{h_j} \in \Sigma_{h_j}(C)$. It has been shown in Section 3.7 that the parallel branches in $L(\sigma_{h_{j-1}}, \sigma_{h_j})$ form a coset in the partition

$$p_{h_{j-1},h_j}(C)/C^{\text{tr}}_{h_{j-1},h_j}.$$

Therefore, the number of parallel branches between two adjacent states $\sigma_{h_{j-1}}$ and σ_{h_j} in the j-th section of $T(U)$ is

$$\left|L(\sigma_{h_{j-1}}, \sigma_{h_j})\right| = 2^{k(C_{h_{j-1},h_j})}. \tag{6.4}$$

In Section 3.7, we have also shown that

$$L(\sigma_0, \sigma_{h_{j-1}}) \in p_{0,h_{j-1}}(C)/C^{\text{tr}}_{0,h_{j-1}} \tag{6.5}$$

and

$$L(\sigma_0, \sigma_{h_j}) \in p_{0,h_j}(C)/C^{\text{tr}}_{0,h_j}. \tag{6.6}$$

If $\sigma_{h_{j-1}}$ and σ_{h_j} are adjacent, then

$$\begin{aligned} L(\sigma_0, \sigma_{h_{j-1}}) \circ L(\sigma_{h_{j-1}}, \sigma_{h_j}) &\triangleq \\ \{\boldsymbol{u} \circ \boldsymbol{v} : \boldsymbol{u} \in L(\sigma_0, \sigma_{h_{j-1}}) &\text{ and } \boldsymbol{v} \in L(\sigma_{h_{j-1}}, \sigma_{h_j})\}, \end{aligned} \tag{6.7}$$

is the set of paths in $T(U)$ that diverge from the initial state σ_0, converge at the state $\sigma_{h_{j-1}}$, and then transverse the parallel branches in $L(\sigma_{h_{j-1}}, \sigma_{h_j})$ to the state σ_{h_j} as shown in Figure 6.2. It has also been shown in Section 3.7 that $L(\sigma_0, \sigma_{h_{j-1}}) \circ L(\sigma_{h_{j-1}}, \sigma_{h_j})$ is a subcode of a coset in the partition $p_{0,h_j}(C)/C^{\text{tr}}_{0,h_j}$. Let $\Sigma_{h_{j-1}}(\sigma_{h_j})$ denote the set of states in the state space $\Sigma_{h_{j-1}}(C)$ that are adjacent to state σ_{h_j} as shown in Figure 6.2. Then

$$\bigcup_{\sigma_{h_{j-1}} \in \Sigma_{h_{j-1}}(\sigma_{h_j})} L(\sigma_0, \sigma_{h_{j-1}}) \circ L(\sigma_{h_{j-1}}, \sigma_{h_j}) \triangleq L(\sigma_0, \sigma_{h_j}) \tag{6.8}$$

is a coset in the partition $p_{0,h_j}(C)/C^{\text{tr}}_{0,h_j}$.

Note that the dimensions of $L(\sigma_0, \sigma_{h_{j-1}})$, $L(\sigma_{h_{j-1}}, \sigma_{h_j})$ and $L(\sigma_0, \sigma_{h_j})$ are $k(C^{\text{tr}}_{0,h_{j-1}})$, $k(C^{\text{tr}}_{h_{j-1},h_j})$ and $k(C^{\text{tr}}_{0,h_j})$, respectively. It follows from (6.8) that the number of states in $\Sigma_{h_{j-1}}(\sigma_{h_j})$ is given by

$$\begin{aligned} \left|\Sigma_{h_{j-1}}(\sigma_{h_j})\right| &= 2^{k(C^{\text{tr}}_{0,h_j}) - k(C^{\text{tr}}_{0,h_{j-1}}) - k(C^{\text{tr}}_{h_{j-1},h_j})} \\ &= 2^{k(C_{0,h_j}) - k(C_{0,h_{j-1}}) - k(C_{h_{j-1},h_j})}. \end{aligned} \tag{6.9}$$

This implies that the number of composite branches converging into a state $\sigma_{h_j} \in \Sigma_{h_j}(C)$, called the **incoming degree** of σ_{h_j}, is given by

$$\deg(\sigma_{h_j})_{\text{in}} \triangleq 2^{k(C_{0,h_j}) - k(C_{0,h_{j-1}}) - k(C_{h_{j-1},h_j})}. \tag{6.10}$$

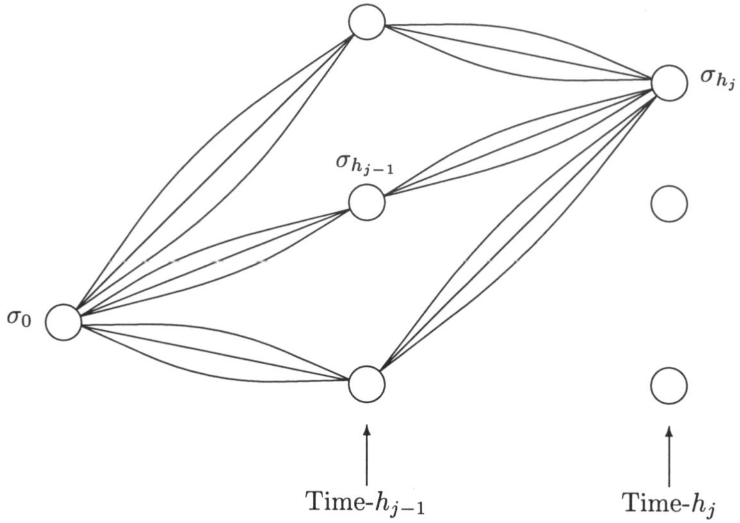

Figure 6.2. State connectivity.

This number is a measure of the **state connectivity** of the sectionalized code trellis $T(U)$. In an IC implementation of a Viterbi decoder, this number is known as the **radix number**, a key design parameter.

Since each composite branch $L(\sigma_{h_{j-1}}, \sigma_{h_j})$ in the j-th section of $T(U)$ is a coset in the partition $p_{h_{j-1},h_j}(C)/C^{\text{tr}}_{h_{j-1},h_j}$, the number of distinct composite branches in the j-th section of $T(U)$ is

$$2^{k(p_{h_{j-1},h_j}(C))-k(C_{h_{j-1},h_j})}. \tag{6.11}$$

It follows from (6.2) and (6.10) that the total number of composite branches in the j-th section of $T(U)$ is given by

$$2^{K-k(C_{0,h_{j-1}})-k(C_{h_j,N})-k(C_{h_{j-1},h_j})}. \tag{6.12}$$

From (6.11) and (6.12), we find that each distinct composite branch appears in the j-th section of $T(U)$

$$2^{K-k(C_{0,h_{j-1}})-k(C_{h_j,N})-k(p_{h_{j-1},h_j}(C))} \tag{6.13}$$

times.

From (6.2) (with h_j replaced by h_{j-1}) and (6.12), we can compute the number of composite branches diverging from a state $\sigma_{h_{j-1}} \in \Sigma_{h_{j-1}}(C)$ at time-h_{j-1} as

$$2^{k(C_{h_{j-1},N})-k(C_{h_j,N})-k(C_{h_{j-1},h_j})}, \tag{6.14}$$

which is called the **outgoing degree** of $\sigma_{h_{j-1}}$, denoted $\deg(\sigma_{h_{j-1}})_{\text{out}}$. Equations (6.4), (6.10)–(6.12) and (6.14) give the branch complexity and state connectivity of the j-th section of a minimal L-section trellis $T(U)$ with section boundary locations in $U = \{h_0, h_1, \ldots, h_L\}$.

Define

$$\delta_j \triangleq \log_2 \ \deg(\sigma_{h_j})_{\text{in}}$$

with $\sigma_{h_j} \in \Sigma_{h_j}(C)$. The ordered sequence $(\delta_1, \delta_2, \ldots, \delta_L)$ is called the **converging branch dimension profile (CBDP)**. Define

$$\lambda_j \triangleq \log_2 \ \deg(\sigma_{h_j})_{\text{out}}.$$

The ordered sequence $(\lambda_0, \lambda_1, \ldots, \lambda_{L-1})$ is called the **diverging branch dimension profile (DBDP)**.

Let M_j denote the total number of composite branches in the j-th trellis section (given by (6.12)) and define

$$\beta_j \triangleq \log_2 M_j.$$

The ordered sequence

$$(\beta_1, \beta_2, \ldots, \beta_L)$$

is called the **branch complexity profile (BCP)**. The branch complexity of the minimal L-section trellis $T(U)$ in terms of the total number of branches in the trellis is given by

$$B = \sum_{j=1}^{L} 2^{\beta_j} \cdot 2^{k(C_{h_{j-1},h_j})}. \tag{6.15}$$

Since each branch in the j-th section of $T(U)$ represents $h_j - h_{j-1}$ code bits, it is equivalent to $h_j - h_{j-1}$ branches in the bit-level N-section trellis T for the code. Therefore, the branch complexity in terms of bit branches is given by

$$E = \sum_{j=1}^{L} (h_j - h_{j-1}) \cdot 2^{\beta_j} \cdot 2^{k(C_{h_{j-1},h_j})}. \tag{6.16}$$

If the section boundary is $U = \{0, 1, 2, \ldots, N\}$, then (6.16) gives the branch (or edge) complexity of the bit-level N-section minimal trellis of the code.

6.3 A PROCEDURE FOR CONSTRUCTING A MINIMAL L-SECTION TRELLIS

A minimal L-section trellis diagram for an (N, K) linear block code C can be constructed directly from the TOGM G. Let $U = \{h_0, h_1, \ldots, h_L\}$ be the set of section boundary locations with $h_0 = 0 < h_1 < \cdots < h_{L-1} < h_L = N$. Again the construction of the minimal L-section trellis diagram $T(U)$ with section boundary set U is carried out serially, section by section. Suppose $T(U)$ has been constructed up to the j-th section (i.e., up to time-h_j) with $1 \leq j < L$. Now we begin to construct the $(j+1)$-th section from time-h_j to time-h_{j+1}. Partition the rows of the TOGM G into three disjoint subsets, $G^p_{h_j}$, $G^f_{h_j}$, and $G^s_{h_j}$ as follows (also shown in Figure 6.3):

(1) $G^p_{h_j}$ consists of those rows in G whose spans are contained in the interval $[1, h_j]$.

(2) $G^f_{h_j}$ consists of those rows in G whose spans are contained in the interval $[h_j + 1, N]$.

(3) $G^s_{h_j}$ consists of those rows in G whose active spans contain the time index h_j.

It is clear that $G^p_{h_j}$ and $G^f_{h_j}$ generate the past and future codes, C_{0,h_j} and $C_{h_j,N}$, respectively. Let $A^s_{h_j}$ be the set of information bits that correspond to the rows of $G^s_{h_j}$. Then the bits in $A^s_{h_j}$ define the state space $\Sigma_{h_j}(C)$ at time-h_j. That is, for any binary $\rho_{h_j} = |A^s_{h_j}|$-tuple, which represents values of information bits in $A^s_{h_j}$, there is a corresponding state in $\Sigma_{h_j}(C)$.

To determine the composite branches between states in $\Sigma_{h_j}(C)$ and states in $\Sigma_{h_{j+1}}(C)$ and the parallel branches between two adjacent states, we further partition the rows of $G^f_{h_j}$ into three subsets, $G^{f,p}_{h_j,h_{j+1}}$, $G^{f,s}_{h_j,h_{j+1}}$ and $G^f_{h_{j+1}}$ as follows (see Figure 6.3):

(1) $G^{f,p}_{h_j,h_{j+1}}$ consists of those rows of $G^f_{h_j}$ whose spans are contained in the interval $[h_j + 1, h_{j+1}]$.

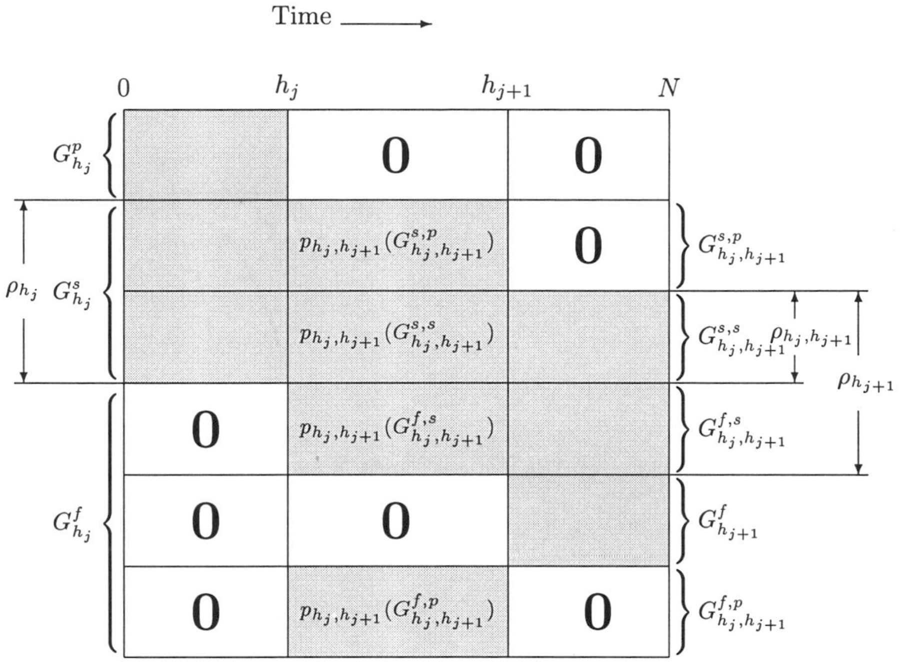

Figure 6.3. Partition of the TOGM G.

(2) $G^{f,s}_{h_j,h_{j+1}}$ consists of those rows of $G^f_{h_j}$ whose active spans contain the time index h_{j+1}.

(3) The remaining rows in $G^f_{h_j}$ form $G^f_{h_{j+1}}$.

Let $A^{f,p}_{h_j,h_{j+1}}$ and $A^{f,s}_{h_j,h_{j+1}}$ denote the subsets of information bits that correspond to the rows in $G^{f,p}_{h_j,h_{j+1}}$ and $G^{f,s}_{h_j,h_{j+1}}$, respectively. Then the information bits in these two sets may be regarded as the current input information bits. These input bits together with the state of the encoder at time-h_j uniquely determine the output code bits between time-h_j and time-h_{j+1}. Note that the information bits in $A^{f,p}_{h_j,h_{j+1}}$ only affect the output during the interval between time-h_j and time-h_{j+1}. Therefore, they determine the parallel branches between two adjacent states. The information bits in $A^{f,s}_{h_j,h_{j+1}}$ determine the diverging composite branches from a state in $\Sigma_{h_j}(C)$.

Let $p_{h_j,h_{j+1}}(G^{f,p}_{h_j,h_{j+1}})$, $p_{h_j,h_{j+1}}(G^{f,s}_{h_j,h_{j+1}})$ and $p_{h_j,h_{j+1}}(G^{s}_{h_j})$ denote the truncations of $G^{f,p}_{h_j,h_{j+1}}$, $G^{f,s}_{h_j,h_{j+1}}$ and $G^{s}_{h_j}$ from time-h_j to time-h_{j+1}. The rows in $p_{h_j,h_{j+1}}(G^{f,p}_{h_j,h_{j+1}})$ span the code $C^{\text{tr}}_{h_j,h_{j+1}}$, and the rows in $p_{h_j,h_{j+1}}(G^{f,p}_{h_j,h_{j+1}})$, $p_{h_j,h_{j+1}}(G^{f,s}_{h_j,h_{j+1}})$ and $p_{h_j,h_{j+1}}(G^{s}_{h_j})$ span the truncated code $p_{h_j,h_{j+1}}(C)$. Then every composite branch between a state $\sigma_{h_j} \in \Sigma_{h_j}(C)$ and a state $\sigma_{h_{j+1}} \in \Sigma_{h_{j+1}}(C)$ is a coset in the partition $p_{h_j,h_{j+1}}(C)/C^{\text{tr}}_{h_j,h_{j+1}}$. The number of parallel branches between two adjacent states is therefore $|C^{\text{tr}}_{h_j,h_{j+1}}|$.

Let σ_{h_j} be the state at time-h_j defined by the binary ρ_{h_j}-tuple formed by the binary information bits in the set $A^{s}_{h_j}$,

$$(a_1^{(j)}, a_2^{(j)}, \ldots, a_{\rho_{h_j}}^{(j)}), \tag{6.17}$$

where $\rho_{h_j} = \log_2 |\Sigma_{h_j}(C)| = |G^{s}_{h_j}|$. Let $\{\boldsymbol{g}_1^{(j)}, \boldsymbol{g}_2^{(j)}, \ldots, \boldsymbol{g}_{\rho_{h_j}}^{(j)}\}$ denote the rows in $G^{s}_{h_j}$. Then

$$\boldsymbol{u} = a_1^{(j)} \cdot \boldsymbol{g}_1^{(j)} + a_2^{(j)} \cdot \boldsymbol{g}_2^{(j)} + \cdots + a_{\rho_{h_j}}^{(j)} \cdot \boldsymbol{g}_{\rho_{h_j}}^{(j)} \tag{6.18}$$

is a codeword (or path) passing through the state σ_{h_j} at time-h_j. Let $p_{h_j,h_{j+1}}(\boldsymbol{u})$ denote the branch on $\boldsymbol{u}$ from time-h_j to time-h_{j+1}. Let $B_{h_j,h_{j+1}}$ denote the code of length $h_{j+1} - h_j$ generated by $p_{h_j,h_{j+1}}(G^{f,s}_{h_j,h_{j+1}})$. Then for every vector $\boldsymbol{b} \in B_{h_j,h_{j+1}}$, there is a composite branch diverging from the state σ_{h_j} which consists of the following parallel branches,

$$\{p_{h_j,h_{j+1}}(\boldsymbol{u}) + \boldsymbol{b} + \boldsymbol{x} : \boldsymbol{x} \in C^{\text{tr}}_{h_j,h_{j+1}}\}. \tag{6.19}$$

Therefore, the number of composite branches diverging from the state σ_{h_j} is $|B_{h_j,h_{j+1}}|$.

Next, we analyze the state transitions. Partition the matrix $G^{s}_{h_j}$ into two submatrices, $G^{s,p}_{h_j,h_{j+1}}$ and $G^{s,s}_{h_j,h_{j+1}}$, where

(1) $G^{s,p}_{h_j,h_{j+1}}$ consists of those rows in $G^{s}_{h_j}$ whose active spans do not contain the time index h_{j+1}, and

(2) $G^{s,s}_{h_j,h_{j+1}}$ consists of those rows in $G^{s}_{h_j}$ whose active spans contain the time index h_{j+1}.

Let $A^{s,p}_{h_j,h_{j+1}}$ and $A^{s,s}_{h_j,h_{j+1}}$ denote the sets of information bits corresponding to $G^{s,p}_{h_j,h_{j+1}}$ and $G^{s,s}_{h_j,h_{j+1}}$, respectively. Then the set of information bits that defines the state space $\Sigma_{h_{j+1}}(C)$ at time-h_{j+1} is given by

$$\begin{aligned} A^s_{h_{j+1}} &= (A^s_{h_j} \setminus A^{s,p}_{h_j,h_{j+1}}) \cup A^{f,s}_{h_j,h_{j+1}} \\ &= A^{s,s}_{h_j,h_{j+1}} \cup A^{f,s}_{h_j,h_{j+1}}. \end{aligned} \tag{6.20}$$

Therefore, the state transitions from time-h_j to time-h_{j+1} are completely specified by the change from $A^s_{h_j}$ to $A^s_{h_{j+1}}$.

Define

$$\rho_{h_j,h_{j+1}} \triangleq |A^{s,s}_{h_j,h_{j+1}}| = |G^{s,s}_{h_j,h_{j+1}}|. \tag{6.21}$$

Then it follows from (6.20) (also Figure 6.3) that

$$|A^{s,p}_{h_j,h_{j+1}}| = \rho_{h_j} - \rho_{h_j,h_{j+1}}, \tag{6.22}$$

and

$$|A^{f,s}_{h_j,h_{j+1}}| = \rho_{h_{j+1}} - \rho_{h_j,h_{j+1}}. \tag{6.23}$$

Let $\boldsymbol{a}^0_{h_j}$ be the binary $(\rho_{h_j} - \rho_{h_j,h_{j+1}})$-tuple formed by the binary information bits in the set $A^{s,p}_{h_j,h_{j+1}}$, $\boldsymbol{a}_{h_j}$ be the binary $\rho_{h_j,h_{j+1}}$-tuple formed by the information bits in the set $A^{s,s}_{h_j,h_{j+1}}$, and $\boldsymbol{a}^*_{h_j}$ be the binary $(\rho_{h_{j+1}} - \rho_{h_j,h_{j+1}})$-tuple formed by the information bits in the set $A^{f,s}_{h_j,h_{j+1}}$. Then $(\boldsymbol{a}^0_{h_j}, \boldsymbol{a}_{h_j})$ defines a state, denoted $\sigma(\boldsymbol{a}^0_{h_j}, \boldsymbol{a}_{h_j})$, in the state space $\Sigma_{h_j}(C)$ at time-h_j, and $(\boldsymbol{a}_{h_j}, \boldsymbol{a}^*_{h_j})$ defines a state, denoted $\sigma(\boldsymbol{a}_{h_j}, \boldsymbol{a}^*_{h_j})$, in the state space $\Sigma_{h_{j+1}}(C)$ at time-h_{j+1}. State $\sigma(\boldsymbol{a}^0_{h_j}, \boldsymbol{a}_{h_j})$ is adjacent to state $\sigma(\boldsymbol{a}_{h_j}, \boldsymbol{a}^*_{h_j})$. The composite branch that connects these two states in the trellis is given by (6.19) with

$$\boldsymbol{b} = \boldsymbol{a}^*_{h_j} \cdot p_{h_j,h_{j+1}}(G^{f,s}_{h_j,h_{j+1}}). \tag{6.24}$$

Note that these two states share a common $\boldsymbol{a}_{h_j}$. For $\boldsymbol{a}'_{h_j} \neq \boldsymbol{a}_{h_j}$, the state $\sigma(\boldsymbol{a}^0_{h_j}, \boldsymbol{a}_{h_j})$ at time-h_j is not adjacent to the state $\sigma(\boldsymbol{a}'_{h_j}, \boldsymbol{a}^*_{h_j})$ at time-h_{j+1}. Therefore, from each state $\sigma(\boldsymbol{a}^0_{h_j}, \boldsymbol{a}_{h_j})$ in $\Sigma_{h_j}(C)$, there are $2^{\rho_{h_{j+1}} - \rho_{h_j,h_{j+1}}}$ possible transitions to the states $\sigma(\boldsymbol{a}_{h_j}, \boldsymbol{a}^*_{h_j})$ in $\Sigma_{h_{j+1}}(C)$ with $\boldsymbol{a}^*_{h_j} \in \{0,1\}^{\rho_{h_{j+1}} - \rho_{h_j,h_{j+1}}}$. This completely specifies the state transitions from time-h_j to time-h_{j+1}.

State labeling based on the state defining information set $A^s_{h_j}$ with $0 \leq j \leq L$ is exactly the same as described in Section 4.1. We may use either a K-tuple or a $\rho_{L,\max}(C)$-tuple to label a state. In general, $\rho_{L,\max}(C)$ is much smaller than K, and hence using a $\rho_{L,\max}(C)$-bit label for a state is more efficient.

Suppose an L-section trellis diagram $T(U)$ with boundary location set $U = \{h_0, h_1, \ldots, h_L\}$ has been constructed up to time-h_j. The trellis section from time-h_j to time-h_{j+1} can be constructed by the following procedure:

(1) Form and label the states in $\Sigma_{h_{j+1}}(C)$ based on $A^s_{h_{j+1}}$.

(2) For each state in $\Sigma_{h_j}(C)$, determine its transitions to the states in $\Sigma_{h_{j+1}}(C)$ based on the state transition rules described above.

(3) For two adjacent states, $\sigma(\boldsymbol{a}^0_{h_j}, \boldsymbol{a}_{h_j})$ and $\sigma(\boldsymbol{a}_{h_j}, \boldsymbol{a}^*_{h_j})$, at time-$h_j$ and time-h_{j+1}, connect them by parallel branches given by (6.19).

Repeat the above procedure until the L-section trellis $T(U)$ is completed.

Example 6.2 Consider the $(8,4,4)$ RM code with the TOGM G as

$$G = \begin{bmatrix} 1 & 1 & 1 & 1 & 0 & 0 & 0 & 0 \\ 0 & 1 & 0 & 1 & 1 & 0 & 1 & 0 \\ 0 & 0 & 1 & 1 & 1 & 1 & 0 & 0 \\ 0 & 0 & 0 & 0 & 1 & 1 & 1 & 1 \end{bmatrix}.$$

Suppose we want to construct a 4-section trellis for this code with boundary locations in $U = \{0,2,4,6,8\}$. First, we find from G that the state space dimension profile is $(0,2,2,2,0)$. Therefore, $\rho_{4,\max}(C) = 2$. We also find that

$$C^{tr}_{0,2} = C^{tr}_{2,4} = C^{tr}_{4,6} = C^{tr}_{6,8} = \{\mathbf{0}\},$$

$$p_{0,2}(G^{f,s}_{0,2}) = \begin{bmatrix} 1 & 1 \\ 0 & 1 \end{bmatrix}, \qquad p_{2,4}(G^{f,s}_{2,4}) = \begin{bmatrix} 1 & 1 \end{bmatrix},$$

$$p_{4,6}(G^{f,s}_{4,6}) = \begin{bmatrix} 1 & 1 \end{bmatrix}, \qquad p_{6,8}(G^{f,s}_{6,8}) = \emptyset.$$

The state defining information sets at the boundary locations, $0, 2, 4, 6$, and 8, are given in Table 6.1. Following the constructing procedure given above, we obtain the 4-section trellis diagram for the $(8,4,4)$ RM code as shown in Figure 6.4, where the states are labeled based on the information defining sets using $\rho_{4,\max}(C)$-tuples with $\rho_{4,\max}(C) = 2$.

$\triangle\triangle$

Table 6.1. State defining information sets for a 4-section trellis for the $(8,4,4)$ RM code.

	Time $\longrightarrow$				
	0	2	4	6	8
$A^s_{h_j}$	$\emptyset$	$\{a_1, a_2\}$	$\{a_2, a_3\}$	$\{a_2, a_4\}$	$\emptyset$
$A^{s,p}_{h_j,h_{j+1}}$	$\emptyset$	$\{a_1\}$	$\{a_3\}$	$\{a_2, a_4\}$	$\emptyset$
$A^{f,s}_{h_j,h_{j+1}}$	$\{a_1, a_2\}$	$\{a_3\}$	$\{a_4\}$	$\emptyset$	$\emptyset$

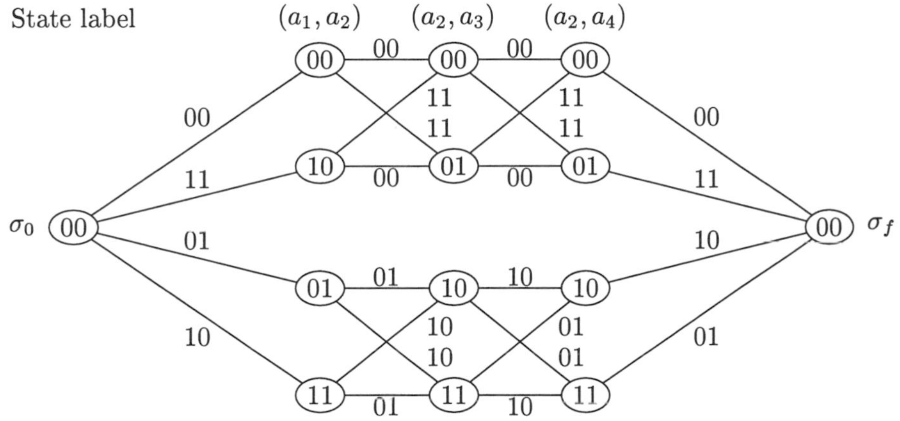

Figure 6.4. A minimal 4-section trellis diagram for the $(8,4,4)$ RM code with 2-bit state labels.

Construction of $T(U)$ can be achieved by using the state labeling based on a parity-check matrix H for C [101]. Let

$$H_{h_j,h_{j+1}} = [\boldsymbol{h}_{h_j+1}, \boldsymbol{h}_{h_j+2}, \ldots, \boldsymbol{h}_{h_{j+1}}] \tag{6.25}$$

denote the submatrix of the parity-check matrix H of C that consists of columns from $\boldsymbol{h}_{h_j+1}$ to $\boldsymbol{h}_{h_{j+1}}$. Let $l(\sigma_{h_j})$ be the label for the state σ_{h_j}. Then the composite branch given by (6.19) connects the state σ_{h_j} to the state $\sigma_{h_{j+1}} \in \Sigma_{h_{j+1}}(C)$ at time-h_{j+1} that is labeled by

$$l(\sigma_{h_{j+1}}) = l(\sigma_{h_j}) + (p_{h_j,h_{j+1}}(\boldsymbol{u}) + \boldsymbol{b}) \cdot H^T_{h_j,h_{j+1}}. \tag{6.26}$$

Eq.(6.26) gives the connection from a starting state σ_{h_j} at time-h_j to a destination state $\sigma_{h_{j+1}}$ at time-h_{j+1}.

To facilitate the construction of the $(j+1)$-th trellis section between time-h_j and time-h_{j+1}, we form a table, denoted Q_{h_j}, at the completion of the construction of the j-th section. Each entry in Q_{h_j} is a triplet,

$$(\boldsymbol{f}, l(\sigma), \boldsymbol{c}).$$

The first component $\boldsymbol{f}$ is a binary ρ_{h_j}-tuple formed by a specific combination of the ρ_{h_j} information bits in $A^s_{h_j}$. This ρ_{h_j}-tuple defines a specific state σ in $\Sigma_{h_j}(C)$. The second component $l(\sigma)$ is simply the label of state σ. The third component is given by

$$\begin{aligned} \boldsymbol{c} &= p_{h_j,h_{j+1}}(\boldsymbol{u}) \\ &= p_{h_j,h_{j+1}}(\boldsymbol{f} \cdot G^s_{h_j}), \end{aligned} \tag{6.27}$$

where $\boldsymbol{u} = \boldsymbol{f} \cdot G^s_{h_j}$ is given by (6.18).

Construction of the $(j+1)$-th section of $T(U)$ is carried out as follows:

(1) Form $C^{\rm tr}_{h_j,h_{j+1}}$ and $B_{h_j,h_{j+1}}$.

(2) For every entry $(\boldsymbol{f}, l(\sigma), \boldsymbol{c}) \in Q_{h_j}$ and every $\boldsymbol{b} \in B_{h_j,h_{j+1}}$, form the composite branch,

$$\{\boldsymbol{b} + \boldsymbol{c} + \boldsymbol{x} : \boldsymbol{x} \in C^{\rm tr}_{h_j,h_{j+1}}\}, \tag{6.28}$$

(3) For every starting state $\sigma \in \Sigma_{h_j}(C)$, and $\boldsymbol{b} \in B_{h_j,h_{j+1}}$, the destination state $\sigma' \in \Sigma_{h_{j+1}}(C)$ is labeled with

$$l(\sigma') = l(\sigma) + (\boldsymbol{b} + \boldsymbol{c}) \cdot H^T_{h_j,h_{j+1}}. \tag{6.29}$$

Repeat the above process until the L-section trellis $T(U)$ is completed.

The trellis construction procedures presented in Section 4.1, Section 4.2, and this section only provide the general steps of construction. A detail and efficient trellis construction procedure is given in Appendix A.

6.4 PARALLEL STRUCTURE

Consider the trellis section from time-h_j to time-h_{j+1}. For a given $\rho_{h_j,h_{j+1}}$-tuple $\boldsymbol{a}_{h_j}$, define the following two sets of states at time-h_j and time-h_{j+1}, respectively:

$$S_L(\boldsymbol{a}_{h_j}) \triangleq \left\{\sigma(\boldsymbol{a}^0_{h_j}, \boldsymbol{a}_{h_j}) : \boldsymbol{a}^0_{h_j} \in \{0,1\}^{\rho_{h_j} - \rho_{h_j,h_{j+1}}}\right\} \tag{6.30}$$

and

$$S_R(\boldsymbol{a}_{h_j}) \triangleq \left\{\sigma(\boldsymbol{a}_{h_j}, \boldsymbol{a}^*_{h_j}) : \boldsymbol{a}^*_{h_j} \in \{0,1\}^{\rho_{h_{j+1}} - \rho_{h_j,h_{j+1}}}\right\}. \tag{6.31}$$

Then $S_L(\boldsymbol{a}_{h_j})$ and $S_R(\boldsymbol{a}_{h_j})$ are subspaces of the state spaces, $\Sigma_{h_j}(C)$ and $\Sigma_{h_{j+1}}(C)$, respectively. From the state transition analysis given in the previous section, we observe the following:

(1) Every state in $S_L(\boldsymbol{a}_{h_j})$ is adjacent to all the $2^{\rho_{h_{j+1}} - \rho_{h_j,h_{j+1}}}$ states in $S_R(\boldsymbol{a}_{h_j})$ and is not adjacent to any other state in $\Sigma_{h_{j+1}}(C)$.

(2) Every state in $S_R(\boldsymbol{a}_{h_j})$ is adjacent from all the $2^{\rho_{h_j} - \rho_{h_j,h_{j+1}}}$ states in $S_L(\boldsymbol{a}_{h_j})$ and is not adjacent from any other state in $\Sigma_{h_j}(C)$.

Therefore, the states in $S_L(\boldsymbol{a}_{h_j})$, the states in $S_R(\boldsymbol{a}_{h_j})$, and the composite branches connecting them form a **completely connected subtrellis** (known as a **complete bipartite graph**). Since there are $2^{\rho_{h_j,h_{j+1}}}$ possible $\rho_{h_j,h_{j+1}}$-tuple $\boldsymbol{a}_{h_j}$, there are $2^{\rho_{h_j,h_{j+1}}}$ such completely connected subtrellises in the trellis section time-h_j and time-h_{j+1}. All these subtrellises are **structurally identical (isomorphic)**, and there are no cross connections between them. These subtrellises are called **parallel components**. The parallel structure of a trellis section is shown in Figure 6.5.

It follows from the definition of $\rho_{h_j,h_{j+1}}$ given by (6.21) and the partition of the TOGM G shown in Figure 6.3 that

$$\rho_{h_j,h_{j+1}} = K - k(C_{h_j,N}) - k(C_{0,h_{j+1}}) + k(C_{h_j,h_{j+1}}). \tag{6.32}$$

Therefore, the total number of parallel components in the trellis section from time-h_j to time-h_{j+1} is

$$2^{K - k(C_{h_j,N}) - k(C_{0,h_{j+1}}) + k(C_{h_j,h_{j+1}})}. \tag{6.33}$$

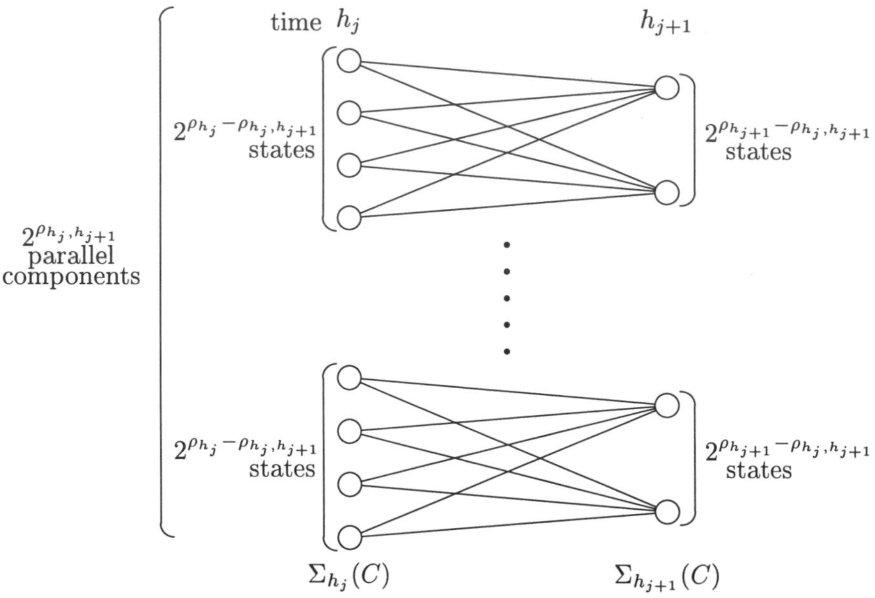

Figure 6.5. Parallel structure in a trellis section.

From (6.2) and (6.33), we find that the numbers of states in $S_L(\boldsymbol{a}_{h_j})$ and $S_R(\boldsymbol{a}_{h_j})$ are:

$$2^{k(C_{0,h_{j+1}})-k(C_{0,h_j})-k(C_{h_j,h_{j+1}})} \tag{6.34}$$

and

$$2^{k(C_{h_j,N})-k(C_{h_{j+1},N})-k(C_{h_j,h_{j+1}})}, \tag{6.35}$$

respectively. Equations (6.30) to (6.35) completely characterize the parallel components in the $(j+1)$-th trellis section of $T(U)$.

Consider the 4-section trellis diagram for the $(8,4)$ RM code shown in Figure 6.4. There are two parallel components in both the second and third sections of the trellis. Each component has two states at each end.

Analysis of the parallel structure of a sectionalized code trellis is presented in the Appendix A.

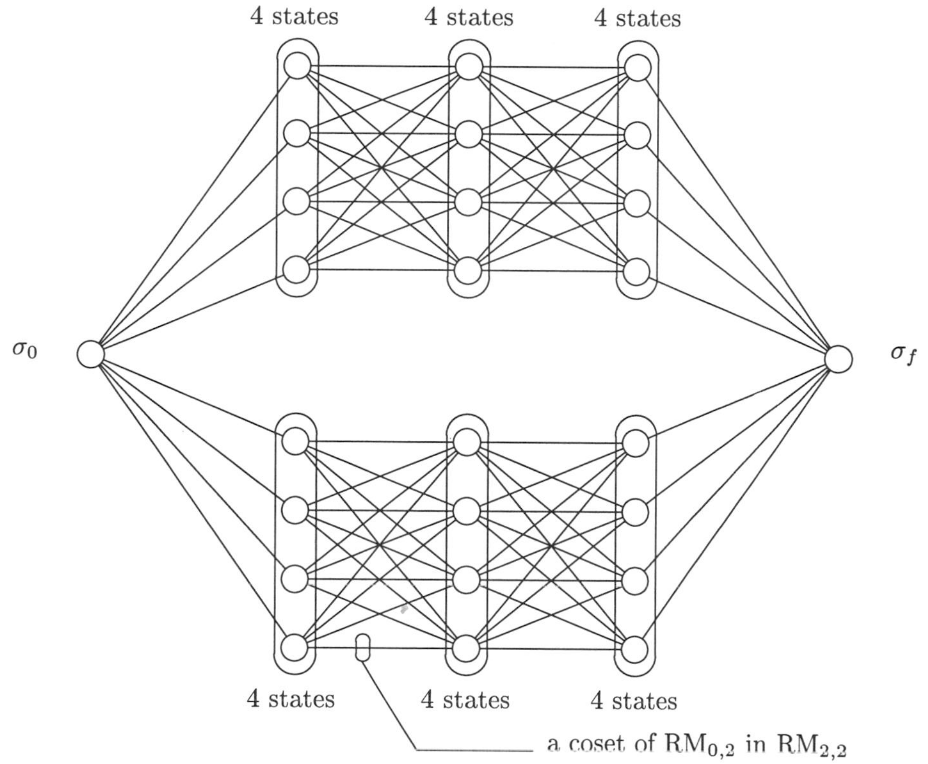

Figure 6.6. The 4-section minimal trellis diagram $T(\{0,4,8,12,16\})$ for $\mathrm{RM}_{2,4}$.

Example 6.3 Consider the $\mathrm{RM}_{2,4}$ code which is a (16,11) code. The 4-section minimal trellis diagram $T(U)$ with section boundary locations in $U = \{0,4,8,12,16\}$ is depicted in Figure 6.6. There are two parallel and structural identical components in both the second and third sections of the trellis, and each component is a complete bipartite graph. Each component has 4 states at each end. Each state at boundary location 8 has 4 composite branches diverging from it. For $1 \leq j \leq 4$, $p_{4(j-1),4j}(\mathrm{RM}_{2,4}) = \mathrm{RM}_{2,2}$ and $C^{\mathrm{tr}}_{4(j-1),4j} = \mathrm{RM}_{0,2}$. Therefore, there are 2 parallel branches between any two adjacent states whose 4-bit label sequences form a coset of $\mathrm{RM}_{0,2}$ in $\mathrm{RM}_{2,2}$. In both the second and third sections of the trellis, each coset in $\mathrm{RM}_{2,2}/\mathrm{RM}_{0,2}$ appears 4 times as the composite branch label. In fact, the entire trellis consists of two 4-section parallel and structurally identical subtrellises without cross connections between them. The maximum state complexity is 8. Therefore it is possible to devise two identical trellis-based decoders, say Viterbi decoders, to process the entire trellis in parallel. This not only simplifies the decoding complexity but also speeds up the decoding process.

△△

The parallel components in a trellis section can be partitioned into groups of the same size in such a way that [44]: (1) two parallel components in the same group are **identical up to path labeling**, and (2) if there is a common label sequence in two parallel components, then they are in the same group. Since all the parallel components in a group have the same label set, in a trellis-based decoding algorithm, only the metrics of the branches in one of the parallel components need to be computed. This results in a reduction of branch metric computation.

Let $\bar{C}_{h_j,h_{j+1}}$ denote the subcode of C that consists of those codewords whose components from the $(h_j + 1)$-th bit to the h_{j+1}-th bit positions are all zero. Then each group consists of $2^{\lambda_{h_j,h_{j+1}}(C)}$ identical parallel components where [44]

$$\lambda_{h_j,h_{j+1}}(C) = k(\bar{C}_{h_j,h_{j+1}}) - k(p_{0,h_j}(C_{0,h_{j+1}}) \cap p_{0,h_j}(\bar{C}_{h_j,h_{j+1}})) - k(C_{h_{j+1},N}). \tag{6.36}$$

Each parallel component can be decomposed into subtrellises with simple **uniform structure** [44] as shown in Figure 6.7. Consider a parallel component, denoted Λ. Let $\Sigma_{h_j}(\Lambda)$ and $\Sigma_{h_{j+1}}(\Lambda)$ denote the state spaces at two ends of Λ. We first partition $\Sigma_{h_j}(\Lambda)$ into blocks, called **left U-blocks**, which satisfy the following condition:

(B1) If two states σ_{h_j} and σ'_{h_j} in $\Sigma_{h_j}(\Lambda)$ are in the same left U-block, then they have the same set of diverging composite branches, i.e.,

$$\begin{aligned} &\{L(\sigma_{h_j}, \sigma_{h_{j+1}}) : \sigma_{h_{j+1}} \in \Sigma_{h_{j+1}}(\Lambda)\} \\ &= \{L(\sigma'_{h_j}, \sigma_{h_{j+1}}) : \sigma_{h_{j+1}} \in \Sigma_{h_{j+1}}(\Lambda)\} \end{aligned} \tag{6.37}$$

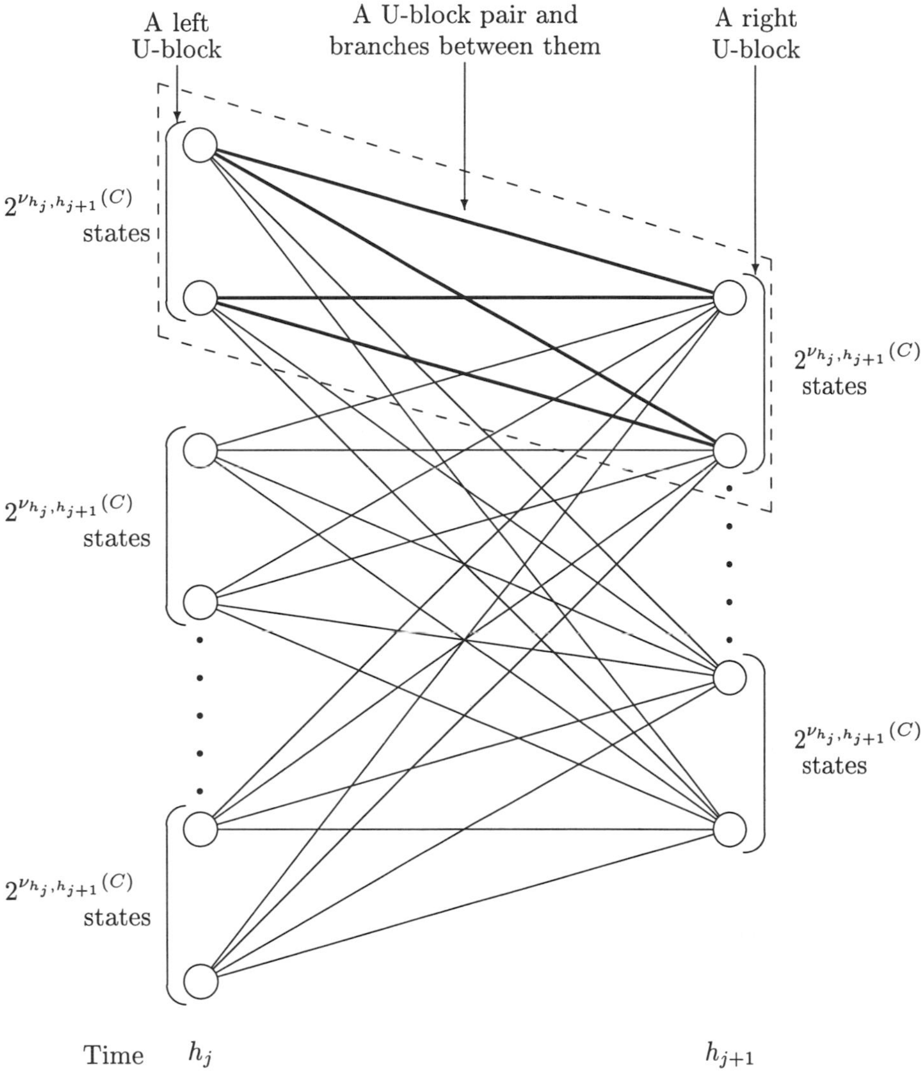

Figure 6.7. Partition of a parallel component.

and otherwise

$$\{L(\sigma_{h_j}, \sigma_{h_{j+1}}) : \sigma_{h_{j+1}} \in \Sigma_{h_{j+1}}(\Lambda)\} \\ \cap \{L(\sigma'_{h_j}, \sigma_{h_{j+1}}) : \sigma_{h_{j+1}} \in \Sigma_{h_{j+1}}(\Lambda)\} = \emptyset. \tag{6.38}$$

We next partition $\Sigma_{h_{j+1}}(\Lambda)$ into blocks, called **right U-blocks**, which satisfy the following conditions:

(B2) If two states $\sigma_{h_{j+1}}$ and $\sigma'_{h_{j+1}}$ in $\Sigma_{h_{j+1}}(\Lambda)$ are in the same right U-block, then they have the same set of converging composite branches, i.e.,

$$\{L(\sigma_{h_j}, \sigma_{h_{j+1}}) : \sigma_{h_j} \in \Sigma_{h_j}(\Lambda)\} = \{L(\sigma_{h_j}, \sigma'_{h_{j+1}}) : \sigma_{h_j} \in \Sigma_{h_j}(\Lambda)\} \tag{6.39}$$

and otherwise

$$\{L(\sigma_{h_j}, \sigma_{h_{j+1}}) : \sigma_{h_j} \in \Sigma_{h_j}(\Lambda)\} \cap \{L(\sigma_{h_j}, \sigma'_{h_{j+1}}) : \sigma_{h_j} \in \Sigma_{h_j}(\Lambda)\} = \emptyset. \tag{6.40}$$

Each left U-block (or right U-block) consists of $2^{\nu_{h_j,h_{j+1}}(C)}$ states [44], where

$$\nu_{h_j,h_{j+1}}(C) \triangleq k(p_{0,h_{j+1}}(\bar{C}_{h_j,h_{j+1}} \cap (C_{0,h_{j+1}} \oplus C_{h_j,N}))) - k(C_{0,h_j}). \tag{6.41}$$

A pair of a left U-block and a right U-block is called a **U-block pair**. It follows from the conditions (B1) and (B2), that each U-block pair $(B_h, B_{h_{j+1}})$ has the following **uniform** properties:

(1) For any two states $\sigma_{h_{j+1}}$ and $\sigma'_{h_{j+1}}$ in $B_{h_{j+1}}$,

$$\{L(\sigma_{h_j}, \sigma_{h_{j+1}}) : \sigma_{h_j} \in B_{h_j}\} = \{L(\sigma_{h_j}, \sigma'_{h_{j+1}}) : \sigma_{h_j} \in B_{h_j}\}. \tag{6.42}$$

(2) For any two states σ_{h_j} and σ'_{h_j} in B_{h_j},

$$\{L(\sigma_{h_j}, \sigma_{h_{j+1}}) : \sigma_{h_{j+1}} \in B_{h_{j+1}}\} = \{L(\sigma'_{h_j}, \sigma_{h_{j+1}}) : \sigma_{h_{j+1}} \in B_{h_{j+1}}\}. \tag{6.43}$$

The first property simply says that for a U-block pair $(B_{h_j}, B_{h_{j+1}})$, the set of composite branches from states in the left U-block B_{h_j} converging to any state in the right U-block $B_{h_{j+1}}$ is the same. The second property simply says that the set of composite branches diverging from any state in B_{h_j} to states in $B_{h_{j+1}}$ is the same. Two different U-block pairs have mutually disjoint composite branch sets.

7 PARALLEL DECOMPOSITION AND LOW-WEIGHT SUBTRELLISES

A minimal code trellis has the least overall state and branch complexity but is, in general, densely connected. For long codes with large dimensions, it is very difficult to implement any trellis-based MLD algorithm based on a full code trellis on IC (integrated circuit) chip(s) for hardware implementation. To overcome this implementation difficulty, one possible approach is to decompose the minimal trellis diagram of a code into parallel and structurally identical subtrellises of smaller dimensions without cross connections between them so that each subtrellis with reasonable complexity can be put on a single IC chip of reasonable size. This approach is called **parallel decomposition**.

Parallel decomposition should be done in such a way that the maximum state space dimension of the minimal trellis of a code is not exceeded. Parallel decomposition has other advantages. Since all the subtrellises are structurally identical, we can devise identical decoders of much smaller complexity to process these subtrellises in parallel. This not only simplifies the decoding complexity but also speeds up the decoding process. Parallel decomposition also resolves wire routing problem in IC implementation and reduces internal communications. Wire routing is a major problem in IC implementation of a trellis-based decoder.

Another approach to overcome decoding complexity difficulty for long codes is to devise suboptimal decoding algorithms to provide an effective trade-off between error performance and decoding complexity. Trellis-based suboptimal decoding algorithms have recently been devised. These decoding algorithms are based on a purged trellis which consists of only low weight codewords (or paths). Such a low-weight subtrellis has much smaller state and branch complexities than the full code trellis.

In this chapter, we first present a method for parallel decomposition of a minimal trellis without exceeding the maximum state space dimension and then present a method for purging a code trellis to obtain low-weight subtrellises. The structure and complexity of low-weight subtrellises are also discussed.

7.1 PARALLEL DECOMPOSITION OF A TRELLIS DIAGRAM

Consider the minimal N-section trellis T for a binary (N, K) linear block code C. Let $\rho_{\max}(C)$ denote the maximum state space dimension of T, i.e.,

$$\rho_{\max}(C) \triangleq \max_{0 \leq i \leq N} \rho_i(C)$$

where $\rho_i(C) \triangleq \log_2 |\Sigma_i(C)|$. The objective is to construct a trellis for C which is a **disjoint union** of a certain desired number of parallel isomorphic subtrellises under the constraint that its state space dimension at every time-i for $0 \leq i \leq N$ is less than or equal to the maximum state space dimension $\rho_{\max}(C)$ of the minimal N-section trellis T. This trellis may not be a minimal trellis for C.

Define the non-empty index set

$$I_{\max}(C) = \{i : \rho_i(C) = \rho_{\max}(C)\}. \tag{7.1}$$

Suppose we choose a subcode C' of C such that $\dim(C') = \dim(C) - 1$ by removing a row $\boldsymbol{g}$ from the TOGM G of C. The set of coset representatives $[C/C']$ is generated by $\boldsymbol{g}$. Recall that $\rho_i(C)$ is equal to the number of rows of G whose active spans contain i. Then we have

$$\rho_i(C') = \rho_i(C) - 1 \tag{7.2}$$

for $i \in \{\text{aspan}(\boldsymbol{g})\}$ and

$$\rho_i(C') = \rho_i(C) \tag{7.3}$$

for $i \notin \{\text{aspan}(\boldsymbol{g})\}$. It follows from (7.1) to (7.3) that we obtain Theorem 7.1 [73].

Theorem 7.1 If there exists a row $\boldsymbol{g}$ in the TOGM G for the code C such that $\text{aspan}(\boldsymbol{g}) \supseteq I_{\max}(C)$, then we can form a subcode C' of C generated by $G \setminus \{\boldsymbol{g}\}$ such that $\rho_{\max}(C') = \rho_{\max}(C) - 1$ and

$$I_{\max}(C') = I_{\max}(C) \cup \{i : \rho_i(C) = \rho_{\max}(C) - 1, i \notin \text{aspan}(\boldsymbol{g})\}.$$

$\triangle\triangle$

Since G is a TOGM, $G' = G \setminus \{g\}$ is also a TOGM. If the condition of Theorem 7.1 is satisfied, then it is possible to construct a trellis for C which consists of two parallel isomorphic subtrellises, one for C' and the other for the coset of C'. Each subtrellis is a minimal trellis and has maximum state space dimension equal to $\rho_{\max}(C) - 1$. Therefore, the maximum state space dimension of the resultant trellis is still $\rho_{\max}(C)$.

We can apply Theorem 7.1 again to C' if there exists a row $\boldsymbol{g}' \in G'$ such that $\text{aspan}(\boldsymbol{g}') \supseteq I_{\max}(C')$. This results in a subcode C'' with $\dim(C'') = \dim(C') - 1$ and $\rho_{\max}(C'') = \rho_{\max}(C') - 1$. In this case, we are able to construct a trellis for C with four parallel isomorphic subtrellises, one for each coset in C/C''. The maximum state space dimension of each subtrellis is $\rho_{\max}(C) - 2$ and each subtrellis is minimal. Therefore the maximum state space dimension is still $\rho_{\max}(C)$. Theorem 7.1 can be applied repeatedly until either the desired level of decomposition is achieved or no row in the generator matrix can be found to satisfy the condition in Theorem 7.1.

Now we put Theorem 7.1 in a general form. Let $R(C)$ be the following subset of rows of the TOGM G,

$$R(C) \triangleq \{\boldsymbol{g} \in G : \text{aspan}(\boldsymbol{g}) \supseteq I_{\max}(C)\}. \tag{7.4}$$

Let $\rho = |R(C)|$.

Theorem 7.2 [73] With $R(C)$ defined as above and $\rho = |R(C)|$, let $1 \leq \rho' \leq \rho$. There exists a subcode of C' of C such that $\rho_{\max}(C') = \rho_{\max}(C) - \rho'$ and

$\dim(C') = \dim(C) - \rho'$ if and only if there exists a subset $R' \subseteq R(C)$ consisting of ρ' rows of $R(C)$ such that for every i with $\rho_i(C) > \rho_{\max}(C) - \rho'$, there exist at least $\rho_i(C) - (\rho_{\max}(C) - \rho')$ rows in R' whose active spans contain i. The set of coset representatives for C/C' is generated by R'.

Proof: Suppose there exists a subset $R' = \{\boldsymbol{g}'_1, \boldsymbol{g}'_2, \ldots, \boldsymbol{g}'_{\rho'}\}$ consisting of ρ' rows of $R(C)$ which satisfies the condition of the theorem. Since $R' \subseteq R(C)$, $I_{\max}(C) \subseteq \text{aspan}(\boldsymbol{g}'_l)$ for $1 \leq l \leq \rho'$. Consider the subcode C' generated by $G \setminus R'$. It follows from the definition of $R(C)$ that $\dim(C') = \dim(C) - \rho'$. For those $i \in I_{\max}(C)$, we can determine $\rho_i(C')$ by counting the number of rows $\boldsymbol{g}$ in $G \setminus R'$ that are active at position i. This number is exactly $\rho_{\max}(C) - \rho'$. For those $i \notin I_{\max}(C)$ and $\rho_i(C) > \rho_{\max}(C) - \rho'$, we are assured by the condition that $\rho_i(C)$ will be reduced by at least $\rho_i(C) - (\rho_{\max}(C) - \rho')$. This guarantees that $\rho_{\max}(C') = \rho_{\max}(C) - \rho'$.

To prove the converse, let C' be a subcode of C of dimension $\dim(C) - \rho'$ such that $\rho_{\max}(C') = \rho_{\max}(C) - \rho'$. Without loss of generality, we let C' be generated by $G \setminus R'$ where R' is some subset of rows of the TOGM G with $|R'| = \rho'$. Let $N_i(R')$ be the number of rows $\boldsymbol{g}'$ in R' such that $i \in \text{aspan}(\boldsymbol{g}')$. Then, for every $i \in \{0, 1, \ldots, N\}$, we have

$$\rho_i(C') + N_i(R') = \rho_i(C). \tag{7.5}$$

Therefore,

$$\begin{aligned} N_i(R') &= \rho_i(C) - \rho_i(C') \\ &\geq \rho_i(C) - \rho_{\max}(C'). \end{aligned} \tag{7.6}$$

This says that for every $i \in \{0, 1, \ldots, N\}$, there exist at least $\rho_i(C) - \rho_{\max}(C') = \rho_i(C) - (\rho_{\max}(C) - \rho')$ rows in R' that are active. For $i \in I_{\max}(C)$, we have

$$N_i(R') \geq \rho_{\max}(C) - \rho_{\max}(C') = \rho'. \tag{7.7}$$

Since $|R'| = \rho'$, the equality of (7.7) must hold. Therefore, the active spans of all the rows in R' contain $I_{\max}(C)$. As a result, $R' \subseteq R(C)$.

△△

Two direct consequences of Theorem 7.2 are given in Corollaries 7.1 and 7.2.

Corollary 7.1 In parallel decomposition of a minimal trellis diagram for a linear block code C, the maximum number of parallel isomorphic subtrellises one can obtain such that the total state space dimension at any time never exceeds $\rho_{\max}(C)$ is upper bounded by $2^{|R(C)|}$.

△△

Corollary 7.2 The logarithm base-2 of the number of parallel isomorphic subtrellises in a minimal L-section trellis with section boundary locations in $\{h_0, h_1, \ldots, h_L\}$ for a binary linear block code is given by the number of rows in its trellis oriented generator matrix whose active spans contain the section boundary locations, $\{h_1, h_2, \ldots, h_{L-1}\}$.

△△

Theorem 7.2 actually provides a method to choose a subcode C' of a given code with $\rho_{\max}(C') = \rho_{\max}(C) - \dim([C/C'])$, such that a trellis T for C can be built with the following properties:

(1) The maximum state space dimension of T is $\rho_{\max}(C)$ (for minimal trellis).

(2) T is the union of $2^{\dim[C/C']}$ parallel isomorphic subtrellises T_i with each T_i being isomorphic to the minimal trellis for C'.

Example 7.1 Again we consider the $(8,4)$ RM code given in Example 3.1, whose minimal 4-section trellis diagram with section boundary locations in $U = \{0,2,4,6,8\}$ is shown in Figure 6.1. By examining the TOGM G of the code (the last matrix given in Example 3.1), we find only one row, the second row, whose active span contains $\{2,4,6\}$. Therefore, the minimal 4-section trellis consists of 2 parallel isomorphic subtrellises without cross connections between them.

△△

Example 7.2 Consider the extended and permuted $(32,21)$ BCH code. A parity check matrix for this code is shown in Figure 7.1. The set of spans of any TOGM and the state space dimension profile of the 32-section minimal trellis for this code are given in Tables 7.1 and 7.2, respectively. The 4-section minimal trellis with section boundaries in $\{0,8,16,24,32\}$ has the state space dimension profile, $\{0,7,9,7,0\}$, which gives $\rho_{4,\max}(C) = 9$. There is only one

$$
\begin{bmatrix}
1&1&1&1&1&1&1&1&1&1&1&1&1&1&1&1&0&0&0&0&0&0&0&0&0&0&0&0&0&0&0&0\\
0&1&0&1&1&0&1&0&1&0&0&1&0&1&1&0&1&1&0&0&1&1&0&0&0&0&0&0&0&0&0&0\\
0&0&1&1&1&0&0&1&1&1&0&0&1&0&0&1&0&1&0&1&0&0&0&0&1&0&1&0&0&0&0&0\\
0&0&0&1&1&1&1&0&0&0&1&0&1&1&0&1&0&1&0&0&0&1&0&0&1&0&0&0&1&0&0&0\\
0&0&0&0&1&1&1&1&1&0&1&0&1&0&1&0&0&1&0&1&1&0&1&0&0&0&0&0&0&0&0&0\\
0&0&0&0&0&1&1&0&0&0&0&0&0&1&1&0&1&0&1&0&1&1&0&0&1&0&1&0&1&1&0&0\\
0&0&0&0&0&0&1&1&0&1&0&1&0&1&1&0&0&1&1&0&1&0&1&0&1&1&0&0&0&0&0&0\\
0&0&0&0&0&0&0&0&1&1&1&1&1&1&1&1&1&1&1&1&1&1&1&1&0&0&0&0&0&0&0&0\\
0&0&0&0&0&0&0&0&0&1&0&1&1&0&1&0&0&1&0&1&0&1&0&1&1&1&1&1&0&0&0&0\\
0&0&0&0&0&0&0&0&0&0&1&1&0&0&1&1&0&1&1&0&1&0&0&1&0&1&0&1&1&0&1&0\\
0&0&0&0&0&0&0&0&0&0&0&0&0&0&0&0&1&1&1&1&1&1&1&1&1&1&1&1&1&1&1&1
\end{bmatrix}
$$

Figure 7.1. Parity check matrix for the extended and permuted $(32, 21)$ BCH code.

row, row-8, in G whose active span contains $\{8, 16, 24\}$. Therefore, this 4-section minimal trellis consists of only 2 parallel isomorphic subtrellises.

First we consider the decomposition of the 4-section minimal trellis with section boundary locations in $U = \{0, 8, 16, 24, 32\}$. Note that $I_{\max}(C) = \{16\}$ and from Table 7.1, we find that there are 9 rows in G with active spans containing $\{16\}$ and hence $|R(C)| = 9$. In an attempt to construct a trellis consisting of 64 parallel isomorphic subtrellises while satisfying the upper bound of 9 on the maximum state space dimension, we let $\rho' = 6$. So $\rho_{4,\max}(C) - \rho' = \rho_{4,\max}(C') = 3$. The set $\{i : \rho_i(C) > \rho_{4,\max}(C')\} = \{8, 16, 24\}$. However, we find that no subset R' of $R(C)$ exists satisfying the condition in Theorem 7.2. Hence we can not construct a trellis that consists of 64 parallel isomorphic subtrellises for this code without violating the constraint on the maximum state space dimension. If we chose $\rho' = 5$, then we can find a subset $R' = \{g_6, g_7, g_8, g_{12}, g_{15}\} \subseteq R(C)$ that satisfies the condition of Theorem 7.2. Hence choosing the subcode C' generated by $G \setminus R'$, we obtain a trellis for C consisting of 32 parallel isomorphic subtrellises. Each subtrellis is isomorphic to the minimal trellis for C' which has $\rho_{4,\max}(C') = 4$.

Table 7.1. Set of row spans of the trellis oriented generator matrix of the extended and permuted $(32, 21)$ primitive BCH code.

row-#	span	row-#	span	row-#	span
1	[1,8]	8	[8,25]	15	[15,27]
2	[2,15]	9	[9,16]	16	[17,24]
3	[3,13]	10	[10,23]	17	[18,31]
4	[4,14]	11	[11,19]	18	[19,29]
5	[5,12]	12	[12,26]	19	[20,30]
6	[6,18]	13	[13,20]	20	[21,28]
7	[7,21]	14	[14,22]	21	[25,32]

Table 7.2. The state space dimension profile of the 32-section minimal trellis for the extended and permuted $(32, 21)$ primitive BCH code.

Time i	Dimension ρ_i	Time i	Dimension ρ_i	Time i	Dimension ρ_i
0	0	11	10	22	9
1	1	12	10	23	8
2	2	13	10	24	7
3	3	14	10	25	7
4	4	15	10	26	6
5	5	16	9	27	5
6	6	17	10	28	4
7	7	18	10	29	3
8	7	19	10	30	2
9	8	20	10	31	1
10	9	21	10	32	0

For the same code, the 32-section minimal trellis with section boundary set $U = \{2i : 0 \leq i \leq 16\}$ has $\rho_{32,\max}(C) = 10$ and $I_{\max}(C) = \{12, 14, 18, 20\}$. Using Table 7.1, we find that $R(C) = \{\boldsymbol{g}_7, \boldsymbol{g}_8, \boldsymbol{g}_{10}, \boldsymbol{g}_{12}\}$. In an attempt to build a trellis for C consisting of 4 parallel isomorphic subtrellises while satisfying the upper bound of 10 on the maximum state space dimension, we let $\rho' = 2$. So $\rho_{32,\max}(C') = 8$. The set

$$\{i : \text{even } i \text{ and } \rho_i(C) > \rho_{32,\max}(C')\} = \{10, 12, 14, 16, 18, 20, 22\}.$$

From Table 7.1, we find $R' = \{\boldsymbol{g}_8, \boldsymbol{g}_{10}\}$ satisfying the condition in Theorem 7.2. Therefore, a trellis with 4 parallel isomorphic subtrellises with total state space dimension not exceeding $\rho_{32,\max}(C) = 10$ can be built.

△△

Basically, Theorem 7.2 provides a method for decomposing a complex minimal trellis into subtrellises without exceeding the maximum state complexity. This decomposition of a trellis into parallel and structurally identical subtrellises of smaller state complexity without cross connections between them has significant advantages for IC implementation of a trellis-based decoding algorithm, such as the Viterbi algorithm. Identical decoders of **much simpler complexity** can be devised to process the subtrellises independently in parallel without internal communications (or information transfer) between them. Internal information transfer limits the decoding speed. Furthermore, the number of computations to be carried out per subtrellis is much smaller than that of a fully connected trellis. As a result, the parallel structure not only simplifies the decoding complexity but also speeds up the decoding process.

7.2 LOW-WEIGHT SUBTRELLIS DIAGRAMS

Consider a binary (N, K) linear block code C with **weight profile** $W = \{0, w_1, w_2, \ldots, w_m\}$ where w_1 is the minimum weight and $w_1 < w_2 < \ldots < w_m \leq N$. Let $T(U)$ be the L-section trellis (minimal or non-minimal) with section boundary locations in $U = \{h_0, h_1, \ldots, h_L\}$. For two adjacent states σ and σ', with $\sigma \in \Sigma_{h_j}(C)$ and $\sigma' \in \Sigma_{h_{j+1}}(C)$, let $L(\sigma, \sigma')$ denote the set of parallel branches connecting the states σ and σ'. Let $\boldsymbol{b} \in L(\sigma, \sigma')$ and $w(\boldsymbol{b})$ denote the Hamming weight of $\boldsymbol{b}$. Let $\alpha(L(\sigma, \sigma'))$ denote the minimum weight of the parallel branches in $L(\sigma, \sigma')$.

In the following, we consider purging $T(U)$ to obtain a subtrellis diagram, denoted $T_{\min}(\mathbf{0})$, which consists of only the all-zero codeword and the minimum weight codewords of C. Such a trellis is called the **minimum-weight trellis** diagram for C. We say that this minimum weight subtrellis is **centered around** the all-zero codeword $\mathbf{0}$.

For any state $\sigma \in \Sigma_{h_j}(C)$ with $0 \le j \le L$, let (σ) denote the minimum of the weights of all paths from the initial state σ_0 to σ. We call (σ) the **minimum path weight to** the state σ. For every state in $T(U)$, (σ) can be computed by an algorithm very similar to the well known Viterbi algorithm as follows: Assume that (σ) is known for every state $\sigma \in \Sigma_{h_i}(C)$ for $0 \le i \le j$. Let $\sigma' \in \Sigma_{h_{j+1}}(C)$. Then (σ') is given by

$$(\sigma') = \min_{\sigma \in F(\sigma')} \{(\sigma) + \alpha(L(\sigma, \sigma'))\}, \tag{7.8}$$

where

$$F(\sigma') \triangleq \{\sigma \in \Sigma_{h_j}(C) : L(\sigma, \sigma') \neq \emptyset\}. \tag{7.9}$$

The process begins with $(\sigma_0) = 0$. Once (σ) is determined for every state $\sigma \in \Sigma_{h_j}(C)$, the states in $\Sigma_{h_{j+1}}(C)$ can be processed. This computation is repeated until $(\sigma_f) = 0$ is determined.

For any state $\sigma \in \Sigma_{h_j}(C)$ with $0 \le j \le L$, let $[\sigma]$ denote the **minimum of weights** of all paths from σ to the final state σ_f. We call $[\sigma]$ the **minimum path weight from** σ. For every state $\sigma \in T(U)$, $[\sigma]$ can be computed recursively from σ_f as follows: Assume that $[\sigma]$ is known for every state $\sigma \in \Sigma_{h_i}(C)$ for $j < i \le L$. Let σ' be a state in $\Sigma_{h_j}(C)$. Then $[\sigma']$ is given by

$$[\sigma'] = \min_{\sigma \in G(\sigma')} \{[\sigma] + \alpha(L(\sigma', \sigma))\}, \tag{7.10}$$

where

$$G(\sigma') \triangleq \{\sigma \in \Sigma_{h_{j+1}}(C) : L(\sigma', \sigma) \neq \emptyset\}. \tag{7.11}$$

Note that among all the paths passing through a given state σ, there is **at least** one path with weight $(\sigma) + [\sigma]$ and no path passing through that state has weight less than $(\sigma) + [\sigma]$. Let the all-zero path $\mathbf{0}$ be denoted by the state sequence

$$\sigma_0 = \sigma_0^{(\mathbf{0})}, \sigma_{h_1}^{(\mathbf{0})}, \ldots, \sigma_{h_L}^{(\mathbf{0})} = \sigma_f.$$

It is clear that $(\sigma_{h_j}^{(0)}) = [\sigma_{h_j}^{(0)}] = 0$ for $0 \leq j \leq L$ and for any state σ not in this sequence $(\sigma) + [\sigma] > 0$.

Now we describe a systematic method for deleting states and branches in $T(U)$ to obtain a subtrellis which contains all the codewords with weights up to and including d, where d is a nonzero weight in the weight profile W of C. This method consists of the following rules [75]:

Rule 1: If for every state $\sigma \in T(U)$, $(\sigma) + [\sigma] > d$, delete the state σ and all branches from and to σ.

Rule 2: Let σ and σ' be any two adjacent states in $\Sigma_{h_j}(C)$ and $\Sigma_{h_{j+1}}(C)$ with $0 \leq j < L$, respectively. Let $\boldsymbol{b} \in L(\sigma, \sigma')$. If $(\sigma) + [\sigma'] + w(\boldsymbol{b}) > d$, delete the branch $\boldsymbol{b}$.

Rule 3: If as a result of applications of Rules 1 and 2 to all the states and branches in $T(U)$, any state $\sigma(\neq \sigma_0)$ has no incoming branches, then σ and all its outgoing branches are deleted. Any state $\sigma(\sigma \neq \sigma_f)$ has no outgoing branches, then σ and all its incoming branches are deleted.

The above purging rules are applied repeatedly until further application leaves the trellis unaltered. Let $T_d(\mathbf{0})$ denote the resultant purged trellis. $T_d(\mathbf{0})$ contains the following codewords of C: (1) all codewords of weights up to and including d; and (2) possibly some codewords of weights greater than d which correspond to nonzero paths in $T(U)$ that diverge from and remerge with the all-zero path more than once. For each of these paths, the weight of a partial path between the adjacent diverging and remerging states, called **side-loop**, is d or less but not zero.

If the **purging parameter** d is equal to the minimum weight of C, (i.e., $d = w_1$), then $T_{w_1}(\mathbf{0})$ contains: (1) the all-zero codeword $\mathbf{0}$; (2) all the codewords of minimum weights w_1; and (3) those codewords of weights greater than w_1 which correspond to nonzero paths that diverge from and remerge with the all-zero path more than once. For each of these nonzero paths, the weight of each side-loop is w_1.

Next we modify the subtrellis diagram $T_{w_1}(\mathbf{0})$ so that the resultant trellis contains only the all-zero path and all the paths which correspond to the minimum-weight codewords of the code. The modification of $T_{w_1}(\mathbf{0})$ is done as follows:

Step 1: Create $L-1$ new states, denoted $\tilde{\sigma}_{h_1}^{(0)}, \tilde{\sigma}_{h_2}^{(0)}, \ldots, \tilde{\sigma}_{h_{L-1}}^{(0)}$, one at each section boundary location except $h_0 = 0$ and $h_L = N$;

Step 2: For $1 \leq j < L-1$, connect the state $\tilde{\sigma}_{h_j}^{(0)}$ to state $\tilde{\sigma}_{h_{j+1}}^{(0)}$ by a single branch with the all-zero $(h_{j+1} - h_j)$-tuple label. Also connect $\tilde{\sigma}_{h_{L-1}}^{(0)}$ to the final state σ_f by a single all-zero label; and

Step 3: For every branch with label $\boldsymbol{b} \neq \mathbf{0}$ in $T_{w_1}(\mathbf{0})$ that merges into one of the states, $\sigma_{h_j}^{(0)}$, on the all-zero path, $\sigma_0^{(0)}, \sigma_{h_1}^{(0)}, \ldots, \sigma_{h_L}^{(0)}$, from any state σ, delete this branch and create a new branch with the same label from σ to the state $\tilde{\sigma}_{h_j}^{(0)}$.

Steps 1-3 ensure that there is no path in the new trellis that after merging with the all-zero path diverges from it again before terminating at the final state σ_f. Consequently, the new trellis contains only the all-zero path $\mathbf{0}$ and the minimum weight codewords of C.

This new subtrellis is called the **minimum-weight trellis diagram** for C, denoted $T_{\min}(C)$ (or simply $T_{\min}(\mathbf{0})$). Let $\boldsymbol{c}$ be a nonzero codeword in C. The subtrellis that contains $\boldsymbol{c}$ and all the codewords in C at minimum distance from $\boldsymbol{c}$ can be obtained by adding $\boldsymbol{c}$ to all the codewords in $T_{\min}(\mathbf{0})$. This subtrellis, denoted $T_{\min}(\boldsymbol{c})$, is called the minimum-weight (or **minimum distance**) trellis centered around $\boldsymbol{c}$.

The minimum-weight trellis $T_{\min}(\mathbf{0})$ is **sparsely** connected and generally has much smaller state and branch complexities than that of the full trellis for the code. For example, the state space complexity profile for the 4-section minimum-weight trellis for the $(64, 42)$ RM code is $(1, 157, 157, 157, 1)$ as compared to $(1, 1024, 1024, 1024, 1)$ for the 4-section full trellis of the code. For the same code, the 8-section minimum-weight trellis has the state complexity profile

$$(1, 45, 157, 717, 157, 717, 157, 45, 1)$$

as compared to

$$(1, 128, 1024, 8192, 1024, 8192, 1024, 128, 1)$$

for the 8-section full trellis of the code. For this code, the 8-section minimum-weight trellis has a total of $4,524$ branches as compared to a total of $278,784$ branches for the 8-section full trellis.

For any weight w_i in the weight profile W, the w_i-weight subtrellis can be obtained in a similar way by setting the purging parameter $d = w_i$.

7.3 STRUCTURE ANALYSIS AND CONSTRUCTION OF LOW-WEIGHT SUBTRELLISES

As shown in the previous section, a low-weight subtrellis for a code can be obtained by purging the full code trellis. However, for long codes, constructing a large full code trellis and then purging it can be very time consuming if not impossible. Therefore, it is desirable to construct a low-weight subtrellis directly from the low-weight codewords of a code if they are known, e.g., RM codes. In this section, we study the structure and complexity of subtrellises for low-weight codewords of binary linear block codes.

Consider a binary (N, K) linear block code C with **weight profile** $W = \{0, w_1, w_2, \ldots w_m\}$ where w_1 is the minimum weight and $w_1 < w_2 < \cdots < w_m \leq N$. For a subcode C' of C, let $T_{C'}$ denote a minimal trellis for C', and let $\Sigma_h(C')$ denote the state space of $T_{C'}$ at time-h for $0 \leq h \leq N$. Let σ_0 and σ_f denote the initial state and the final state of $T_{C'}$, respectively, for any subcode C' of C. In the following, we develop some structural properties of a low-weight subtrellis of a code.

Theorem 7.3 Let C' be a subcode (**linear or nonlinear**) of C. Then there is a **unique mapping** φ from $\Sigma(C')$ to $\Sigma(C)$ such that for each state $\sigma' \in \Sigma_h(C')$ with $0 \leq h \leq N$, $\varphi(\sigma') \in \Sigma_h(C)$ and

$$L(\sigma_0, \sigma') \subseteq L(\sigma_0, \varphi(\sigma')). \tag{7.12}$$

Proof: For $\boldsymbol{u}_1 \in L(\sigma_0, \sigma')$, there is a unique state $\sigma \in \Sigma_h(C)$ such that $\boldsymbol{u}_1 \in L(\sigma_0, \sigma)$. For $\boldsymbol{u}_2 \in L(\sigma_0, \sigma')$, since $\boldsymbol{u}_1 + \boldsymbol{u}_2 \in C^{\text{tr}}_{0,h}$, then $\boldsymbol{u}_2$ must be in $L(\sigma_0, \sigma)$. This implies Eq.(7.12).

$\triangle\triangle$

Let φ^{-1} denote the inverse mapping from $\Sigma(C)$ to the family of $2^{|\Sigma(C')|}$ subsets of $\Sigma(C')$. For $\sigma \in \Sigma(C)$, if $\varphi^{-1}(\sigma)$ is empty, then σ is said to be deleted from T_C to obtain $T_{C'}$, and if $|\varphi^{-1}(\sigma)| \geq 2$, then σ is said to be split into states in $T_{C'}$.

For a nonzero weight $w \in W$, let C_w be defined as

$$C_w \triangleq \{\boldsymbol{u} \in C : w(\boldsymbol{u}) = w\}, \tag{7.13}$$

which is simply the set of codewords in C with weight w. For $0 \leq h < h' \leq N$, let $w_{h,h'}(\boldsymbol{u})$ denote the weight of $p_{h.h'}(\boldsymbol{u})$. In the following, we derive some structural properties of the minimal trellis for the subcode C_w, denoted T_{C_w}.

Theorem 7.4 Let w be a nonzero weight of C. For a codeword $\boldsymbol{u}$ in C_w and $0 \leq h < h' \leq N$, suppose the path labeled $\boldsymbol{u}$ in T_{C_w} from the initial state σ_0 to the final state σ_f pass through two states $\sigma \in \Sigma_h(C_w)$ and $\sigma' \in \Sigma_{h'}(C_w)$. For any $\boldsymbol{v} \in p_{h,h'}(C)$, $\boldsymbol{v} \in L(\sigma, \sigma')$ if and only if the following conditions hold:

$$\boldsymbol{v} + p_{h,h'}(\boldsymbol{u}) \in C^{\text{tr}}_{h,h'}, \tag{7.14}$$

and

$$w(\boldsymbol{v}) = w_{h,h'}(\boldsymbol{u}). \tag{7.15}$$

Proof: We first prove the only if (necessary) conditions. Suppose $\boldsymbol{v} \in L(\sigma, \sigma')$. Since both $\boldsymbol{v}$ and $p_{h,h'}(\boldsymbol{u})$ are in $L(\sigma, \sigma')$, the condition of (7.14) follows from the fact that C is linear. Define

$$\boldsymbol{v}' \triangleq p_{0,h}(\boldsymbol{u}) \circ \boldsymbol{v} \circ p_{h',N}(\boldsymbol{u}),$$

where $\circ$ denote the concatenation operation. Then $\boldsymbol{v}'$ is also a path in T_{C_w} from the initial state σ_0 to the final state σ_f passing through the states $\sigma \in \Sigma_h(C_w)$ and $\sigma' \in \Sigma_{h'}(C_w)$. Note that $\boldsymbol{u}$ and $\boldsymbol{v}'$ are identical from σ_0 to σ and from σ' to σ_f. Since both are in C_w and have weight w, the condition of (7.15) must hold.

Next we prove that Equations (7.14) and (7.15) are sufficient conditions for a vector $\boldsymbol{v} \in p_{h,h'}(C)$ to be in $L(\sigma, \sigma')$. Let $\boldsymbol{v}$ be a vector in $p_{h.h'}(C)$. Suppose $\boldsymbol{v}$ satisfies the conditions of (7.14) and (7.15). Define the following vector:

$$\boldsymbol{v}' \triangleq p_{0,h}(\boldsymbol{u}) \circ \boldsymbol{v} \circ p_{h',N}(\boldsymbol{u}). \tag{7.16}$$

It follows from (7.14) that $\boldsymbol{v}'$ is a codeword in C. Note that $\boldsymbol{v}'$ and $\boldsymbol{u}$ are identical in first h and last $N - h'$ components. Then it follows from (7.15) that

$$w(\boldsymbol{v}') = w(\boldsymbol{u}) = w,$$

and hence $\boldsymbol{v}' \in C_w$. From (7.14), (7.15) and (7.16), we find that

$$p_{0,h'}(\boldsymbol{v}') + p_{0,h'}(\boldsymbol{u}) \in C_{0,h'}^{\mathrm{tr}},$$

and

$$w_{0,h'}(\boldsymbol{v}') = w_{0,h'}(\boldsymbol{u}).$$

For any binary $N-h'$ tuple $\boldsymbol{x}$, $p_{0,h'}(\boldsymbol{v}') \circ \boldsymbol{x} \in C_w$ if and only if $p_{0,h'}(\boldsymbol{u}) \circ \boldsymbol{x} \in C_w$. Since T_{C_w} is minimal, $p_{0,h'}(\boldsymbol{v}')$ and $p_{0,h'}(\boldsymbol{u})$ are in $L(\sigma_0, \sigma')$. Since $p_{0,h}(\boldsymbol{v}') = p_{0,h}(\boldsymbol{u}) \in L(\sigma_0, \sigma)$, we conclude that $\boldsymbol{v} \in L(\sigma, \sigma')$.

△△

Corollary 7.3 For $0 \leq h < h' \leq N$, $\sigma \in \Sigma_h(C_w)$, $\sigma' \in \Sigma_{h'}(C_w)$ and different $\boldsymbol{u}$ and $\boldsymbol{v}$ in $L(\sigma, \sigma')$, the following inequalities hold:

$$w(\boldsymbol{u} + \boldsymbol{v}) \geq w_1, \tag{7.17}$$

and

$$w(\boldsymbol{u}) = w(\boldsymbol{v}) \geq w_1/2. \tag{7.18}$$

Proof: In equality (7.17) simply follows from (7.14). From (7.15) and (7.17), we have

$$\begin{aligned} 2w(\boldsymbol{u}) = 2w(\boldsymbol{v}) &= w(\boldsymbol{u}) + w(\boldsymbol{v}) \\ &\geq w(\boldsymbol{u} + \boldsymbol{v}) \geq w_1. \end{aligned} \tag{7.19}$$

Eq.(7.19) implies (7.18).

△△

For $0 \leq h \leq N$, let σ be a state in $\Sigma_h(C_w)$. Then it follows from Theorem 7.4 (Eq.(7.15)) that all the paths in T_{C_w} from the initial state σ_0 to state σ have the same weight, denoted $W_L(\sigma)$, which is called the **left weight** of state σ. Also the paths in T_{C_w} from state σ to the find state σ_f have the same weight $w - W_L(\sigma)$, denoted $W_R(\sigma)$. We call $W_R(\sigma) \triangleq w - W_L(\sigma)$ the **right weight** of state σ.

A direct consequence of Theorem 7.4 and Corollary 7.3 is Corollary 7.4.

Corollary 7.4 For a state σ in the minimal trellis for C_w, the following conditions hold:

(i) If $0 \leq W_L(\sigma) < w_1/2$, then $|L(\sigma_0, \sigma)| = 1$ (i.e., there is only one path in T_{C_w} connecting the initial state σ_0 to state σ);

(ii) if $w - w_1/2 < W_L(\sigma) \leq w$, then $|L(\sigma, \sigma_f)| = 1$ (i.e., there is only one path in T_{C_w} connecting state σ to the final state σ_f); and

(iii) if $|L(\sigma_0, \sigma)| \geq 2$ and $|L(\sigma, \sigma_f)| \geq 2$, then $w_1/2 \leq W_L(\sigma) \leq w - w_1/2$.

△△

For the case (i) (or (ii)) of the above corollary, the analysis of T_{C_w} is equivalent to the analysis of a coset in $p_{h,N}(C)/C^{\rm tr}_{h,N}$ (or $p_{0,h}(C)/C^{\rm tr}_{0,h}$) with the restriction of weight $w - W_L(\sigma)$ (or $W_L(\sigma)$). Case (iii) is of our main concern.

Theorem 7.5 Let C' be a subcode of a binary (N, K) linear block code C. For $0 \leq h_1 < h_2 < h_3 \leq n$, let $\boldsymbol{u}$ and $\boldsymbol{v}$ be two different paths in $T_{C'}$ from a state in $\Sigma_{h_1}(C')$ to a state in $\Sigma_{h_3}(C')$ such that $w(\boldsymbol{u}) = w(\boldsymbol{v}) = w_1/2$ and both paths pass through a common state in $\Sigma_{h_2}(C)$. Then, $\boldsymbol{u}$ and $\boldsymbol{v}$ have the following structures:

(i) Either

$$p_{h_1,h_2}(\boldsymbol{u}) = p_{h_1,h_2}(\boldsymbol{v}) = \mathbf{0}_{h_2-h_1}$$

or

$$p_{h_2,h_3}(\boldsymbol{u}) = p_{h_2,h_3}(\boldsymbol{v}) = \mathbf{0}_{h_3-h_2},$$

where $\mathbf{0}_{h_2-h_1}$ and $\mathbf{0}_{h_3-h_2}$ denote the all-zero $(h_2 - h_1)$-tuple and the all-zero $(h_3 - h_2)$-tuple, respectively, and

(ii) For any i with $h_1 \leq i \leq h_3$, the pair of the i-th components of $\boldsymbol{u}$ and $\boldsymbol{v}$ is either $(0,0), (0,1)$ or $(1,0)$.

Proof: Corollary 7.3 implies part (i) of the theorem. Since

$$\begin{aligned} w_1 &= w(\boldsymbol{u}) + w(\boldsymbol{v}) \\ &\geq w(\boldsymbol{u} + \boldsymbol{v}) \\ &\geq w_1, \end{aligned}$$

then $w(\boldsymbol{u}) + w(\boldsymbol{v}) = w(\boldsymbol{u} + \boldsymbol{v})$. This implies (ii) of the theorem.

△△

For a block code B of length N, and integers h, w and w' such that $0 \leq w \leq h$ and $0 \leq w' \leq N - h$, let $B_{(w,w'),h}$ denote the set of those codewords in B whose weight of the first h components is w and whose weight of the last $N-h$ components is w'.

It follows from Theorem 7.5 that the truncated subtrellis of $T_{C_{(w_1/2, w-w_1/2),h}}$ from time-0 to time-h has a very simple structure. Recall that there is a **one-to-one** mapping λ from the cosets in $C/(C_{0,h} \oplus C_{h,N})$ to the states in the state space $\Sigma_h(C)$ of the minimal trellis for C at time-h. Let B be a coset in $C/(C_{0,h} \oplus C_{h,N})$ and $\lambda(B)$ denote the state at time-h that corresponds to B. Then

$$L(\sigma_0, \lambda(B)) \circ L(\lambda(B), \sigma_f) = B. \tag{7.20}$$

From (7.15), the relation between $\Sigma_h(C_w)$ and $\Sigma_h(C)$ (Theorem 7.3) and (7.20), we have the following theorem.

Theorem 7.6 There is a one-to-one mapping λ_w from

$$\begin{aligned}[C/(C_{0,h} \oplus C_{h,N})]_w \triangleq \{ & B_{(w_L, w-w_L),h} \neq \emptyset : \\ & B \in C/(C_{0,h} \oplus C_{h,N}), 0 \leq w_L \leq w\}\end{aligned}$$

to $\Sigma_h(C_w)$ such that for $D \in [C/(C_{0,h} \oplus C_{h,N})]_w$,

$$L(\sigma_0, \lambda_w(D)) \circ L(\lambda_w(D), \sigma_f) = D, \tag{7.21}$$

i.e.,

$$L(\sigma_0, \lambda_w(D)) = p_{0,h}(D), \tag{7.22}$$

and

$$L(\lambda_w(D), \sigma_f) = p_{h,N}(D). \tag{7.23}$$

△△

For $B \in C/(C_{0,h} \oplus C_{h,N})$, if $B_w = \emptyset$, then the state $\lambda(B)$ is deleted in T_{C_w}, and if there are two or more w_L such that $0 \leq w_L \leq w$ and $B_{(w_L, w-w_L),h} \in [C/(C_{0,h} \oplus C_{h,N})]_w$, then the state $\lambda(B)$ is split into two or more states in T_{C_w}. If some states in $\Sigma_h(C)$ are split into several states in $\Sigma_h(C_w)$, then strictly speaking, T_{C_w} is not a subtrellis of T_C. However, we simply call T_{C_w} as w-weight subtrellis of T_C.

Table 7.3. The Minimum Weight Subtrellis of $\mathrm{RM}_{r,m}$.

Type of States at time $N/2$				N_{ST}
W_L	W_R	N_L	N_R	
0	2^{m-r}	1	$2^{r-1}\begin{bmatrix} m-1 \\ r-1 \end{bmatrix}$	1
2^{m-r}	0	$2^{r-1}\begin{bmatrix} m-1 \\ r-1 \end{bmatrix}$	1	1
2^{m-r-1}	2^{m-r-1}	2^r	2^r	$\begin{bmatrix} m-1 \\ r \end{bmatrix}$
Total		$1+2^r+2^{r-1}\begin{bmatrix} m-1 \\ r-1 \end{bmatrix}$	$1+2^r+2^{r-1}\begin{bmatrix} m-1 \\ r-1 \end{bmatrix}$	$2+\begin{bmatrix} m-1 \\ r \end{bmatrix}$
Full Trellis		$2^{\binom{m-1}{r}+\sum_{i=0}^{r-1}\binom{m-1}{i}}$	$2^{\binom{m-1}{r}+\sum_{i=0}^{r-1}\binom{m-1}{i}}$	$2^{\binom{m-1}{r}}$

Theorem 7.6 simply says that for a state σ with left weight w_L in $\Sigma_h(C_w)$, the set of all paths from the initial state to the final state that pass through the state σ is the (nonempty) set of all codewords in a coset of $C_{0,h} \oplus C_{h,N}$ in C whose left and right weights are w_L and $w - w_L$ with respect to h, respectively.

Let t be a small integer greater than or equal to one, and let w_t be the t-th smallest nonzero weight in the weight profile W of C. By merging states in $T_{C_{w_i}}$ with $1 \leq i \leq t$ properly, we can construct a trellis with the smallest state complexity for a subcode C' of C [48] such that

(i)

$$\bigcup_{i=1}^{t} C_{w_i} \subseteq C' \subseteq C - \{\mathbf{0}\}, \tag{7.24}$$

where $\mathbf{0}$ is the all-zero codeword in C, and

(ii) C' has the smallest trellis state complexity, and no branch can be deleted from $T_{C'}$ without violating the above condition (i).

The trellis $T_{C'}$ is called **weight-($w_1 \sim w_t$) subtrellis** of C. For simplicity, the weight-($w_1 \sim w_1$) subtrellis is called the weight-w_1 (or minimum-weight) subtrellis of C.

Table 7.4. The Weight $w_1 \sim w_2$ Subtrellises of $\mathrm{RM}_{2,6}$ and $\mathrm{RM}_{3,6}$, where $w_1 = 2^{m-r}$ and $w_2 = 2^{m-r} + 2^{m-r-1}$.

<table>
<tr><th rowspan="2">Code</th><th colspan="4">Type of States at time $n/2$</th><th rowspan="2">N_{ST}</th></tr>
<tr><th>W_L</th><th>W_R</th><th>N_L</th><th>N_R</th></tr>
<tr><td rowspan="8">$\mathrm{RM}_{6,2}$</td><td>0</td><td>16</td><td>1</td><td>62</td><td>1</td></tr>
<tr><td>16</td><td>0</td><td>62</td><td>1</td><td>1</td></tr>
<tr><td>8</td><td>16</td><td rowspan="2">4</td><td>56</td><td rowspan="3">155</td></tr>
<tr><td>8</td><td>8</td><td rowspan="2">4</td></tr>
<tr><td>16</td><td>8</td><td>56</td></tr>
<tr><td>12</td><td>12</td><td>16</td><td>16</td><td>868</td></tr>
<tr><td colspan="2">Total</td><td>$23,251$</td><td>$23,251$</td><td>$1,025$</td></tr>
<tr><td colspan="2">Full Trellis</td><td>2^{16}</td><td>2^{16}</td><td>$1,024$</td></tr>
<tr><td rowspan="10">$\mathrm{RM}_{6,3}$</td><td>0</td><td>8</td><td>1</td><td>620</td><td rowspan="2">1</td></tr>
<tr><td>0</td><td>12</td><td>1</td><td>$13,888$</td></tr>
<tr><td>8</td><td>0</td><td>620</td><td>1</td><td rowspan="2">1</td></tr>
<tr><td>12</td><td>0</td><td>$13,888$</td><td>1</td></tr>
<tr><td>4</td><td>8</td><td rowspan="2">8</td><td>336</td><td rowspan="3">155</td></tr>
<tr><td>4</td><td>4</td><td rowspan="2">8</td></tr>
<tr><td>8</td><td>4</td><td>336</td></tr>
<tr><td>6</td><td>6</td><td>32</td><td>32</td><td>868</td></tr>
<tr><td colspan="2">Total</td><td>$95,606$</td><td>$95,606$</td><td>$1,025$</td></tr>
<tr><td colspan="2">Full Trellis</td><td>2^{26}</td><td>2^{26}</td><td>$1,024$</td></tr>
</table>

Example 7.3 Consider $\mathrm{RM}_{r,m}$ as C. Then, $N = 2^m$ and $w_1 = 2^{m-r}$. For $h = N/2$, $C^{\mathrm{tr}}_{0,N/2} = C^{\mathrm{tr}}_{N/2,N} = \mathrm{RM}_{r-1,m-1}$. **The split weight distributions** of cosets of $\mathrm{RM}_{r-1,m-1} \circ \mathrm{RM}_{r-1,m-1}$ in $\mathrm{RM}_{r,m}$ have been obtained for $m \leq 9$ except for $\mathrm{RM}_{4,9}$. The states σ's in $\Sigma_h(C_w)$ can be classified by the values of the left weight $W_L(\sigma)$, the right weight $W_R(\sigma)$, the number $N_L(\sigma)$ of left paths from the initial state σ_0 to σ, $|L(\sigma_0, \sigma)|$, and the number $N_R(\sigma)$ of right paths from σ to the final state σ_f, $|L(\sigma, \sigma_f)|$. **The structure** of $T_{[\mathrm{RM}_{r,m}]_{2^{m-r}}}$ at time-$N/2$, in terms of the above classification of states, and the number N_{ST} of states of each class are given in Table 7.3 [48, 50], where for $0 \leq j \leq i$, $\left[{i \atop j} \right]$

Table 7.5. The Weight $w_1 \sim w_2$ Subtrellis of EBCH$(64, 24)$ where $w_1 = 16$ and $w_2 = 18$.

Type of States at time $n/2$				N_{ST}	Subcode
W_L	W_R	N_L	N_R		
0	16	1	62	1	$\mathrm{RM}_{2,6}$
16	0	62	1	1	$\mathrm{RM}_{2,6}$
6	12	1	16	96	
12	6	16	1	96	
8	8	4	4	155	$\mathrm{RM}_{2,6}$
8	10	1	4	960	
10	8	4	1	960	
Total		7,115	7,115	2,269	
Full Trellis		2^{18}	2^{18}	4,096	

denotes Gaussian binomial coefficient [66] defined as

$$\begin{bmatrix} i \\ j \end{bmatrix} \triangleq \begin{cases} \prod_{p=0}^{j-1} \frac{2^{i-p}-1}{2^{j-p}-1}, & \text{for } j \geq 1, \\ 1, & \text{for } j = 0. \end{cases}$$

Since $\begin{bmatrix} m-1 \\ r \end{bmatrix} = \begin{bmatrix} m-1 \\ m-r-1 \end{bmatrix}$ [66], the minimum weight subtrellises of $\mathrm{RM}_{r,m}$ and its dual code $\mathrm{RM}_{m-r-1,m}$ have the same number of states at time-$N/2$. The number of states of $T_{[\mathrm{RM}_{r,m}]_{2^{m-r}}}$ at time-$N/4$ and time-$3N/4$ is also $\begin{bmatrix} m-1 \\ r \end{bmatrix}$ and the detailed structure of $T_{[\mathrm{RM}_{r,m}]_{2^{m-r}}}$ is given in [50]. The structure of $T_{[\mathrm{RM}_{r,m}]_w}$ at time-$N/2$, where $2^{m-r} \leq w < 2^{m-r+1}$, has been determined and is presented in [49]. Table 7.4 shows the structures of 2-section trellises $T_{[\mathrm{RM}_{2,6}]_{16\sim 24}}$ and $T_{[\mathrm{RM}_{3,6}]_{8\sim 12}}$ with section boundaries $\{0, 32, 64\}$. These subtrellises have one additional state to the trellises of $\mathrm{RM}_{2,6}$ and $\mathrm{RM}_{3,6}$ at time 32, because one state is split into two states to exclude the zero codeword.

△△

Example 7.4 Let $N = 2^m$. Let EBCH(N, K) denote the extended code of a permuted binary primitive BCH code of length $N - 1$ with K information

Table 7.6. The Weight $w_1 \sim w_2$ Subtrellis of EBCH$(64, 45)$ where $w_1 = 8$ and $w_2 = 10$.

Type of States at time $N/2$				N_{ST}	Subcode
W_L	W_R	N_L	N_R		
0	8	1	620	1	$\text{RM}_{3,6}$
8	0	620	1	1	$\text{RM}_{3,6}$
2	6	1	32	112	
2	8		320		
6	2	32	1	112	
8	2	320			
4	4	8	8	155	$\text{RM}_{3,6}$
4	6	2	24	2,240	
4	4		2		
6	4	24			
4	6	2	32	1,680	
6	4	32	2	1,680	
Total		156,757	156,757	5,981	
Full Trellis		2^{39}	2^{39}	8192	

bits. The permutation is chosen in such a way that the resultant code has a RM code as a large subcode [17, 36, 37]. Choose EBCH(N, K) as C and $N/2$ as h. Then, the following symmetric properties hold [17]:

$$C^{\text{tr}}_{0,N/2} = C^{\text{tr}}_{N/2,N}.$$

and for any state σ in $\Sigma_{N/2}(C_w)$, there is a state σ' in $\Sigma_{N/2}(C_w)$, such that $W_L(\sigma') = W_R(\sigma)$ and $W_R(\sigma') = W_L(\sigma)$. The split weight distributions of cosets of $C^{\text{tr}}_{0,N/2} \circ C^{\text{tr}}_{N/2,N}$ in C have been computed for all EBCH codes of length 64 and EBCH$(128, 29)$ and EBCH(128, 36) codes. Tables 7.5 to 7.7 show the structures of 2-section trellises with section boundaries $\{0, N/2, N\}$, $T_{[\text{EBCH}(64,24)]_{w_1 \sim w_2}}$, where $w_1 = 16$ and $w_2 = 18$, and $T_{[\text{EBCH}(64,45)]_{w_1 \sim w_2}}$, where $w_1 = 8$ and $w_2 = 10$, and $T_{[\text{EBCH}(128,36)]_{w_1 \sim w_2}}$, where $w_1 = 32$ and $w_2 = 36$,

Table 7.7. The Weight $w_1 \sim w_2$ Subtrellis of EBCH$(123, 36)$ where $w_1 = 32$ and $w_2 = 36$.

Type of States at time $N/2$				N_{ST}	Subcode
W_L	W_R	N_L	N_R		
0	32	1	126	1	$RM_{2,7}$
32	0	126	1	1	$RM_{2,7}$
14	22	1	1	2304	
22	14	1	1	2304	
16	16	4	4	651	$RM_{2,7}$
16	20	1	1	4032	
20	16	1	1	4032	
18	18	1	1	3584	
Total		18987	18987	16909	
Full Trellis		2^{29}	2^{29}	2^{22}	

respectively. Since the set of minimum weight codewords of EBCH$(64, 24)$ (or EBCH(128,36)) spans the subcode $RM_{2,6}$ (or $RM_{2,7}$), $T_{[\text{EBCH}(64,24)]_{w_1}}$ (or $T_{[\text{EBCH}(128,36)]_{w_1}}$) is isomorphic to $T_{[RM_{2,6}]_{w_1}}$ (or $T_{[RM_{2,7}]_{w_1}}$), where $w_1 = 16$ (or $w_1 = 32$).

△△

8 METHODS FOR CONSTRUCTING CODES AND TRELLISES

This chapter is concerned with construction of codes and their trellises. Methods for constructing long powerful codes from short component codes or component codes of smaller dimensions are presented. These methods include squaring construction, direct-sum, interleaving, product, concatenation and multilevel concatenation. Also presented in this chapter are techniques for constructing trellises for codes from their component codes. One such technique is the product of directed graphs.

8.1 SQUARING CONSTRUCTION OF CODES

The **squaring construction** is a method for constructing long codes from a sequence of subcodes of a given short code [24]. The codes constructed by this method have simple trellis structure and can be decoded with multistage decoding.

Let C_0 be a binary (n, k_0) linear block code with minimum Hamming distance d_0. Let $C_1, C_2, \ldots, C_m$ be a sequence of subcodes of C_0 such that

$$C_0 \supset C_1 \supset C_2 \supset \cdots \supset C_m. \tag{8.1}$$

For $0 \le i \le m$, let G_i, k_i and d_i be the generator matrix, the dimension and the minimum distance of the subcode C_i, respectively. Form a chain of partitions as follows:

$$C_0/C_1, C_0/C_1/C_2, \ldots, C_0/C_1/\cdots/C_m.$$

For $0 \le i < m$, let $G_{i/i+1}$ denote the generator matrix for the coset representative space $[C_i/C_{i+1}]$. The rank of $G_{i/i+1}$ is

$$\text{Rank}(G_{i/i+1}) = \text{Rank}(G_i) - \text{Rank}(G_{i+1}). \tag{8.2}$$

Without loss of generality, we assume that $G_0 \supset G_1 \supset G_2 \supset \cdots \supset G_m$. Then, for $0 \le i < m$,

$$G_{i/i+1} = G_i \backslash G_{i+1}. \tag{8.3}$$

One-level squaring construction is based on C_1 and the partition C_0/C_1. Let $\boldsymbol{a} = (a_1, a_2, \ldots, a_n)$ and $\boldsymbol{b} = (b_1, b_2, \ldots, b_n)$ be two binary n-tuples and $(\boldsymbol{a}, \boldsymbol{b})$ denote the $2n$-tuple, $(a_1, a_2, \ldots, a_n, b_1, b_2, \ldots, b_n)$. Form the following set of $2n$-tuples:

$$|C_0/C_1|^2 \triangleq \{(\boldsymbol{a} + \boldsymbol{x}, \boldsymbol{b} + \boldsymbol{x}) : \boldsymbol{a}, \boldsymbol{b} \in C_1 \text{ and } \boldsymbol{x} \in [C_0/C_1]\}. \tag{8.4}$$

Then $|C_0/C_1|^2$ is an $(2n, k_0 + k_1)$ linear block code with minimum Hamming distance,

$$D_1 \triangleq \min\{2d_0, d_1\}. \tag{8.5}$$

The generator matrix for $|C_0/C_1|^2$ is given by

$$G = \begin{bmatrix} G_1 & 0 \\ 0 & G_1 \\ G_{0/1} & G_{0/1} \end{bmatrix} \tag{8.6}$$

This matrix can be expressed in the following form:

$$G = I_2 \otimes G_1 \oplus (1, 1) \otimes G_{0/1}, \tag{8.7}$$

where $\otimes$ denotes the **Kronecker product**, $\oplus$ the **direct sum** and I_2 the **identity matrix** of dimension 2.

Now we extend the 1-level squaring construction to two-level squaring construction. First we form two codes, $U \triangleq |C_0/C_1|^2$ and $V \triangleq |C_1/C_2|^2$, using 1-level squaring construction. It is easy to see that V is a subcode of U. The 2-level squaring construction based on the partitions, C_0/C_1, C_1/C_2, and $C_0/C_1/C_2$ gives the following code:

$$\begin{aligned} |C_0/C_1/C_2|^4 &\triangleq \{(\boldsymbol{a}+\boldsymbol{x}, \boldsymbol{b}+\boldsymbol{x}) : \boldsymbol{a}, \boldsymbol{b} \in V \text{ and } \boldsymbol{x} \in [U/V]\} \\ &= \{(\boldsymbol{a}+\boldsymbol{x}, \boldsymbol{b}+\boldsymbol{x}) : \boldsymbol{a}, \boldsymbol{b} \in |C_1/C_2|^2 \\ &\qquad \text{and } \boldsymbol{x} \in [|C_0/C_1|^2 \,/\, |C_1/C_2|^2]\}, \end{aligned} \tag{8.8}$$

which is simply the code obtained by 1-level squaring construction based on V and U/V. This code is an $(4n, k_0 + 2k_1 + k_2)$ linear block code with minimum Hamming distance,

$$D_2 \triangleq \min\{4d_0, 2d_1, d_2\}. \tag{8.9}$$

Let G_U, G_V and $G_{U/V}$ denote the generator matrices for U, V and $[U/V]$, respectively. Then the generator matrix for $|U/V|^2$ is

$$G = \begin{bmatrix} G_V & 0 \\ 0 & G_V \\ G_{U/V} & G_{U/V} \end{bmatrix}. \tag{8.10}$$

Put G_V and G_U in the form of (8.6) and note that $G_{U/V} = G_U \backslash G_V$. Then the generator matrix G of $|C_0/C_1/C_2|^4 = |U/V|^2$ given by (8.10) can be put in the following form:

$$G = \begin{bmatrix} G_2 & 0 & 0 & 0 \\ 0 & G_2 & 0 & 0 \\ 0 & 0 & G_2 & 0 \\ 0 & 0 & 0 & G_2 \\ G_{0/1} & G_{0/1} & G_{0/1} & G_{0/1} \\ G_{1/2} & G_{1/2} & G_{1/2} & G_{1/2} \\ 0 & 0 & G_{1/2} & G_{1/2} \\ 0 & G_{1/2} & 0 & G_{1/2} \end{bmatrix}. \tag{8.11}$$

This matrix can be expressed in the following compact form:

$$G = I_4 \otimes G_2 \oplus (1111) \otimes G_{0/1} \oplus \begin{bmatrix} 1 & 1 & 1 & 1 \\ 0 & 0 & 1 & 1 \\ 0 & 1 & 0 & 1 \end{bmatrix} \otimes G_{1/2}. \tag{8.12}$$

Note that

$$(1111) \text{ and } \begin{bmatrix} 1 & 1 & 1 & 1 \\ 0 & 0 & 1 & 1 \\ 0 & 1 & 0 & 1 \end{bmatrix}$$

are the generator matrices of the zero-th and first-order RM codes of length 4.

Higher level squaring construction can be carried out recursively in a similar manner. For $m \geq 2$, let

$$U_m \triangleq |C_0/C_1/\cdots/C_{m-1}|^{2^{m-1}}$$

and

$$V_m \triangleq |C_1/C_2/\cdots/C_m|^{2^{m-1}}$$

denote the two codes obtained by $(m-1)$-level squaring construction. The code obtained by m-level squaring construction is given by

$$|C_0/C_1/\cdots/C_m|^{2^m} \triangleq \{(\boldsymbol{a}+\boldsymbol{x}, \boldsymbol{b}+\boldsymbol{x}) : \boldsymbol{a}, \boldsymbol{b} \in V_m \text{ and } \boldsymbol{x} \in [U_m/V_m]\}. \tag{8.13}$$

The generator matrix is given by

$$G = I_{2^m} \otimes G_m \oplus \sum_{0 \leq r < m} G_{\mathrm{RM}}(r,m) \otimes G_{r/r+1}, \tag{8.14}$$

where I_{2^m} denotes the identity matrix of dimension 2^m and $G_{\mathrm{RM}}(r,m)$ is the generator matrix of the r-th order RM code, $\mathrm{RM}_{r,m}$, of length 2^m.

RM codes are good examples of the squaring construction. Long RM codes can be constructed from short RM codes iteratively using the squaring construction. From the construction of RM codes given in Section 2.5, we find that for $0 < i \leq r$,

$$\mathrm{RM}_{r,m} \supset \mathrm{RM}_{r-1,m} \supset \cdots \supset \mathrm{RM}_{r-i,m}. \tag{8.15}$$

Consider the RM code, $\mathrm{RM}_{r,m}$. As shown in Section 2.5, this code can be obtained from $\mathrm{RM}_{r,m-1}$ and $\mathrm{RM}_{r-1,m-1}$ codes using the $|\boldsymbol{u}|\boldsymbol{u}+\boldsymbol{v}|$-construction.

It follows from (2.40) that the generator matrix for the $\mathrm{RM}_{r,m}$ code can be expressed as follows:

$$G_{\mathrm{RM}}(r,m) = \begin{bmatrix} G_{\mathrm{RM}}(r,m-1) & G_{\mathrm{RM}}(r,m-1) \\ 0 & G_{\mathrm{RM}}(r-1,m-1) \end{bmatrix}. \tag{8.16}$$

Define

$$\Delta_{\mathrm{RM}}(r/r-1,m-1) \triangleq G_{\mathrm{RM}}(r,m-1) \backslash G_{\mathrm{RM}}(r-1,m-1). \tag{8.17}$$

Note that $\Delta_{\mathrm{RM}}(r/r-1,m-1)$ consists of those rows in $G_{\mathrm{RM}}(r,m-1)$ but not in $G_{\mathrm{RM}}(r-1,m-1)$ and it spans the coset representative space $[\mathrm{RM}_{r,m-1}/\mathrm{RM}_{r-1,m-1}]$. Now we can put $G_{\mathrm{RM}}(r,m-1)$ in the following form:

$$G_{\mathrm{RM}}(r,m-1) = \begin{bmatrix} G_{\mathrm{RM}}(r-1,m-1) \\ \Delta_{\mathrm{RM}}(r/r-1,m-1) \end{bmatrix}. \tag{8.18}$$

Replacing $G_{\mathrm{RM}}(r,m-1)$ in (8.16) with the expression of (8.18) and performing row operations, we put $G_{\mathrm{RM}}(r,m)$ in the following form:

$$G_{\mathrm{RM}}(r,m) = \begin{bmatrix} G_{\mathrm{RM}}(r-1,m-1) & 0 \\ 0 & G_{\mathrm{RM}}(r-1,m-1) \\ \Delta_{\mathrm{RM}}(r/r-1,m-1) & \Delta_{\mathrm{RM}}(r/r-1,m-1) \end{bmatrix}. \tag{8.19}$$

This is exactly the generator matrix form of 1-level squaring construction. Therefore, the r-th order RM code of length 2^m can be constructed from the r-th order and $(r-1)$-th order RM codes of length 2^{m-1}, i.e.,

$$\mathrm{RM}_{r,m} = |\mathrm{RM}_{r,m-1}/\mathrm{RM}_{r-1,m-1}|^2. \tag{8.20}$$

Since

$$\mathrm{RM}_{r,m-1} = |\mathrm{RM}_{r,m-2}/\mathrm{RM}_{r-1,m-2}|^2$$

and

$$\mathrm{RM}_{r-1,m-1} = |\mathrm{RM}_{r-1,m-2}/\mathrm{RM}_{r-2,m-2}|^2,$$

then $\mathrm{RM}_{r,m}$ can be constructed from $\mathrm{RM}_{r,m-2}$, $\mathrm{RM}_{r-1,m-2}$ and $\mathrm{RM}_{r-2,m-2}$ using two-level squaring construction, i.e.,

$$\mathrm{RM}_{r,m} = |\mathrm{RM}_{r,m-2}/\mathrm{RM}_{r-1,m-2}/\mathrm{RM}_{r-2,m-2}|^4. \tag{8.21}$$

Repeating the above process, we find that for $1 \leq \mu \leq r$, the $\text{RM}_{r,m}$ code can be expressed as a μ-level squaring construction code as follows:

$$\text{RM}_{r,m} = |\text{RM}_{r,m-\mu}/\text{RM}_{r-1,m-\mu}/\cdots/\text{RM}_{r-\mu,m-\mu}|^{2^\mu}. \qquad (8.22)$$

A problem which is related to the construction of codes from component codes is code **decomposition**. A code is said to be **decomposable** if it can be expressed in terms of component codes. A code is said to be μ-level decomposable if it can be expressed as an μ-level squaring construction code from a sequence of subcodes of a given code as shown in (8.13) or as a μ-level concatenated code [24]. From (8.22), we see that a RM code is μ-level decomposable.

A μ-level decomposable code can be decoded in multiple stages, component codes are decoded sequentially one at a time and decoded information is passed from one stage to the next stage. This multistage decoding provides a good trade-off between error performance and decoding complexity. A trellis-based decoding can be used at each stage of decoding.

8.2 TRELLISES FOR CODES BY SQUARING CONSTRUCTION

Codes constructed by the squaring construction have simple trellis structure and their trellis diagrams can be constructed easily. We will use the 2-level squaring construction for illustration.

Consider the code

$$C = |C_0/C_1/C_2|^4$$

given by (8.8) whose generator matrix is given by (8.11) or (8.12). This code has length $N = 4n$. We consider the construction of a minimal 4-section trellis for this code with section boundary locations in $\{0, n, 2n, 3n, 4n\}$. In order to obtain a TOGM for C, we must use trellis oriented form for all the matrices in (8.12). Note that if G_0 is in TOF, then $G_{0/1}$, $G_{1/2}$ and G_2 are in TOF. By simple row operations, the generator matrix of the 1st-order RM codes of length 4 can be put in trellis oriented form as follows:

$$\begin{bmatrix} 1 & 1 & 0 & 0 \\ 0 & 1 & 1 & 0 \\ 0 & 0 & 1 & 1 \end{bmatrix}.$$

The generator matrix for $C = |C_0/C_1/C_2|^4$ in trellis oriented form is given below,

$$G = (1111) \otimes G_{0/1} \oplus \begin{bmatrix} 1 & 1 & 0 & 0 \\ 0 & 1 & 1 & 0 \\ 0 & 0 & 1 & 1 \end{bmatrix} \otimes G_{1/2} \oplus \begin{bmatrix} 1 & 0 & 0 & 0 \\ 0 & 1 & 0 & 0 \\ 0 & 0 & 1 & 0 \\ 0 & 0 & 0 & 1 \end{bmatrix} \otimes G_2. \quad (8.23)$$

The trellis construction then follows from the generator matrix given in (8.23) and the construction procedure given in Section 6.3. First, we notice that the compositions of first and fourth columns of G are the same, and also the compositions of the second and third columns are the same. Furthermore we find that:

(1)

$$\begin{aligned} G_n^s &= (1111) \otimes G_{0/1} \oplus (1100) \otimes G_{1/2}, \\ G_{2n}^s &= (1111) \otimes G_{0/1} \oplus (0110) \otimes G_{1/2}, \\ G_{3n}^s &= (1111) \otimes G_{0/1} \oplus (0011) \otimes G_{1/2}. \end{aligned}$$

(2)

$$\begin{aligned} G_{0,n}^{f,p} &= (1000) \otimes G_2, & G_{n,2n}^{f,p} &= (0100) \otimes G_2, \\ G_{2n,3n}^{f,p} &= (0010) \otimes G_2, & G_{3n,4n}^{f,p} &= (0001) \otimes G_2. \end{aligned}$$

(3)

$$\begin{aligned} G_{0,n}^{f,s} &= (1111) \otimes G_{0/1} \oplus (1100) \otimes G_{1/2}, \\ G_{n,2n}^{f,s} &= (0110) \otimes G_{1/2}, \\ G_{2n,3n}^{f,s} &= (0011) \otimes G_{1/2}, \\ G_{3n,4n}^{f,s} &= \emptyset. \end{aligned}$$

From the above observations, we can say that the trellis has the following general structure:

(1) The state spaces, $\Sigma_n(C)$, $\Sigma_{2n}(C)$ and $\Sigma_{3n}(C)$, have the same number of states, $2^{k_0-k_2}$.

(2) For all four sections, each composite branch consists of 2^{k_2} parallel branches. These 2^{k_2} parallel branches form the codewords in C_2 or a coset in $C_0/C_1/C_2$.

(3) There are $2^{k_0-k_2}$ composite branches diverging from the initial state σ_0. There are $2^{k_1-k_2}$ composite branches diverging from each state in $\Sigma_n(C)$ or $\Sigma_{2n}(C)$. There is only one composite branch diverging from each state in $\Sigma_{3n}(C)$. All the states in $\Sigma_{3n}(C)$ converge to the final state σ_f.

(4) The total number of composite branches in either section-2 or section-3 is $2^{k_0+k_1-2k_2}$. There are $2^{k_1-k_2}$ composite branches converging into a state in either $\Sigma_{2n}(C)$ or $\Sigma_{3n}(C)$.

(5) All the branches (including the parallel branches) diverging from a state in $\Sigma_n(C)$ or in $\Sigma_{2n}(C)$ form the codewords in C_1 or a coset in C_0/C_1. All the branches converging into a state in $\Sigma_{2n}(C)$ or $\Sigma_{3n}(C)$ form the codewords in C_1 or a coset in C_0/C_1.

From the above properties, we see that section-2 and section-3 are structurally identical. The third and fourth sections form the mirror image of the first and second sections. Since the active spans of all the rows in $(1111) \otimes G_{0/1}$ contain the section boundaries n, $2n$ and $3n$, it follows from Corollary 7.2 that the overall trellis consists of $2^{k_0-k_1}$ parallel isomorphic subtrellises without cross-connections between them.

Following the procedure given in Section 6.3, we can construct the minimal 4-section trellis diagram for $C = |C_0/C_1/C_2|^4$.

The 2^l-section minimal trellis for an l-level squaring construction code $|C/C_1/\cdots/C_l|^{2^l}$ can be analyzed in the same way. Therefore, trellises for these codes have simple structural properties, such as regularity, parallel structure and mirror symmetry. Since RM codes can be constructed from RM codes of shorter lengths using squaring construction, their trellises can be constructed and analyzed easily.

Example 8.1 Consider the three RM codes, $\text{RM}_{2,3} = (8,7,2) = C_0$, $\text{RM}_{1,3} = (8,4,4) = C_1$ and $\text{RM}_{0,3} = (8,1,8) = C_2$. The 2-level squaring construction based on these 3 codes yields the $\mathrm{RM}_{2,5} = (32, 16, 8)$ code, i.e.,

$$\mathrm{RM}_{2,5} = |\mathrm{RM}_{2,3}/\mathrm{RM}_{1,3}/\mathrm{RM}_{0,3}|^4 .$$

It follows from the above analysis of a 2-level squaring construction code that each state space at the boundaries, n, $2n$ and $3n$, consists of 64 states. There are 8 composite branches diverging from each state in $\Sigma_8(C)$ or $\Sigma_{16}(C)$, and there are 8 composite branches converging into each state in $\Sigma_{16}(C)$ or $\Sigma_{24}(C)$. Each composite branch consists of two parallel branches, which are the two codewords in a coset in the partition $\mathrm{RM}_{2,3}/\mathrm{RM}_{1,3}/\mathrm{RM}_{0,3}$. The 16 branches diverging from any state in $\Sigma_8(C)$ or $\Sigma_{16}(C)$ form the codewords in $\mathrm{RM}_{1,3}$ or a coset in $\mathrm{RM}_{2,3}/\mathrm{RM}_{1,3}$, and the 16 branches converging into any state in $\Sigma_{16}(C)$ or $\Sigma_{24}(C)$ also form the codewords in $\mathrm{RM}_{1,3}$ or a coset in $\mathrm{RM}_{2,3}/\mathrm{RM}_{1,3}$. The overall trellis consists of 8 parallel isomorphic subtrellises without cross connections between them. Each subtrellis has 8 states at time-8, -16 or -24. Therefore, 8 identical trellis-based decoders can be devised to process the overall trellis in parallel.

$\triangle\triangle$

The next example analyzes the structure of the uniform 8-section minimal trellis diagram for a specific RM code. Analysis of uniform 8-section trellises for all other RM codes are the same.

Example 8.2 Let $m = 6$ and $r = 3$. The third-order RM code, $\mathrm{RM}_{3,6}$, is a $(64, 42)$ linear block code with minimum distance 8. From (8.22), this code can be constructed by 3-level squaring construction using $C_0 = \mathrm{RM}_{3,3}$, $C_1 = \mathrm{RM}_{2,3}$, $C_2 = \mathrm{RM}_{1,3}$ and $C_3 = \mathrm{RM}_{0,3}$ codes as component codes. Then

$$\mathrm{RM}_{3,6} = |\mathrm{RM}_{3,3}/\mathrm{RM}_{2,3}/\mathrm{RM}_{1,3}/\mathrm{RM}_{0,3}|^{2^3} .$$

It follows from (8.14) that the generator matrix of the $\mathrm{RM}_{3,6}$ code is given by

$$\begin{aligned} G &= I_8 \otimes G_3 \oplus G_{\mathrm{RM}}(0,3) \otimes G_{0/1} \\ &\quad \oplus G_{\mathrm{RM}}(1,3) \otimes G_{1/2} \oplus G_{\mathrm{RM}}(2,3) \otimes G_{2/3} \end{aligned}$$

where G_3, $G_{0/1}$, $G_{1/2}$ and $G_{2/3}$ are generator matrices for $\mathrm{RM}_{0,3}$, $[\mathrm{RM}_{3,3}/\mathrm{RM}_{2,3}]$, $[\mathrm{RM}_{2,3}/\mathrm{RM}_{1,3}]$ and $[\mathrm{RM}_{1,3}/\mathrm{RM}_{0,3}]$ respectively, and $\mathrm{Rank}(G_3) = 1$,

$\text{Rank}(G_{0/1}) = 1$, $\text{Rank}(G_{1/2}) = 3$ and $\text{Rank}(G_{2/3}) = 3$. Put all the matrices in trellis oriented form. Then the TOGM G for the $\text{RM}_{3,6}$ code is given by

$$\begin{aligned} G &= (11111111) \otimes G_{0/1} \\ &\oplus \begin{bmatrix} 1 & 1 & 1 & 1 & 0 & 0 & 0 & 0 \\ 0 & 1 & 0 & 1 & 1 & 0 & 1 & 0 \\ 0 & 0 & 1 & 1 & 1 & 1 & 0 & 0 \\ 0 & 0 & 0 & 0 & 1 & 1 & 1 & 1 \end{bmatrix} \otimes G_{1/2} \\ &\oplus \begin{bmatrix} 1 & 1 & 0 & 0 & 0 & 0 & 0 & 0 \\ 0 & 1 & 1 & 0 & 0 & 0 & 0 & 0 \\ 0 & 0 & 1 & 1 & 0 & 0 & 0 & 0 \\ 0 & 0 & 0 & 1 & 1 & 0 & 0 & 0 \\ 0 & 0 & 0 & 0 & 1 & 1 & 0 & 0 \\ 0 & 0 & 0 & 0 & 0 & 1 & 1 & 0 \\ 0 & 0 & 0 & 0 & 0 & 0 & 1 & 1 \end{bmatrix} \otimes G_{2/3} \oplus I_3 \otimes G_3. \end{aligned}$$

From this matrix, we notice that the compositions of the first and 8-th columns are identical, the compositions of 2nd, 3rd, 6-th and 7-th columns are identical, and the compositions of 4-th and 5-th columns are identical. Furthermore, we find that:

(1)

$$\begin{aligned} G_n^s &= (11111111) \otimes G_{0/1} \oplus (11110000) \otimes G_{1/2} \\ &\qquad \oplus (11000000) \otimes G_{2/3}, \\ G_{2n}^s &= (11111111) \otimes G_{0/1} \oplus \begin{bmatrix} 1 & 1 & 1 & 1 & 0 & 0 & 0 & 0 \\ 0 & 1 & 0 & 1 & 1 & 0 & 1 & 0 \end{bmatrix} \otimes G_{1/2} \\ &\qquad \oplus (01100000) \otimes G_{2/3}, \\ G_{3n}^s &= (11111111) \otimes G_{0/1} \oplus \begin{bmatrix} 1 & 1 & 1 & 1 & 0 & 0 & 0 & 0 \\ 0 & 1 & 0 & 1 & 1 & 0 & 1 & 0 \\ 0 & 0 & 1 & 1 & 1 & 1 & 0 & 0 \end{bmatrix} \otimes G_{1/2} \\ &\qquad \oplus (00110000) \otimes G_{2/3}, \\ G_{4n}^s &= (11111111) \otimes G_{0,1} \oplus \begin{bmatrix} 0 & 1 & 0 & 1 & 1 & 0 & 1 & 0 \\ 0 & 0 & 1 & 1 & 1 & 1 & 0 & 0 \end{bmatrix} \otimes G_{1/2} \\ &\qquad \oplus (00011000) \otimes G_{2/3}, \end{aligned}$$

$$G^s_{5n} = (11111111) \otimes G_{0/1} \oplus \begin{bmatrix} 0 & 1 & 0 & 1 & 1 & 0 & 1 & 0 \\ 0 & 0 & 1 & 1 & 1 & 1 & 0 & 0 \\ 0 & 0 & 0 & 0 & 1 & 1 & 1 & 1 \end{bmatrix} \otimes G_{1/2} \oplus (00001100) \otimes G_{2/3},$$

$$G^s_{6n} = (11111111) \otimes G_{0/1} \oplus \begin{bmatrix} 0 & 1 & 0 & 1 & 1 & 0 & 1 & 0 \\ 0 & 0 & 0 & 0 & 1 & 1 & 1 & 1 \end{bmatrix} \otimes G_{1/2} \oplus (00000110) \otimes G_{2/3},$$

$$G^s_{7n} = (11111111) \otimes G_{0/1} \oplus (00001111) \otimes G_{1/2} \oplus (00000011) \otimes G_{2/3}.$$

(2) For $1 \leq i \leq 8$, let $\boldsymbol{u}_i$ denote the i-th row of I_8. Then, $G^{f,p}_{(i-1)n,in} = \boldsymbol{u}_i \otimes G_3$ for $1 \leq i \leq 8$.

(3)

$$\begin{aligned} G^{f,s}_{0,n} &= (11111111) \otimes G_3 \oplus (11110000) \otimes G_{1/2} \oplus (11000000) \otimes G_{2/3}, \\ G^{f,s}_{n,2n} &= (01011010) \otimes G_{1/2} \oplus (01100000) \otimes G_{2/3}, \\ G^{f,s}_{2n,3n} &= (00111100) \otimes G_{1/2} \oplus (00110000) \otimes G_{2/3}, \\ G^{f,s}_{3n,4n} &= (00011000) \otimes G_{2/3}, \\ G^{f,s}_{4n,5n} &= (00001111) \otimes G_{1/2} \oplus (00001100) \otimes G_{2/3}, \\ G^{f,s}_{5n,6n} &= (00000110) \otimes G_{2/3}, \\ G^{f,s}_{6n,7n} &= (00000011) \otimes G_{2/3}, \\ G^{f,s}_{7n,8n} &= \emptyset. \end{aligned}$$

(4) For $1 \leq i \leq 8$, $p_{(i-1)n,in}(\text{RM}_{3,6}) = \text{RM}_{3,3}$.

From (1), we find that the state space dimension profile (SSDP) of the 8-section minimal trellis for the $\text{RM}_{3,6}$ code is $(0, 7, 10, 13, 10, 13, 10, 7, 0)$ and $\rho_{8,\max}(\text{RM}_{3,6}) = 13$. From (2) and (4), we find that for $1 \leq i \leq 8$, the partition

$$p_{(i-1)n,in}(C)/C^{\text{tr}}_{(i-1)n,in} = \text{RM}_{3,3}/\text{RM}_{0,3}$$

consists of 2^7 cosets. Therefore, each section of the trellis consists of $2^7 = 128$ distinct composite branches, and each composite branch consists two parallel branches which are codewords of a coset in $\text{RM}_{3,3}/\text{RM}_{0,3}$.

From (3), we find that the diverging branch dimension profile (DBDP) of the trellis is $(7, 6, 6, 3, 6, 3, 3, 0)$. From SSDP and DBDP, we can compute the branch complexity profile (BCP) and the converging branch dimension profile (CBDP) which are $(7, 13, 16, 16, 16, 16, 13, 7)$ and $(0, 3, 3, 6, 3, 6, 6, 7)$, respectively. DBDP and CBDP give the state connectivity at each section boundary location. For example, at boundary location $3n$, there are $2^3 = 8$ composite branches converging into each state and $2^3 = 8$ composite branches diverging from each state. From BCP and (6.15), we find the total number of 8-bit branches in the trellis is

$$\begin{aligned} B &= 2^7 \times 2 + 2^{13} \times 2 + 2^{16} \times 2 + 2^{16} \times 2 + 2^{16} \times 2 \\ &\quad + 2^{16} \times 2 + 2^{13} \times 2 + 2^7 \times 2 \\ &= 557,248. \end{aligned}$$

The total number of bit branches is then $557,248 \times 8 = 4,457,984$.

There is only one row in G whose active span contains the section boundary locations, $n, 2n, 3n, 4n, 5n, 6n$ and $7n$, which is $(11111111) \otimes G_{0/1}$. Therefore, there are only two parallel isomorphic subtrellises in the 8-section minimal trellis of the $\mathrm{RM}_{3,6}$ code. From the above analysis, we find that the trellis has the mirror symmetry structure. The last four sections form the mirror image of the first four sections.

Note that $I_{\max}(\mathrm{RM}_{3,6}) = \{3n, 5n\}$. By inspection of the TOGM G, we find that (see (7.4))

$$R(\mathrm{RM}_{3,6}) = (11111111) \otimes G_{0/1} \oplus \begin{bmatrix} 0 & 1 & 0 & 1 & 1 & 0 & 1 & 0 \\ 0 & 0 & 1 & 1 & 1 & 1 & 0 & 0 \end{bmatrix} \otimes G_{1/2}.$$

From $R(\mathrm{RM}_{3,6})$ and Theorem 7.2, we can decompose the 8-section minimal trellis for $\mathrm{RM}_{3,6}$ into parallel isomorphic subtrellises in various ways without exceeding $\rho_{8,\max}(\mathrm{RM}_{3,6}) = 13$. For example, by deleting the row $(11111111) \otimes G_{0/1}$, and any one among the 3 rows in $(01011010) \otimes G_{1/2}$, we obtain a $(64, 40)$ subcode C' of $\mathrm{RM}_{3,6}$ with SSDP $\{0, 6, 8, 11, 8, 11, 8, 6, 0\}$. Therefore, $\mathrm{RM}_{3,6}$ can be decomposed into 4 parallel isomorphic subtrellises, each with SSDP $\{0, 6, 8, 11, 8, 11, 8, 6, 0\}$. The maximum state space dimension of the resultant decomposed trellis is still $\rho_{8,\max}(\mathrm{RM}_{3,6}) = 13$. In fact, the entire set $R(\mathrm{RM}_{3,6})$ satisfies the condition of Theorem 7.2. Therefore, the 8-section minimal trellis for $\mathrm{RM}_{3,6}$ can be decomposed into 128 parallel isomorphic subtrellises, each has maximum space dimension equal to 64. The maximum state space dimension of the resultant decomposed trellis is still equal to $\rho_{8,\max}(\mathrm{RM}_{3,6}) = 13$ for the 8-section minimal trellis for $\mathrm{RM}_{3,6}$.

$\triangle\triangle$

Consider the 2^μ-section minimal trellis for the $\mathrm{RM}_{r,m}$ code with $0 \leq \mu \leq r$. From (8.14) and (8.22), we find that $C_0 = \mathrm{RM}_{r,m-\mu}$, $C_1 = \mathrm{RM}_{r-1,m-\mu}$ and

$$G_{0/1} = G_{\mathrm{RM}}(r, m-\mu) \backslash G_{\mathrm{RM}}(r-1, m-\mu). \tag{8.24}$$

Then the rows in the TOGM of the $\mathrm{RM}_{r,m}$ code whose active spans contain the section boundary locations, $n, 2n, \ldots, (2^\mu - 1)n$, are the rows in

$$(11\cdots 1) \otimes G_{0/1} = (11\cdots 1) \otimes [G_{\mathrm{RM}}(r, m-\mu) \backslash G_{\mathrm{RM}}(r-1, m-\mu)]. \tag{8.25}$$

Since $|G_{\mathrm{RM}}(r, m-\mu) \backslash G_{\mathrm{RM}}(r-1, m-\mu)| = \binom{m-\mu}{r}$, the 2^μ-section minimal trellis for the $\mathrm{RM}_{r,m}$ consists of $2^{\binom{m-\mu}{r}}$ parallel isomorphic subtrellises without cross connections between them.

8.3 SHANNON AND CARTESIAN PRODUCTS

Let C_1 and C_2 be an (N, K_1, d_1) and an (N, K_2, d_2) binary linear block codes with generator matrices, G_1 and G_2, respectively. Suppose C_1 and C_2 have only the all-zero codeword $\mathbf{0}$ in common, i.e., $C_1 \cap C_2 = \{\mathbf{0}\}$. Their direct-sum, denoted $C_1 \oplus C_2$, is defined as follows:

$$C \triangleq C_1 \oplus C_2 \triangleq \{\boldsymbol{u} + \boldsymbol{v} : \boldsymbol{u} \in C_1, \boldsymbol{v} \in C_2\}. \tag{8.26}$$

Then $C = C_1 \oplus C_2$ is an $(N, K_1 + K_2, d)$ linear block code with minimum distance $d \leq \min\{d_1, d_2\}$ and generator matrix

$$G = \begin{bmatrix} G_1 \\ G_2 \end{bmatrix}.$$

Let T_1 and T_2 be N-section trellis diagrams (minimal or nominimal) for C_1 and C_2, respectively. Then an N-section trellis T for the direct-sum code $C = C_1 \oplus C_2$ can be constructed by taking the **Shannon product** of T_1 and T_2 [41, 90]. Let $T_1 \times T_2$ denote the Shannon product of T_1 and T_2. For

$0 \leq i \leq N$, let $\Sigma_i(C_1)$ and $\Sigma_i(C_2)$ denote the state spaces of T_1 and T_2 at time-i, respectively. The construction of $T_1 \times T_2$ is carried out as follows:

(1) For $0 \leq i \leq N$, form the Cartesian product of $\Sigma_i(C_1)$ and $\Sigma_i(C_2)$,

$$\Sigma_i(C_1) \times \Sigma_i(C_2) \triangleq \{(\sigma_i^{(1)}, \sigma_i^{(2)}) : \sigma_i^{(1)} \in \Sigma_i(C_1), \sigma_i^{(2)} \in \Sigma_i(C_2)\}. \quad (8.27)$$

Then $\Sigma_i(C_1) \times \Sigma_i(C_2)$ forms the state space of $T_1 \times T_2$ at time-i, i.e., two-tuples in $\Sigma_i(C_1) \times \Sigma_i(C_2)$ form the nodes of $T_1 \times T_2$ at level-i.

(2) A state $(\sigma_i^{(1)}, \sigma_i^{(2)}) \in \Sigma_i(C_1) \times \Sigma_i(C_2)$ is adjacent to a state $(\sigma_{i+1}^{(1)}, \sigma_{i+1}^{(2)}) \in \Sigma_{i+1}(C_1) \times \Sigma_{i+1}(C_2)$ if and only if $\sigma_i^{(1)}$ is adjacent to $\sigma_{i+1}^{(1)}$ and $\sigma_i^{(2)}$ is adjacent to $\sigma_{i+1}^{(2)}$. Let $l(\sigma_i^{(1)}, \sigma_{i+1}^{(1)})$ denote the label of the branch that connects $\sigma_i^{(1)}$ to $\sigma_{i+1}^{(1)}$ in trellis T_1 and $l(\sigma_i^{(2)}, \sigma_{i+1}^{(2)})$ denote the label of the branch that connects $\sigma_i^{(2)}$ to $\sigma_{i+1}^{(2)}$ in the trellis T_2 . Then two adjacent states, $(\sigma_i^{(1)}, \sigma_i^{(2)})$ and $(\sigma_{i+1}^{(1)}, \sigma_{i+1}^{(2)})$, in trellis $T_1 \times T_2$ are connected by a branch with label

$$l(\sigma_i^{(1)}, \sigma_{i+1}^{(1)}) + l(\sigma_i^{(2)}, \sigma_{i+1}^{(2)}).$$

For $0 \leq i \leq N$, let $\rho_i(C_1)$ and $\rho_i(C_2)$ be the dimensions of $\Sigma_i(C_1)$ and $\Sigma_i(C_2)$, respectively. Then the state space dimension profile of $T_1 \times T_2$ is

$$(\rho_0(C_1) + \rho_0(C_2), \rho_1(C_1) + \rho_1(C_2), \ldots, \rho_N(C_1) + \rho_N(C_2)).$$

Example 8.3 Let C_1 and C_2 be two linear block codes of length 8 generated by

$$G_1 = \begin{bmatrix} 1 & 1 & 1 & 1 & 0 & 0 & 0 & 0 \\ 0 & 1 & 0 & 1 & 1 & 0 & 1 & 0 \end{bmatrix}$$

and

$$G_2 = \begin{bmatrix} 0 & 0 & 1 & 1 & 1 & 1 & 0 & 0 \\ 0 & 0 & 0 & 0 & 1 & 1 & 1 & 1 \end{bmatrix},$$

respectively. It is easy to check that $C_1 \cap C_2 = \{\mathbf{0}\}$ The direct-sum $C_1 \oplus C_2$ is generated by

$$G = \begin{bmatrix} G_1 \\ G_2 \end{bmatrix} = \begin{bmatrix} 1 & 1 & 1 & 1 & 0 & 0 & 0 & 0 \\ 0 & 1 & 0 & 1 & 1 & 0 & 1 & 0 \\ 0 & 0 & 1 & 1 & 1 & 1 & 0 & 0 \\ 0 & 0 & 0 & 0 & 1 & 1 & 1 & 1 \end{bmatrix},$$

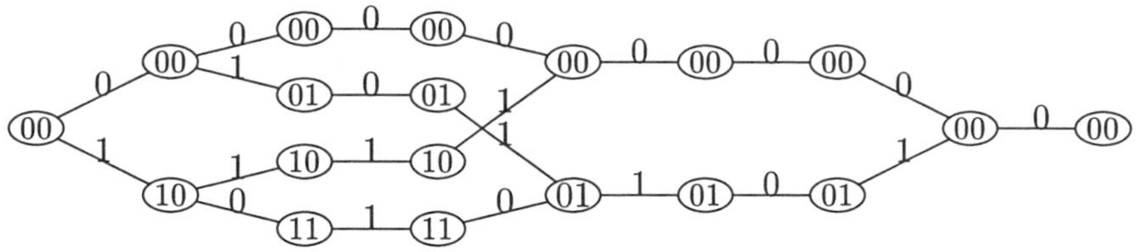

Figure 8.1. The 8-section minimal trellis for the code the code generated by $G_1 = \begin{bmatrix} 1\,1\,1\,1\,0\,0\,0\,0 \\ 0\,1\,0\,1\,1\,0\,1\,0 \end{bmatrix}$.

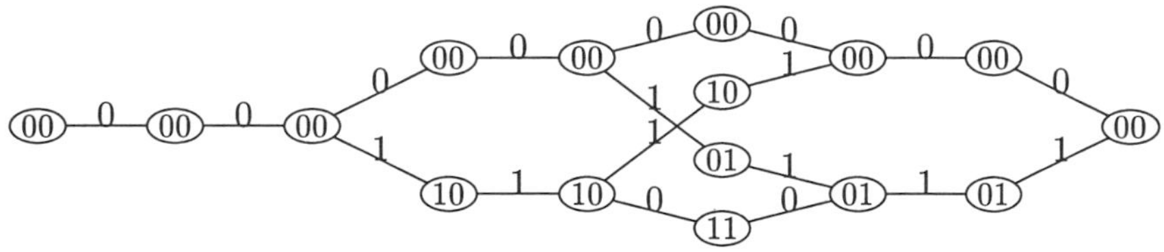

Figure 8.2. The 8-section minimal trellis for the code the code generated by $G_2 = \begin{bmatrix} 0\,0\,1\,1\,1\,1\,0\,0 \\ 0\,0\,0\,0\,1\,1\,1\,1 \end{bmatrix}$.

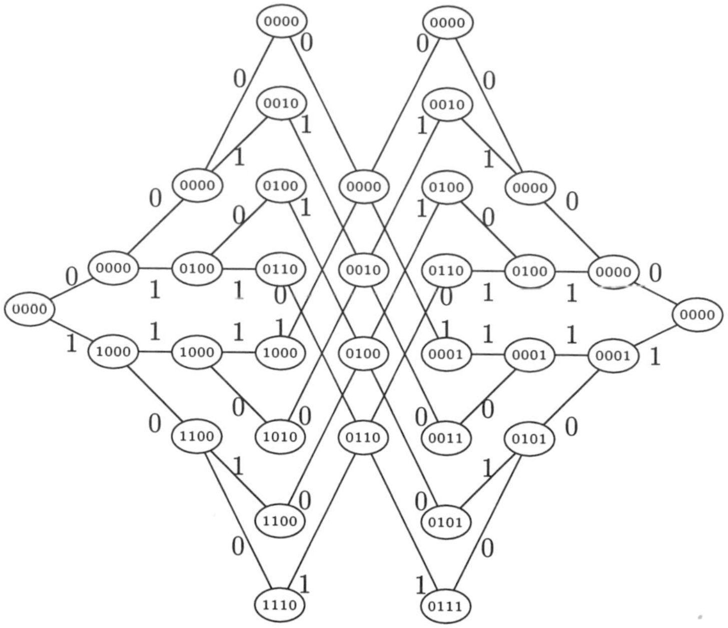

Figure 8.3. Shannon product of the trellises of Figures 8.1 and 8.2.

which is simply the TOGM for the $(8,4,4)$ RM code given in Example 3.1. Both G_1 and G_2 are in TOF. Based on G_1 and G_2, we construct the 8-section trellises T_1 and T_2 for C_1 and C_2 as shown in Figures 8.1 and 8.2, respectively. Taking the Shannon product of T_1 and T_2, we obtain an 8-section trellis $T_1 \times T_2$ for the direct-sum $C_1 \oplus C_2$ as shown in Figure 8.3, which is simply the 8-section minimal trellis for the $(8,4,4)$ RM code as shown in Figure 3.2 (or Figure 4.1). Labeling method by state defining sets of Section 4.1 is used to label the states in T_1, T_2 and $T_1 \times T_2$.

△△

Shannon product can be generalized to construct a trellis for a code which is a direct-sum of m linear block codes. For a positive integer $m \geq 2$ and $1 \leq j \leq m$, let C_j be an (N, K_j, d_j) linear block code. Suppose $C_1, C_2, \ldots, C_m$

satisfying the following condition: For $1 \leq j < j' \leq m$,

$$C_j \cap C_{j'} = \{\mathbf{0}\}. \tag{8.28}$$

This condition simply implies that for $\boldsymbol{v}_j \in C_j$ with $1 \leq j \leq m$,

$$\boldsymbol{v}_1 + \boldsymbol{v}_2 + \cdots + \boldsymbol{v}_m = \mathbf{0}$$

if and only if $\boldsymbol{v}_1 = \boldsymbol{v}_2 = \cdots = \boldsymbol{v}_m = \mathbf{0}$. The direct-sum of $C_1, C_2, \ldots, C_m$ is defined as

$$\begin{aligned} C &\triangleq C_1 \oplus C_2 \oplus \cdots \oplus C_m \\ &= \{\boldsymbol{v}_1 + \boldsymbol{v}_2 + \cdots + \boldsymbol{v}_m : \boldsymbol{v}_j \in C_j, 1 \leq j \leq m\}. \end{aligned} \tag{8.29}$$

Then $C = C_1 \oplus C_2 \oplus \cdots \oplus C_m$ is an (N, K, d) linear block code with

$$\begin{aligned} K &= K_1 + K_2 + \cdots + K_m \\ d &\leq \min_{1 \leq j \leq m} \{d_j\}. \end{aligned}$$

Let G_j be the generator matrix of C_j for $1 \leq j \leq m$. Then $C = C_1 \oplus C_2 \oplus \cdots \oplus C_m$ is generated by the following matrix:

$$G = \begin{bmatrix} G_1 \\ G_2 \\ \vdots \\ G_m \end{bmatrix} \tag{8.30}$$

For $1 \leq j \leq m$, let T_j be an N-section trellis for C_j. Then the Shannon product can be applied to construct an N-section trellis diagram for the direct-sum, $C = C_1 \oplus C_2 \oplus \cdots \oplus C_m$, from the trellises, $T_1, T_2, \ldots, T_m$. For $0 \leq i \leq N$, the state space of the Shannon product $T_1 \times T_2 \times \cdots \times T_m$ at time-i is given by the following set of m-tuples:

$$\begin{aligned} &\Sigma_i(C_1) \times \Sigma_i(C_2) \times \cdots \times \Sigma_i(C_m) \\ &\quad \triangleq \{(\sigma_i^{(1)}, \sigma_i^{(2)}, \ldots, \sigma_i^{(m)}) : \sigma_i^{(j)} \in \Sigma_i(C_j) \text{ for } 1 \leq j \leq m\}. \end{aligned} \tag{8.31}$$

Two states, $\boldsymbol{\sigma}_i \triangleq (\sigma_i^{(1)}, \sigma_i^{(2)}, \ldots, \sigma_i^{(m)})$ and $\boldsymbol{\sigma}_{i+1} \triangleq (\sigma_{i+1}^{(1)}, \sigma_{i+1}^{(2)}, \ldots, \sigma_{i+1}^{(m)})$, in $T_1 \times T_2 \times \cdots \times T_m$ at time-i and time-$(i+1)$ are adjacent if and only if for

$1 \le j \le m$, $\sigma_i^{(j)}$ and $\sigma_{i+1}^{(j)}$ are adjacent. In this case, $\boldsymbol{\sigma}_i$ and $\boldsymbol{\sigma}_{i+1}$ are connected by a branch with label

$$l(\sigma_i^{(1)}, \sigma_{i+1}^{(1)}) + l(\sigma_i^{(2)}, \sigma_{i+1}^{(2)}) + \cdots + l(\sigma_i^{(m)}, \sigma_{i+1}^{(m)}). \tag{8.32}$$

From the above construction of the N-section trellis $T^{\oplus} = T_1 \times T_2 \times \cdots \times T_m$ for the direst-sum code $C = C_1 \oplus C_2 \oplus \cdots \oplus C_m$, we find that the state space dimension of $T^{\oplus}$ at time-i is given by

$$\rho_j(C_1) + \rho_j(C_2) + \cdots + \rho_j(C_m) \tag{8.33}$$

for $0 \le i \le N$, when $T_1, T_2, \ldots, T_m$ are minimal N-section trellises for the component codes $C_1, C_2, \ldots, C_m$, respectively.

Example 8.4 Again we consider the $(8, 4, 4)$ RM code generated by the following TOGM,

$$G = \begin{bmatrix} 1 & 1 & 1 & 1 & 0 & 0 & 0 & 0 \\ 0 & 1 & 0 & 1 & 1 & 0 & 1 & 0 \\ 0 & 0 & 1 & 1 & 1 & 1 & 0 & 0 \\ 0 & 0 & 0 & 0 & 1 & 1 & 1 & 1 \end{bmatrix}.$$

For $1 \le j \le 4$, let C_j be the $(8, 1, 4)$ code generated by the j-th row of G. Then the direct-sum, $C_1 \oplus C_2 \oplus C_3 \oplus C_4$, gives the $(8, 4, 4)$ RM code. The 8-section minimal trellises for the four component codes are shown in Figure 8.4. The Shannon products, $T_1 \times T_2$ and $T_3 \times T_4$, generate the trellises for $C_1 \oplus C_2$ and $C_3 \oplus C_4$, respectively as shown in Figures 8.1 and 8.2. The Shannon product $(T_1 \times T_2) \times (T_3 \times T_4)$ results in the overall trellis for the $(8, 4, 4)$ RM code as shown in Figure 8.3.

Suppose $T_1, T_2, \ldots, T_m$ are minimal N-section trellises for the component codes $C_1, C_2, \ldots, C_m$, respectively. The Shannon product $T_1 \times T_2 \times \cdots \times T_m$ is not necessarily the minimal N-section trellis for the direct-sum $C = C_1 \oplus C_2 \oplus \cdots \oplus C_m$. Let T denote the minimal N-section trellis for C and $\rho_i(C)$ denote the state space dimension of T at time-i for $0 \le i \le N$. Then

$$\rho_i(C) \le \sum_{j=1}^{m} \rho_i(C_j). \tag{8.34}$$

If the equality of (8.34) holds for $0 \le i \le N$, then the Shannon product $T_1 \times T_2 \times \cdots \times T_m$ is the minimal N-section trellis for the direct-sum $C =$

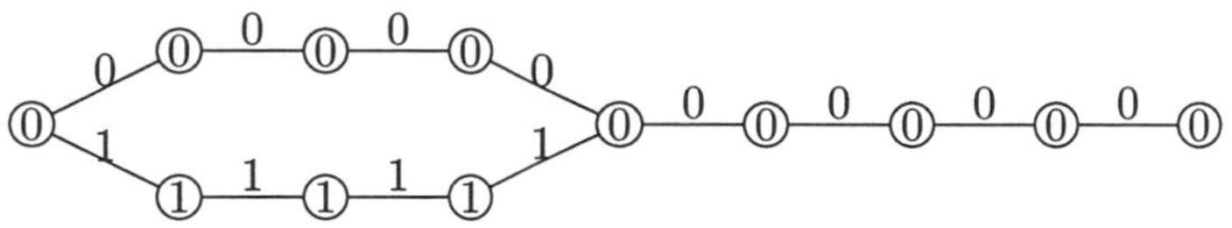

(a) T_1

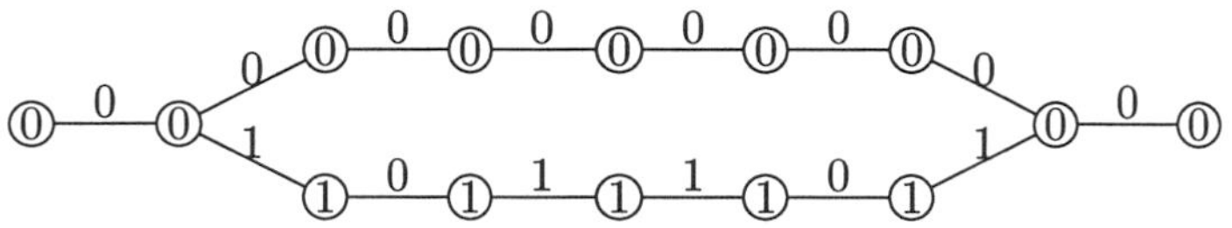

(b) T_2

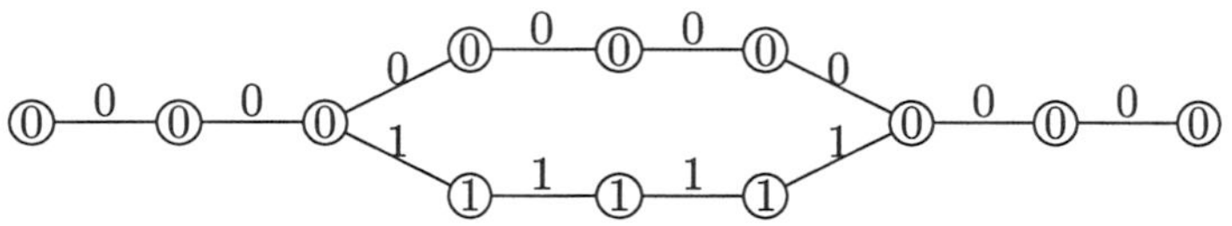

(c) T_3

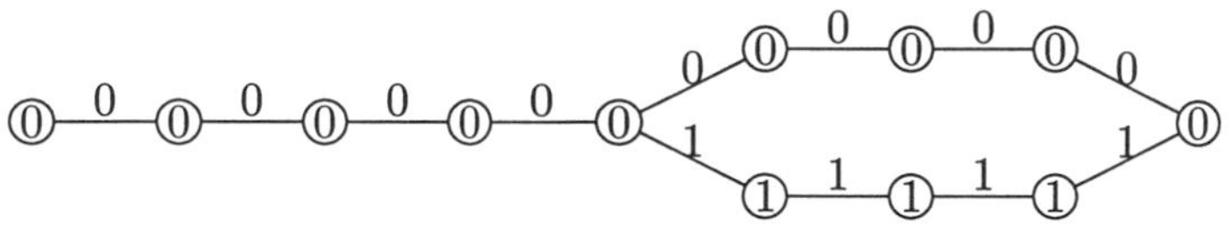

(d) T_4

Figure 8.4. The 8-section minimal trellises for the four component codes.

$C_1 \oplus C_2 \oplus \cdots \oplus C_m$. Theorem 8.1 gives a sufficient condition for the equality of (8.34).

Theorem 8.1 Consider the direct-sum $C = C_1 \oplus C_2 \oplus \cdots \oplus C_m$. For $1 \leq j \leq m$, let T_j be the minimal N-section trellis for the component code C_j. Then, the Shannon product $T_1 \times T_2 \times \cdots \times T_m$ is the minimal N-section trellis for C, if and only if the following condition holds: For $0 \leq i \leq N$, $1 \leq j < j' \leq m$,

$$p_{0,i}(C_j) \cap p_{0,i}(C_{j'}) = \{\mathbf{0}\}. \tag{8.35}$$

Proof: It follows from the condition of (8.35) that we have

$$p_{0,i}(C) = p_{0,i}(C_1) \oplus p_{0,i}(C_2) \oplus \cdots \oplus p_{0,i}(C_m) \tag{8.36}$$

$$C_{0,i} = (C_1)_{0,i} \oplus (C_2)_{0,i} \oplus \cdots \oplus (C_m)_{0,i} \tag{8.37}$$

for $0 \leq i \leq N$. From (8.36) and (8.37), we have

$$k(p_{0,i}(C)) = \sum_{j=1}^{m} k(p_{0,i}(C_j)) \tag{8.38}$$

$$k(C_{0,i}) = \sum_{j=1}^{m} k((C_j)_{0,i}) \tag{8.39}$$

where $k(D)$ denotes the dimension of a linear block code D. For $0 \leq i \leq N$, let $\rho_i(C)$ denote the state space dimension of the minimal N-section trellis for C. From (3.7), we find that for $1 \leq i \leq N$,

$$\rho_i(C) = k(p_{0,i}(C)) - k(C_{0,i}) \tag{8.40}$$

$$\rho_i(C_j) = k(p_{0,i}(C_j) - k((C_j)_{0,i}) \tag{8.41}$$

for $1 \leq j \leq m$. It follows from (8.38) to (8.41) that for $0 \leq i \leq N$,

$$\begin{aligned} \rho_i(C) &= \sum_{j=1}^{m} k(p_{0,i}(C_j)) - \sum_{j=1}^{m} k((C_j)_{0,i}) \\ &= \sum_{j=1}^{m} \rho_i(C_j). \end{aligned} \tag{8.42}$$

This implies that the Shannon product $T_1 \times T_2 \times \cdots \times T_m$ is the minimal N-section trellis for the direct-sum $C = C_1 \oplus C_2 \oplus \cdots \oplus C_m$.

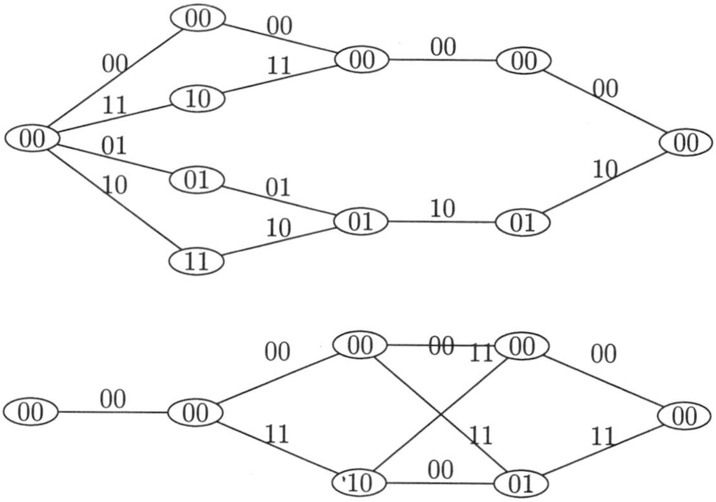

Figure 8.5. 4-section trellises for the trellises of Figures 8.1 and 8.2.

If the condition of (8.35) does not hold for some i, j and j' with $0 \leq i \leq N$ and $1 \leq j < j' \leq m$, then the equality of (8.38) does not hold and therefore, the equality of (8.42) does not hold either.

△△

The Shannon product can be applied to sectionalized trellises. In this case, all the trellises must have the same number of sections and corresponding sections must have the same length.

Example 8.5 Suppose we sectionalize each of the two 8-section trellises of Figure 8.1 and 8.2 into 4 sections, each of length 2. The resultant 4-section trellises are shown in Figure 8.5 and the Shannon product of these two 4-section trellises gives a 4-section trellis as shown in Figure 8.6 which is the same as the 4-section trellis for the $(8, 4, 4)$ RM code as shown in Figure 6.1.

△△

From the above, we see that if a code C can be decomposed as a direct-sum of m component codes, $C_1, C_2, \ldots, C_m$ for which $C_i \cap C_j = \{\mathbf{0}\}$ for $i \neq j$, a

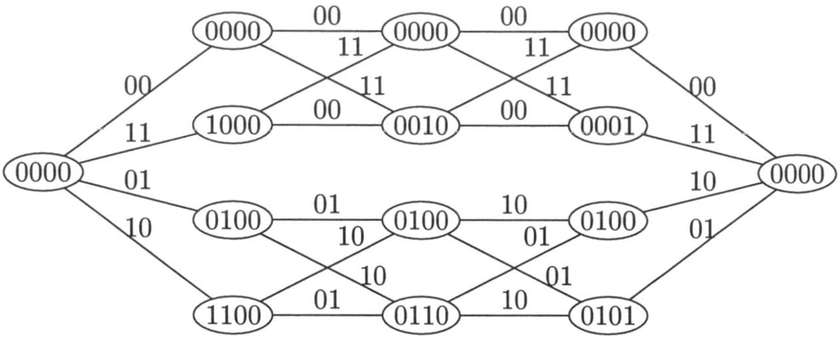

Figure 8.6. A 4-section trellises for the $(8,4,4)$ RM code.

trellis for C can be constructed by taking the Shannon product of the trellises for the component codes. This method is very useful for constructing trellises for long block codes that can be decomposed as direct-sum codes.

For $1 \leq j \leq m$, let C_j be an (N, K_j) linear block code. Take m codewords, one from each code, and arrange them as an array as follows:

$$\begin{aligned} \boldsymbol{v}_1 &= (v_{11}, v_{12} \ldots, v_{1N}) \\ \boldsymbol{v}_2 &= (v_{21}, v_{22} \ldots, v_{2N}) \\ &\vdots \\ \boldsymbol{v}_m &= (v_{m1}, v_{m2} \ldots, v_{mN}). \end{aligned} \tag{8.43}$$

Then we transmit this array column by column. This interleaving of m codes results in an $(mN, K_1 + K_2 + \cdots + K_m)$ linear block code, denoted $C_1 * C_2 * \cdots * C_m$, which is called an interleaved code.

For $1 \leq j \leq m$, let T_j be an N-section trellis for C_j. The **Cartesian product** of $T_1, T_2, \ldots, T_m$ results in an N-section trellis $T^* = T_1 \times T_2 \times \cdots \times T_m$. The formation of state spaces for T^* and condition for state adjacency between states in T^* are the same as for the Shannon product. The difference is

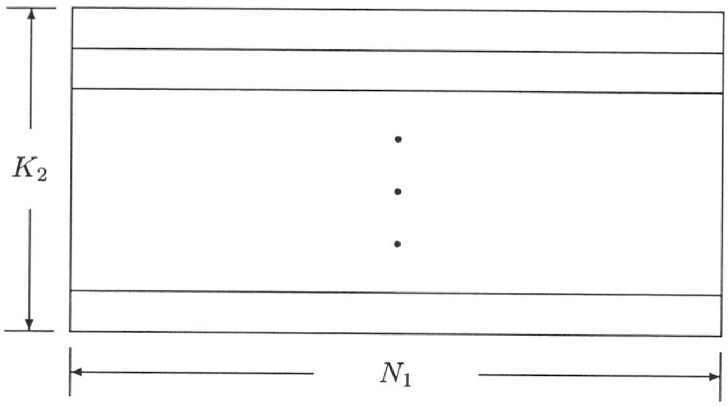

Figure 8.7. An $K_2 \times N_1$ array in which the rows are codewords from C_1.

branch labeling. For two adjacent states, $\boldsymbol{\sigma}_i \triangleq (\sigma_i^{(1)}, \sigma_i^{(2)}, \ldots, \sigma_i^{(m)})$ and $\boldsymbol{\sigma}_{i+1} \triangleq (\sigma_{i+1}^{(1)}, \sigma_{i+1}^{(2)}, \ldots, \sigma_{i+1}^{(m)})$, in T^* at time-i and time-$(i+1)$, let $l_j \triangleq l(\sigma_i^{(j)}, \sigma_{i+1}^{(j)})$ be the label of the branch that connects state $\sigma_i^{(j)}$ and state $\sigma_{i+1}^{(j)}$ for $1 \leq j \leq m$. We connect state $\boldsymbol{\sigma}_i$ and $\boldsymbol{\sigma}_{i+1}$ by a branch which is labeled by the following m-tuple:

$$(l_1, l_2, \ldots, l_m). \tag{8.44}$$

This label is simply a column in the array of (8.43). With this branch labeling, the Cartesian product $T_1 \times T_2 \times \cdots \times T_m$ is an N-section trellis for the interleaved code $C_1 * C_2 * \cdots * C_m$, each branch in the trellis is labeled with m code bits, one from each component code.

In **multilevel coded modulation** [13, 43, 95], each column of the array of (8.43) is mapped into a signal in a signal constellation S with 2^m signal points. This results in a block modulation code over signal space S. With this mapping, $T_1 \times T_2 \times \cdots \times T_m$ becomes a trellis for the modulation code. Modulation codes find applications in bandwidth limited communication systems to achieve reliable data transmission without compromising bandwidth efficiency.

Now we consider **product codes** and their trellis construction. Let C_1 be an (N_1, K_1, d_1) linear block code and C_2 be an (N_2, K_2, d_2) linear block code. Take K_2 codewords from C_1 and form a $K_2 \times N_1$ array of K_2 rows and N_1 columns, each row is a codeword in C_1 as shown in Figure 8.7. Then encode

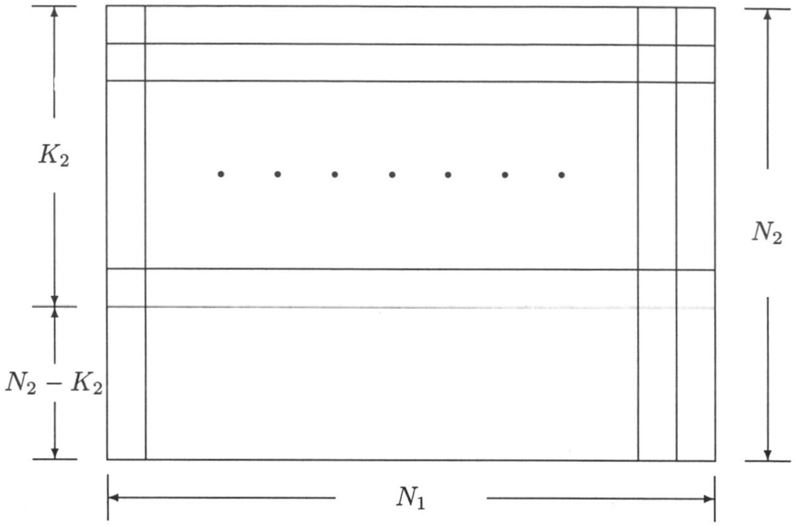

Figure 8.8. A two-dimensional code array.

each column of the array into a codeword in C_2. This results in an $N_2 \times N_1$ code array as shown in Figure 8.8. Every row of this $N_2 \times N_1$ array is a codeword in C_1 and every column is a codeword in C_2. The collection of these $N_2 \times N_1$ code arrays form a two-dimensional product code, denoted $C_1 \times C_2$. If each code array is transmitted column by column (or row by row), we obtain an (N_1N_2, K_1K_2, d_1d_2) linear block code. To construct a trellis for the product $C_1 \times C_2$, we regard the $K_2 \times N_1$ array of Figure 8.7 as an interleaved array with codewords from the same code C_1. We then construct an N-section trellis for the interleaved code

$$\underbrace{C_1 * C_1 * \cdots * C_1}_{K_2} \tag{8.45}$$

using the Cartesian product. Finally each K_2-tuple branch label in the trellis is encoded into a codeword in C_2. This results in an N_1-section trellis for the product code $C_1 \times C_2$, each section is N_2-bit in length.

Product of codes is a powerful technique for constructing long powerful block codes from short component codes.

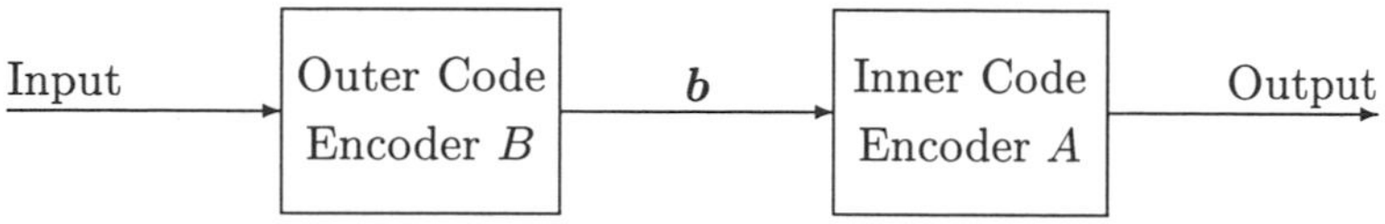

Figure 8.9. A Concatenated encoder.

8.4 MULTILEVEL CONCATENATED CODES AND THEIR TRELLIS CONSTRUCTION

Concatenation is another technique for constructing powerful codes from component codes. Concatenated codes have been widely used in data communication systems to achieve high data reliability with reduced decoding complexity. The concept of concatenated coding was first introduced by Forney [22].

Let A be a binary (n, k, d_A) linear block code and B be an (N, K, d_B) linear block code with symbols from the Galois field GF(2^k). Concatenation of A and B is achieved in two steps (see Figure 8.9):

(1) An information sequence of Kk bits is first divided into K bytes, each byte consists of k bits. Each k-bit byte is regarded as a symbol in GF(2^k). The K-byte information sequence is then encoded into an N-byte codeword

$$\boldsymbol{b} = (b_1, b_2, \ldots, b_N)$$

in B. This is called the **outer encoding** and code B is called the **outer code.**

(2) Each k-bit byte symbol of the outer codeword $\boldsymbol{b} = (b_1, b_2, \ldots, b_N)$ is then encoded into an n-bit codeword in A. This is referred to as the **inner encoding** and A is called the **inner code.**

The above two-step concatenated encoding results in a code sequence of Nn bits. It is a sequence of N codewords in A. The collection of all the possible concatenated code sequences forms an $(Nn, Kk, d_A d_B)$ binary linear block code which is called a **concatenated code**, denoted $B \circ A$.

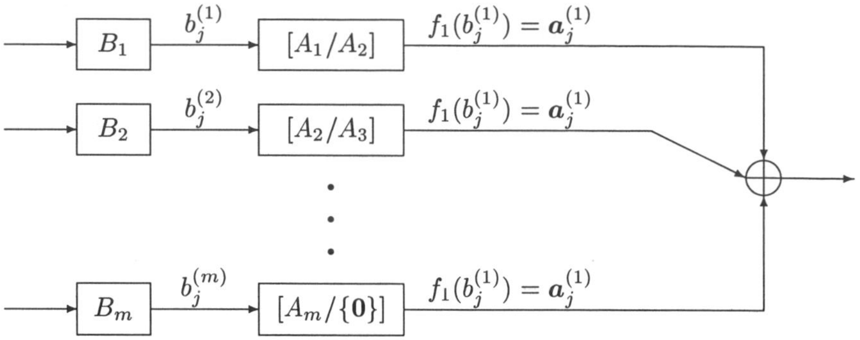

Figure 8.10. An encoder for an m-level concatenated code.

An N-section trellis for the concatenated code $B \circ A$ can be done as follows:

(1) Construct an N-section trellis for the outer code B. Each branch in the trellis is labeled with a k-bit byte.

(2) Each k-bit byte label of a branch in the trellis for the outer code B is encoded into a codeword in the inner code A.

The above construction results in an N-section trellis for the concatenated code $B \circ A$, each branch is labeled by a codeword in the inner code A.

The above one-level concatenated coding scheme can be generalized in multilevel. An m-level concatenated code is formed from a set of m inner codes and a set of m outer codes as shown in Figure 8.10.

The m inner codes are coset codes formed from a binary (n, k_1, d_1) linear block code A_1 and a sequence of m linear subcodes of A_1, denoted $A_2, A_3, \ldots,$ $A_{m+1} = \{\mathbf{0}\}$, such that

$$A_1 \supset A_2 \supset A_3 \supset \cdots \supset A_m \supset A_{m+1} = \{\mathbf{0}\}. \tag{8.46}$$

For $1 \leq i \leq m$, let

(1) k_i and d_{A_i} be the dimension and minimum distance of A_i, respectively;

(2) A_i/A_{i+1} denote the partition of A_i based on its subcode A_{i+1};

(3) $[A_i/A_{i+1}]$ denote the set of representative of cosets in A_i/A_{i+1}.

For $1 \leq i \leq m$, let

$$q_i \triangleq |[A_i/A_{i+1}]| = 2^{k_i - k_{i+1}} \tag{8.47}$$

$$[A_i/A_{i+1}] = \{\boldsymbol{a}_j^{(i)} : 1 \leq j \leq q_i\}. \tag{8.48}$$

Then, for $1 \leq i \leq m$,

$$A_i = \bigcup_{j=1}^{q_i} (\boldsymbol{a}_j^{(i)} + A_{i+1}). \tag{8.49}$$

From (8.49), we see that each codeword in A_1 is a sum of m coset representatives from $[A_1/A_2], [A_2/A_3], \ldots, [A_m/\{\boldsymbol{0}\}]$, respectively. These m coset representative spaces serve as the inner codes for an m-level concatenated code.

For $1 \leq i \leq m$, the i-th outer code, denoted B_i, is an (N, K_i, d_{B_i}) linear block code over GF(q_i) where q_i is defined by (8.47). For $1 \leq i \leq m$, let $f_i(\cdot)$ be a one-to-one mapping from GF(q_i) to $[A_i/A_{i+1}] = \{\boldsymbol{a}_j^{(i)} : 1 \leq j \leq q_i\}$. The mapping $f_i(\cdot)$ is simply an encoding of the symbols from GF(q_i) onto the coset representative space $[A_i/A_{i+1}]$ (inner code encoding). For $1 \leq i \leq m$, let $(b_1^{(i)}, b_2^{(i)}, \ldots, b_N^{(i)})$ be a codeword in the i-th outer code B_i. Then the output sequence of the overall encoder of Figure 8.10 is

$$(\boldsymbol{c}_1, \boldsymbol{c}_2, \ldots, \boldsymbol{c}_N) \tag{8.50}$$

where for $1 \leq j \leq N$,

$$\boldsymbol{c}_j = \sum_{i=1}^{m} f_i(b_j^{(i)}) \tag{8.51}$$

which is a codeword in A_1. Therefore, $(\boldsymbol{c}_1, \boldsymbol{c}_2, \ldots, \boldsymbol{c}_N)$ is a sequence of N codewords in A_1.

The following collection of sequences,

$$C = \{(\boldsymbol{c}_1, \boldsymbol{c}_2, \ldots, \boldsymbol{c}_N) : \boldsymbol{c}_j = \sum_{i=1}^{m} f_i(b_j^{(i)}) \text{ and } (b_1^{(i)}, b_2^{(i)}, \ldots, b_N^{(i)}) \in B_i \text{ for } 1 \leq i \leq m \text{ and } 1 \leq j \leq N\}, \tag{8.52}$$

forms an m-level concatenated code. This multilevel code is denoted as follows:

$$C \triangleq \{B_1, B_2, \ldots, B_m\} \circ \{A_1, A_2, \ldots, A_m\}. \tag{8.53}$$

The minimum distance of C is lower bounded as follows:

$$d_{\min}(C) \geq \min\{d_{A_i} d_{B_i} : 1 \leq i \leq m\}. \tag{8.54}$$

The dimension of C is

$$K = \sum_{i=1}^{m} K_i(k_i - k_{i+1}). \tag{8.55}$$

In an m-level concatenation, the i-th inner and outer codes form the i-th component concatenated code. The output of the i-th concatenated encoder is a sequence of coset representatives from $[A_i/A_{i+1}]$. Note that for $1 \leq i, j \leq m$ and $i \neq j$,

$$[A_i/A_{i+1}] \cap [A_j/A_{j+1}] = \{\mathbf{0}\}. \tag{8.56}$$

It follows from (8.52) and (8.56), C is the direct-sum of m component concatenated codes. Let $B_i \circ [A_i/A_{i+1}]$ denote the i-th component concatenated code for $1 \leq i \leq m$. Then

$$C = B_1 \circ [A_1/A_2] \oplus B_2 \circ [A_2/A_3] \oplus \cdots \oplus B_m \circ [A_m/A_{m+1}]. \tag{8.57}$$

Since C is a direct-sum code, the Shannon product can be used to construct an N-section trellis for C in which the label of each branch is an n-bit codeword in A_1. We first form a trellis diagram for each component concatenated code, denoted T_i, for $1 \leq i \leq m$. Then the Shannon product $T_1 \times T_2 \times \cdots \times T_m$ is an N-section trellis for the multilevel concatenated code C.

The multilevel concatenated coding is a special case of the generalized concatenated coding scheme introduced by Zinoviev [114]. Multilevel concatenated coding together with the Shannon product of trellises can be used to construct good codes with relatively simple trellis complexity.

Example 8.6 Let $n = 7$ and $N = 9$. Let A_1 be the $(7, 7, 1)$ universal code. Choose the following subcodes of A_1:

(1) A_2 is the $(7, 6, 2)$ even parity-check code;

(2) A_3 is a $(7, 3, 4)$ code generated by the following generator matrix:

$$G_3 = \begin{bmatrix} 1 & 0 & 1 & 0 & 1 & 0 & 1 \\ 0 & 1 & 1 & 0 & 0 & 1 & 1 \\ 0 & 0 & 0 & 1 & 1 & 1 & 1 \end{bmatrix};$$

(3) $A_4 = \{(0000000)\}$.

We can readily see that

$$A_1 \supset A_2 \supset A_3 \supset A_4.$$

Form the partitions, A_1/A_2, A_2/A_3 and A_3/A_4. We choose the following coset representative spaces:

(1) $[A_1/A_2] = \{(0000000), (0000001)\}$ which is a $(7, 1, 1)$ code;

(2) $[A_2/A_3]$ is a $(7, 3, 2)$ code generated

$$G_2 = \begin{bmatrix} 0 & 0 & 1 & 0 & 0 & 0 & 1 \\ 0 & 0 & 0 & 0 & 1 & 0 & 1 \\ 0 & 0 & 0 & 0 & 0 & 1 & 1 \end{bmatrix};$$

(3) $[A_3/A_4] = [A_3/\{\mathbf{0}\}]$ is the $(7, 3, 4)$ code generated by G_3.

Then $[A_1/A_2], [A_2/A_3]$ and $[A_3/A_4]$ are used as the inner codes for a 3-level concatenated code.

For the outer codes, the choices are:

(1) B_1 is a binary $(9, 2, 6)$ code which is the product of the $(3, 1, 3)$ repetition code and the $(3, 2, 2)$ even parity-check code;

(2) B_2 is a $(9, 7, 3)$ **maximum distance separable (MDS)** code (an extended RS code) over GF(2^3); and

(3) B_3 is a $(9, 8, 2)$ MDS code (an extended RS code) over GF(2^3).

Then the 3-level concatenated code,

$$C = B_1 \circ [A_1/A_2] \oplus B_2 \circ [A_2/A_3] \oplus B_3 \circ [A_3/\{\mathbf{0}\}]$$

is a $(63, 47, 6)$ binary linear code.

For each component concatenated code, we construct a 9-section minimal trellis, each section is 7-bit long. We find that the maximum state space dimensions of the 9-section minimal trellises for the component concatenated codes are:

$$\rho_{9,\max}(B_1 \circ [A_1/A_2]) = 2,$$
$$\rho_{9,\max}(B_2 \circ [A_2/A_3]) = 6,$$
$$\rho_{9,\max}(B_3 \circ [A_3/\{\mathbf{0}\}]) = 3.$$

The Shannon product of the 3 trellises gives a 9-section trellis for the overall 3-level concatenated code. The maximum state dimension $\rho_{9,\max}(C)$ of the overall trellis is 11.

$\triangle\triangle$

Example 8.7 Let $n = 8$ and $N = 8$. We choose A_1 and its subcodes as the following RM codes of length $2^3 = 8$: $A_1 = \text{RM}_{3,3}$, $A_2 = \text{RM}_{2,3}$, $A_3 = \text{RM}_{1,3}$, $A_4 = \text{RM}_{0,3}$ and $A_5 = \{\mathbf{0}\}$. Then

$$\text{RM}_{3,3} \supset \text{RM}_{2,3} \supset \text{RM}_{1,3} \supset \text{RM}_{0,3} \supset \{\mathbf{0}\},$$

and

$$\begin{aligned}
[A_1/A_2] &= [\text{RM}_{3,3}/\text{RM}_{2,3}], \\
[A_2/A_3] &= [\text{RM}_{2,3}/\text{RM}_{1,3}], \\
[A_3/A_4] &= [\text{RM}_{1,3}/\text{RM}_{0,3}], \\
[A_4/A_5] &= [\text{RM}_{0,3}/\{\mathbf{0}\}].
\end{aligned}$$

For the outer codes, we choose the following:

(1) B_1 is the binary $(8, 1, 8)$ repetition code;

(2) B_2 is the $(8, 5, 4)$ RS code over GF(2^3);

(3) B_3 is the $(8, 7, 2)$ RS code over GF(2^3);

(4) B_4 is the binary $(8, 8, 1)$ universal code.

Then the 4-level concatenated code constructed by using the above inner and outer codes is a $(64, 45, 8)$ binary linear block code.

The maximum state space dimensions of the 8-section minimal trellises for the component concatenated codes are:

$$\begin{aligned}
\rho_{8,\max}(B_1 \circ [\text{RM}_{3,3}/\text{RM}_{2,3}]) &= 1, \\
\rho_{8,\max}(B_2 \circ [\text{RM}_{2,3}/\text{RM}_{1,3}]) &= 9, \\
\rho_{8,\max}(B_3 \circ [\text{RM}_{1,3}/\text{RM}_{0,3}]) &= 3, \\
\rho_{8,\max}(B_4 \circ [\text{RM}_{0,3}/\{\mathbf{0}\}]) &= 0.
\end{aligned}$$

The Shannon product of the trellises of the component concatenated codes gives an 8-section trellis for the overall 4-level concatenated code with maximum state space dimension $\rho_{8,\max}(C) = 13$. The Wolf bound on $\rho_{\max}(C)$ for the code is 19.

The code constructed by this example has the same code parameters as the $(64, 45, 8)$ extended BCH code.

△△

If a code can be decomposed as a multilevel concatenated code, then it can be decoded with multistage decoding to reduce decoding complexity. A code may be decomposed as a multilevel code in many ways. Different decompositions result in different decoding complexities and error performances. Therefore, it is desired to find a decomposition that gives a good trade-off between error performance and decoding complexity.

The most well known example of decomposable codes is the class of RM codes. In Section 8.1, it has been shown that an RM code can be decomposed in many ways using squaring construction (see Eq.(8.22)). For two positive integers r and m such that $0 \leq r < m$, let $K_{r,m}$ and $q_{r,m}$ denote the dimensions of the RM code, $\text{RM}_{r,m}$, and the coset code, $[\text{RM}_{r,m}/\text{RM}_{r-1,m}]$, respectively (Note that for $r = 0$, $\text{RM}_{-1,m} \triangleq \{\mathbf{0}\}$). Let $(n, k, d)^{\rho}$ denote the linear block code over $\text{GF}(2^{\rho})$ obtained by interleaving the binary (n, k, d) code to a depth (or degree) of ρ with each group of ρ interleaved bits regarded as a symbol in $\text{GF}(2^{\rho})$. Then, from (8.22), we can decompose the $\text{RM}_{r,m}$ code as a $(\nu+1)$-level concatenated code as follows [110]:

$$\begin{aligned} \text{RM}_{r,m} \quad = \quad & \{\text{RM}_{0,\mu}^{q_{r,m-\mu}}, \text{RM}_{1,\mu}^{q_{r-1,m-\mu}}, \ldots, \text{RM}_{\nu,\mu}^{K_{r-\nu,m-\mu}}\} \\ & \circ\{\text{RM}_{r,m-\mu}, \text{RM}_{r-1,m-\mu}, \ldots, \text{RM}_{r-\nu,m-\mu}\} \end{aligned} \tag{8.58}$$

where $1 \leq \mu \leq m-1$, $\nu = \mu$ for $r > \mu$ and $\nu = r$ otherwise.

Expression (8.58) displays that an RM code can be decomposed as a multilevel concatenated code in many ways with various number of levels. We also note that in any decomposition of an RM code, all the component inner and outer codes are also RM codes with shorter lengths and smaller dimensions.

Example 8.8 Let $m = 5$ and $r = 2$. Consider the $\text{RM}_{2,5}$ code which is a $(32, 16, 8)$ code. Set $\mu = 2$. Then, it follows from (8.58) that the $\text{RM}_{2,5}$ code can be decomposed into a 3-level concatenated code as follows:

$$\text{RM}_{2,5} = \{\text{RM}_{0,2}^{q_{2,3}}, \text{RM}_{1,2}^{q_{1,3}}, \text{RM}_{2,2}^{K_{0,3}}\} \circ \{\text{RM}_{2,3}, \text{RM}_{1,3}, \text{RM}_{0,3}\}. \tag{8.59}$$

We find that $q_{2,3} = 3, q_{1,3} = 3$ and $K_{0,3} = 1$. Then $\mathrm{RM}_{2,5}$ can be put in the following direct-sum form:

$$\begin{aligned}\mathrm{RM}_{2,5} &= (4,1,4)^3 \circ [(8,7,2)/(8,4,4)] \\ &\oplus (4,3,2)^3 \circ [(8,4,4)/(8,1,8)] \\ &\oplus (4,4,1) \circ [(8,1,8)/\{\mathbf{0}\}] .\end{aligned}$$

Using Cartesian product, we construct 4-section trellises for the outer codes, $(4,1,4)^3$, $(4,3,2)^3$ and $(4,4,1)$ as shown in Figure 8.11. Then we construct the trellises for the 3 component concatenated codes, denoted T_1, T_2 and T_3, using the method described at the beginning of this section. T_1, T_2 and T_3 all have section boundary locations in $\{0, 8, 16, 24, 32\}$. Both T_1 and T_2 have 8 states at time-8, time-16, and time-24. T_3 is a trivial one-state trellis. The Shannon product $T_1 \times T_2 \times T_3$ gives a 4-section trellis for the $\mathrm{RM}_{2,5}$ code with state space dimensional profile,

$$\rho(C) = (0, 6, 6, 6, 0).$$

It has 64 states at time-8, time-16 and time-24. The trellis consists of 8 parallel isomorphic subtrellises without cross connections between them, each subtrellis has a state space dimensional profile $(0, 3, 3, 3, 0)$. Using this trellis, we can devise 8 identical 8-state Viterbi decoders to process the 8 subtrellises in parallel. At the end, there are 8 survivors and the one with the largest metric is the most likely codeword. Therefore, Viterbi decoding of the $\mathrm{RM}_{2,5}$ code is very simple and very high decoding speed can be attained.

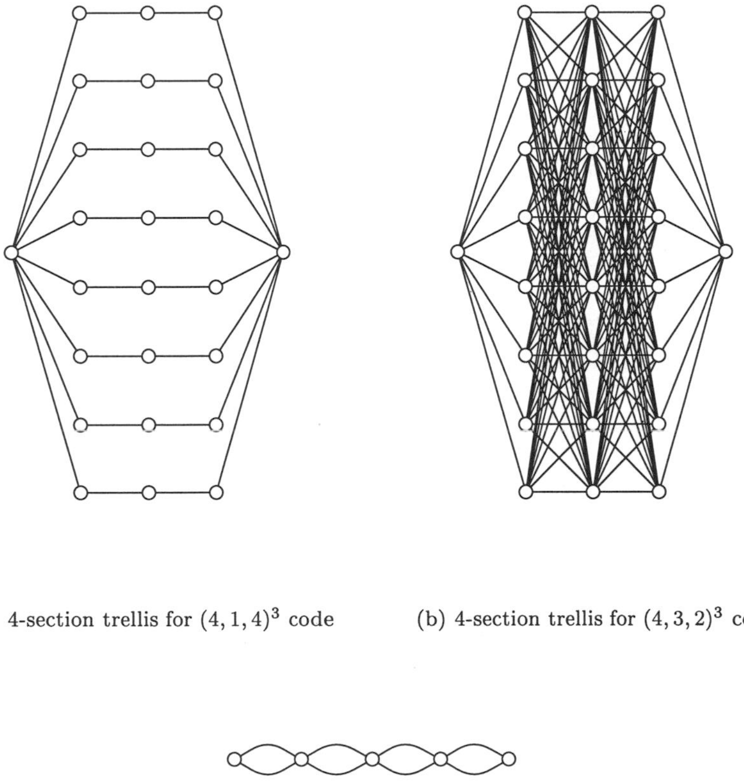

(a) 4-section trellis for $(4,1,4)^3$ code (b) 4-section trellis for $(4,3,2)^3$ code

(c) 4-section trellis for $(4,4,1)$ code

Figure 8.11. Trellises for outer codes.

9 TRELLISES FOR CONVOLUTIONAL CODES AND THEIR RELATED LINEAR BLOCK CODES

The trellis representation of codes was first introduced and used for convolutional codes [23]. This graphical representation, in conjunction with the Viterbi decoding algorithm, has resulted in a wide range of applications of convolutional codes for error control in digital communications over the last two decades. Trellises for convolutional codes are easy to construct. They are time-invariant and have a simple regular structure. In this chapter, we show that the concepts and techniques for constructing trellises for linear block codes developed in the previous chapters can be applied to constructing trellises for convolutional codes with only a slight modification due to the structural difference between these two classes of codes. Also presented in this chapter are trellises for linear block codes obtained by terminating a convolutional code.

9.1 DESCRIPTION OF CONVOLUTIONAL CODES

Consider a rate-k/n convolutional code C. The encoder consists of k inputs, n outputs and k input shift registers for storing information bits. The information sequence $\boldsymbol{a}$ to be encoded is first demultiplexed into k subsequences:

$$\begin{aligned}
\boldsymbol{a}^{(1)} &= (a_0^{(1)}, a_1^{(1)}, \ldots, a_l^{(1)}, \ldots) \\
\boldsymbol{a}^{(2)} &= (a_0^{(2)}, a_1^{(2)}, \ldots, a_l^{(2)}, \ldots) \\
&\vdots \\
\boldsymbol{a}^{(k)} &= (a_0^{(k)}, a_1^{(k)}, \ldots, a_l^{(k)}, \ldots).
\end{aligned} \tag{9.1}$$

These k subsequences are applied to the k inputs of the encoder. At time-l, the k input information bits are

$$a_l^{(1)}, a_l^{(2)}, \ldots, a_l^{(k)}.$$

Let $\boldsymbol{a}_l = (a_l^{(1)}, a_l^{(2)}, \ldots, a_l^{(k)})$. Then the information sequence $\boldsymbol{a}$ can be expressed as

$$\boldsymbol{a} = (\boldsymbol{a}_0, \boldsymbol{a}_1, \ldots, \boldsymbol{a}_l, \ldots). \tag{9.2}$$

The n code sequences generated at the outputs of the encoder are

$$\begin{aligned}
\boldsymbol{u}^{(1)} &= (u_0^{(1)}, u_1^{(1)}, \ldots, u_l^{(1)}, \ldots) \\
\boldsymbol{u}^{(2)} &= (u_0^{(2)}, u_1^{(2)}, \ldots, u_l^{(2)}, \ldots) \\
&\vdots \\
\boldsymbol{u}^{(n)} &= (u_0^{(n)}, u_1^{(n)}, \ldots, u_l^{(n)}, \ldots).
\end{aligned} \tag{9.3}$$

At time-l, the n output code bits are

$$u_l^{(1)}, u_l^{(2)}, \ldots, u_l^{(n)}.$$

These n output code bits depend not only on the k input information bits $(a_l^{(1)}, a_l^{(2)}, \ldots, a_l^{(k)})$ at time-l but also on the state of the encoder, which is defined by the information bits stored in the k input shift registers at time-l. Let $\boldsymbol{u}_l = (u_l^{(1)}, u_l^{(2)}, \ldots, u_l^{(n)})$. The overall code sequence is then represented by

$$\boldsymbol{u} = (\boldsymbol{u}_0, \boldsymbol{u}_1, \ldots, \boldsymbol{u}_l, \ldots), \tag{9.4}$$

which is simply obtained by interleaving the n output code sequences $\boldsymbol{u}^{(1)}$, $\boldsymbol{u}^{(2)}, \ldots, \boldsymbol{u}^{(n)}$. Therefore, every unit of time, k information bits are encoded into n code bits.

For $1 \leq i \leq k$, let m_i be the length of the i-th input register (i.e., the i-th input register consists of m_i memory elements). The total memory of the encoder is then

$$M = \sum_{i=1}^{k} m_i. \tag{9.5}$$

The information bits stored in this memory define the state of the encoder. This state, together with the k input information bits, uniquely determines the n output code bits. Therefore, the encoder can be modeled by a finite state machine (a linear sequential circuit) and its dynamic behavior can be represented by a trellis diagram. The parameter

$$m = \max_{1 \leq i \leq k} \{m_i\} \tag{9.6}$$

is called the **memory order** and $\nu = m + 1$ is called the **constraint length** of the code. A rate-k/n convolutional code of memory order m is commonly called an (n, k, m) convolutional code. Such a code is generated by a generator matrix of the following form [62]:

$$G = \begin{bmatrix} G_0 & G_1 & G_2 & \cdots\cdots & G_m & & & \\ & G_0 & G_1 & \cdots\cdots & G_{m-1} & G_m & & \\ & & G_0 & & G_{m-2} & G_{m-1} & G_m & \\ & & & \ddots & & & & \ddots \end{bmatrix} \tag{9.7}$$

where:

(1) for $0 \leq i \leq m$, G_i is a $k \times n$ submatrix whose entries are

$$G_i = \begin{bmatrix} g_{1,i}^{(1)} & g_{2,i}^{(1)} & \cdots\cdots & g_{n,i}^{(1)} \\ g_{1,i}^{(2)} & g_{2,i}^{(2)} & \cdots\cdots & g_{n,i}^{(2)} \\ \vdots & \vdots & \ddots & \vdots \\ g_{1,i}^{(k)} & g_{2,i}^{(k)} & \cdots\cdots & g_{n,i}^{(k)} \end{bmatrix}, \tag{9.8}$$

(2) both G_0 and G_m are nonzero.

Note that each set of k rows of G is identical to the previous set of k rows but shifted n places to the right. For convenience, we call the matrix $[G_0 G_1 \cdots G_m]$ the **generating pattern**. For an information sequence $\boldsymbol{a} = (\boldsymbol{a}_0, \boldsymbol{a}_1, \ldots, \boldsymbol{a}_l, \ldots)$, the code sequence $\boldsymbol{u} = (\boldsymbol{u}_0, \boldsymbol{u}_1, \ldots, \boldsymbol{u}_l, \ldots)$ is given by

$$\boldsymbol{u} = \boldsymbol{a} \cdot G, \tag{9.9}$$

which is a linear combination of rows of G. For $1 \leq j \leq n$, the j-th output code bit at time-l, with $l \geq 0$, is given by

$$u_l^{(j)} = \sum_{i=1}^{k} a_l^{(i)} g_{j,0}^{(i)} + \sum_{i=1}^{k} a_{l-1}^{(i)} g_{j,1}^{(i)} + \cdots + \sum_{i=1}^{k} a_{l-m}^{(i)} g_{j,m}^{(i)}. \tag{9.10}$$

From (9.10), we see that each block of n code bits not only depends on the current block of k information bits but also depends on the past m blocks of information bits that are stored in the k input shift registers.

9.2 TRELLIS STRUCTURE FOR CONVOLUTIONAL CODES

Since G_0 and G_m are nonzero submatrices, G_0 is the "leading" nonzero submatrix and G_m is the "trailing" nonzero submatrix of each set of k rows of the generator matrix G. Therefore, the span of the l-th set of k rows of G is the interval $[l-1, m+l-1]$ and its active span is the interval $[l-1, m+l-2]$. The above definitions of leading and trailing nonzero submatrices are analogous to the leading and trailing ones of rows in the generator matrix of a linear block code, and thus the generator matrix for an (n, k, m) convolutional code given by (9.7) is already in trellis oriented form. The construction of the code trellis is therefore the same as for linear block codes. At time-l, the generator matrix G can easily be partitioned into three disjoint submatrices, G_l^p, G_l^f, and G_l^s where

(1) G_l^p consists of those rows of G whose spans are contained in the interval $[0, l-1]$,

(2) G_l^f consists of those rows of G whose spans are contained in the interval $[l, \infty)$, and

(3) G_l^s consists of those rows of G whose active spans contain $l-1$.

The information blocks that correspond to the rows of G_l^s are $\boldsymbol{a}_{l-1}, \boldsymbol{a}_{l-2}, \ldots, \boldsymbol{a}_{l-m}$. Among these m information blocks, the information bits that are stored in the k input shift registers are the following M bits:

$$\begin{pmatrix} a_{l-1}^{(1)}, & a_{l-2}^{(1)}, & \cdots & a_{l-m_1}^{(1)} \\ a_{l-1}^{(2)}, & a_{l-2}^{(2)}, & \cdots & a_{l-m_2}^{(2)} \\ \vdots & \vdots & \ddots & \vdots \\ a_{l-1}^{(k)}, & a_{l-2}^{(k)}, & \cdots & a_{l-m_k}^{(k)} \end{pmatrix}. \tag{9.11}$$

These M bits define the state of the encoder at time-l. For $l \geq m$, these bits fill the k input shift registers and the state space $\Sigma_l(C)$ consists of 2^M allowable states. Note that the 2^M combinations of the M bits given in (9.11) do not change with respect to l for $l \geq m$. This says that for $l \geq m$, $\Sigma_l(C) = \Sigma(C)$ is time-invariant. From (9.10), we see that the output function is also time-invariant. For $l < m$, the k input shift registers are not completely filled and the unfilled memory elements are simply set to zero. In this case, the state space $\Sigma_l(C)$ is simply a subspace of $\Sigma(C)$. Therefore, the trellis for a convolutional code is time-invariant. For $l \geq m$, every section is identical.

In the code trellis, the 2^M states are simply labeled by the 2^M combinations of information bits stored in the input shift registers. From (9.10), we see that there are 2^k branches diverging from a state, each branch consists of n code bits and corresponds to a specific combination of the current k information bits, $a_l^{(1)}, a_l^{(2)}, \ldots, a_l^{(k)}$. Given the current state specified by the stored information bits of (9.11), and the current information block $(a_l^{(1)}, a_l^{(2)}, \ldots, a_l^{(k)})$, the next state is then given by

$$\begin{pmatrix} a_{l}^{(1)}, & a_{l-1}^{(1)}, & \cdots & a_{l-m_1+1}^{(1)} \\ a_{l}^{(2)}, & a_{l-1}^{(2)}, & \cdots & a_{l-m_2+1}^{(2)} \\ \vdots & \vdots & \ddots & \vdots \\ a_{l}^{(k)}, & a_{l-1}^{(k)}, & \cdots & a_{l-m_k+1}^{(k)} \end{pmatrix}. \tag{9.12}$$

The branch that connects the two states is labeled by the n code bits given by (9.10). From (9.11) and (9.12), we see that for $l > m$, there are 2^k states at time-$(l-1)$ connecting to the same next state at time-l. This implies that there are 2^k branches merging into each state in the trellis for $l > m$. Due to the regularity and time-invariant structure of the code trellis of a convolutional code, the trellis construction is much simpler than the trellis construction for a linear block code.

Due to the continuous nature of a convolutional code, its code trellis does not terminate and simply repeats indefinitely. If the information sequence $\boldsymbol{a}$ is finite in length, then the code trellis can be forced to terminate by appending m blocks of zero information bits. In this case, the final state is the all-zero state (i.e., the input shift registers contain only zeros). This results in a block code.

Example 9.1 Consider the rate-1/2 convolutional code C of memory order $m = 2$ which is generated by the following generator matrix

$$G = \begin{bmatrix} 1 & 1 & 0 & 1 & 1 & 1 & & & & & & & & \\ & & 1 & 1 & 0 & 1 & 1 & 1 & & & & & & \\ & & & & 1 & 1 & 0 & 1 & 1 & 1 & & & & \\ & & & & & & 1 & 1 & 0 & 1 & 1 & 1 & & \\ & & & & & & & \ddots & & \ddots & & & \ddots & \end{bmatrix}. \tag{9.13}$$

The generating pattern is (110111). The encoder has only one input and two outputs. Let $\boldsymbol{a} = (a_0, a_1, \ldots, a_l, \ldots)$ be the input information sequence. It follows from (9.10) that at time-l, the two output code bits are:

$$\begin{aligned} u_l^{(1)} &= a_l + a_{l-2}, \\ u_l^{(2)} &= a_l + a_{l-1} + a_{l-2}. \end{aligned}$$

From these two equations, we can easily construct the encoder as shown in Figure 9.1(a). The single input register consists of two memory elements. Therefore, for $i \geq 2$, $\Sigma_i(C)$ consists of 4 states labeled by

$$(00), (01), (10), (11).$$

The trellis diagram of this code is shown in Figure 9.1(b) (a reproduction of Figure 3.3).

△△

In general, the trellis of a convolutional code is densely connected and does not consist of parallel subtrellises without cross connections between them. Trellis decomposition is not possible.

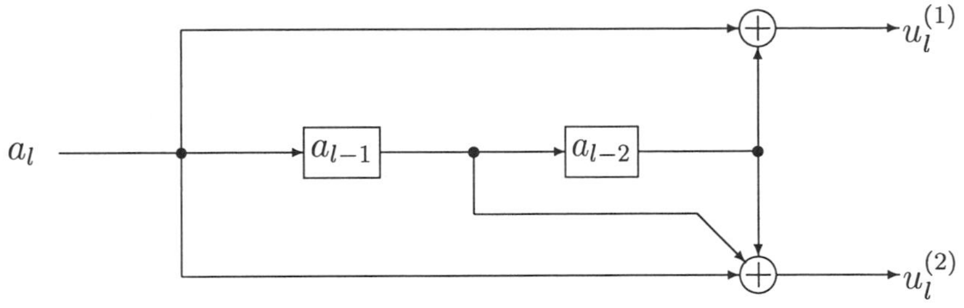

(a) Encoder

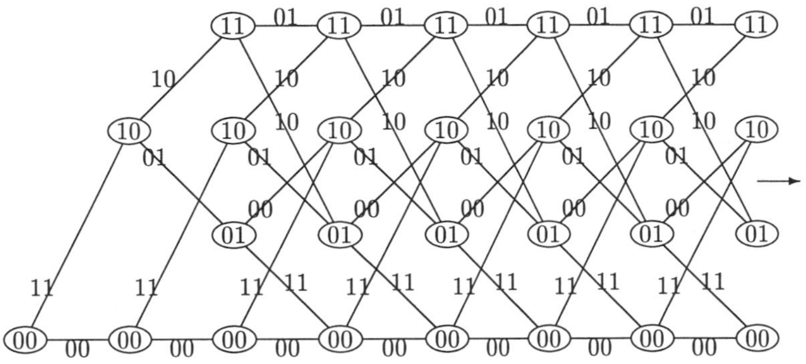

(b) Trellis diagram

Figure 9.1. Encoder and trellis diagram for the $(2, 1, 2)$ convolutional code given in Example 9.1.

9.3 PUNCTURED CONVOLUTIONAL CODES

In the trellis diagram of a rate-k/n convolutional code C with memory order m and total memory M, there are 2^k branches converging into each of the 2^M states in $\Sigma_l(C)$ and 2^k branches diverging from each of these 2^M states for $l > m$. Consequently, the number of branches entering or leaving a state grows exponentially with k. For high rate convolutional codes, large k results in a large radix number and a high degree of state connectivity, which makes the maximum likelihood trellis-based Viterbi decoder difficult to implement. This can be overcome by using a puncturing technique [16] that allows us to construct high rate convolutional codes from a rate-$1/n$ convolutional code. As a result, the trellises of the high rate convolutional codes constructed have the same state complexity and state connectivity as the trellis for the rate-$1/n$ convolutional code. For $l > m$, there are only two branches entering or leaving each state in the trellis. This simplifies the decoder implementation.

For $k \geq 1$ and $\alpha \geq 0$, a rate-$k/(kn - \alpha)$ convolutional code can be obtained by **periodically** deleting (puncturing) α code bits from every k consecutive output code blocks of a rate-$1/n$ convolutional code encoder (no more than $n - 1$ code bits can be deleted from a code block). The punctured (deleted) positions can be specified by an $n \times k$-matrix $P = [p_{i,j}]$ called the **puncturing matrix**, in which $p_{i,j} = 0$ if the i-th bit of the j-th code block is deleted, and $p_{i,j} = 1$ otherwise, for $1 \leq i \leq n$ and $1 \leq j \leq k$. The trellis diagram of the resultant punctured convolutional code is obtained by simply deleting the bit labels corresponding to the punctured positions on each branch of the trellis diagram of the rate-$1/n$ convolutional code. This periodic puncturing results in a non-uniform time-varying trellis with period k. A uniformly sectionalized time-invariant trellis can be obtained by combining every k sections of the time-varying trellis into one section using the technique presented in Section 6.1.

Example 9.2 For the $(2, 1, 2)$ code of Example 9.1, consider the puncturing matrix with $k = 3$ and $\alpha = 2$

$$P = \begin{bmatrix} 1 & 0 & 1 \\ 1 & 1 & 0 \end{bmatrix}.$$

This matrix defines a punctured rate-3/4 convolutional code with generator matrix $\tilde{G}$ obtained from the matrix G given in (9.13) by deleting the third and

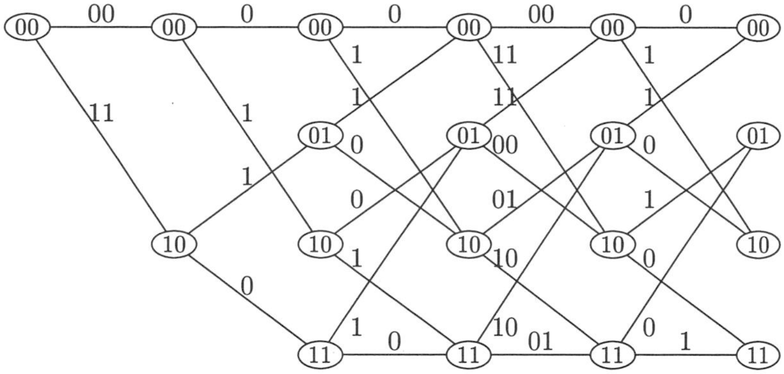

Figure 9.2. The trellis diagram for the rate 3/4 punctured convolutional code given in Example 9.2, obtained by deleting the punctured bits from the original trellis diagram of Example 9.1.

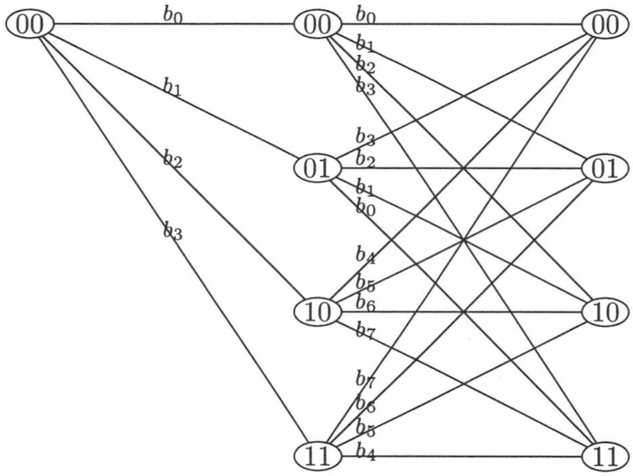

Figure 9.3. The uniformly sectionalized trellis diagram for the rate 3/4 punctured convolutional code given in Example 9.2 with composite branches $b_0 = \{0000, 1111\}, b_1 = \{0010, 1101\}, b_2 = \{0001, 1110\}, b_3 = \{0011, 1100\}, b_4 = \{0110, 1001\}, b_5 = \{0100, 1011\}, b_6 = \{0111, 1000\}, b_7 = \{0101, 1010\}$.

sixth columns in each group of six consecutive columns as

$$
\begin{aligned}
\tilde{G} &= \begin{bmatrix}
1 & 1 & \mathrm{x} & 1 & 1 & \mathrm{x} & & & & & & \\
 & & \mathrm{x} & 1 & 0 & \mathrm{x} & 1 & 1 & \mathrm{x} & & & \\
 & & & & 1 & \mathrm{x} & 0 & 1 & \mathrm{x} & 1 & & \\
 & & & & & \mathrm{x} & 1 & 1 & \mathrm{x} & 1 & 1 & \mathrm{x} \\
 & & & & & & & \ddots & & \ddots & & \ddots
\end{bmatrix} \\
&= \begin{bmatrix}
1 & 1 & 1 & 1 & 0 & 0 & 0 & 0 \\
0 & 0 & 1 & 0 & 1 & 1 & 0 & 0 \\
0 & 0 & 0 & 1 & 0 & 1 & 1 & 0 \\
 & & & & 1 & 1 & 1 & 1 \\
 & & & & & \ddots & & \ddots
\end{bmatrix}. \qquad (9.14)
\end{aligned}
$$

The trellis diagram of the punctured code is shown in Figure 9.2. We see that it has the same structure as the trellis diagram of the $(2,1,2)$ code as shown in Figure 9.1, except for the nonuniform sections. Note that the matrix of (9.14) generates a $(4,3,1)$ convolutional code as described in Section 9.1, with

$$
\begin{aligned}
G_0 &= \begin{bmatrix} 1 & 1 & 1 & 1 \\ 0 & 0 & 1 & 0 \\ 0 & 0 & 0 & 1 \end{bmatrix}, \\
G_1 &= \begin{bmatrix} 0 & 0 & 0 & 0 \\ 1 & 1 & 0 & 0 \\ 0 & 1 & 1 & 0 \end{bmatrix}.
\end{aligned}
$$

The corresponding uniformly sectionalized trellis diagram is shown in Figure 9.3.

△△

Puncturing is a simple method that allows us to construct a rate-k/n_1 convolutional code from a rate-$1/n$ convolutional code ($k > 1, n_1 \geq n$). The trellis diagram for the resultant code has the same structure and state complexity as that of the original rate-$1/n$ convolutional code. As a result, the decoder for the rate-$1/n$ code can be used for decoding the punctured code. In chapter 11, we will see that this property is very interesting for decoding convolutional codes based on their trellises. However, a rate-k/n_1 convolutional code obtained by puncturing generally does not perform as well as the best rate-k/n_1 convolutional code due to a poorer distance profile. On the other hand, the decoding simplicity compensates the degradation in error performance. Furthermore, puncturing allows us to change the code rate depending on the channel conditions without modifying the decoder [16].

9.4 TERMINATION OF CONVOLUTIONAL CODES

From (9.7), we see that the generator matrix of a rate-k/n convolutional code with memory order m has infinite dimension. The encoder encodes the input information sequence continuously and generates a code sequence of infinite length. However, in practical applications, any input information sequence has finite length and consists of a finite number of k-bit information blocks, say x blocks. This finite length input information sequence results in a code sequence of finite length and terminates the encoding process. The termination results in a block code.

Let $\boldsymbol{a} = (\boldsymbol{a}_0, \boldsymbol{a}_1, \ldots, \boldsymbol{a}_{x-1})$ be the information sequence to be encoded. A natural way to terminate the encoding process is to append m zero information blocks to $\boldsymbol{a}$. This results in the information sequence

$$\boldsymbol{a}^* = (\boldsymbol{a}_0, \boldsymbol{a}_1, \ldots, \boldsymbol{a}_{x-1}, \underbrace{\mathbf{0}, \mathbf{0}, \ldots, \mathbf{0}}_{m}).$$

When $\boldsymbol{a}^*$ is completely shifted into the encoder, the encoder returns to the all-zero state, where it started the encoding. The corresponding output code sequence

$$\boldsymbol{u} = (\boldsymbol{u}_0, \boldsymbol{u}_1, \ldots, \boldsymbol{u}_{x-1}, \boldsymbol{u}_x, \boldsymbol{u}_{x+1}, \ldots, \boldsymbol{u}_{x+m-1}),$$

consists of $x+m$ n-bit code blocks where the last m blocks $\boldsymbol{u}_x, \boldsymbol{u}_{x+1}, \ldots, \boldsymbol{u}_{x+m-1}$ correspond to the m zero blocks appended to the information sequence $\boldsymbol{a}$. The above termination process results in an $(n(x+m), kx)$ block code. The generator matrix of this block code simply consists of the first x shifts of the generating pattern $[G_0 G_1 \cdots G_m]$ given in (9.7), i.e.,

$$G = \begin{bmatrix} G_0 & G_1 & G_2 & \cdots & G_m & & & & \\ & G_0 & G_1 & G_2 & \cdots & G_m & & & \\ & & \ddots & & & & \ddots & & \\ & & & G_0 & G_1 & G_2 & \cdots & G_m \end{bmatrix}. \tag{9.15}$$

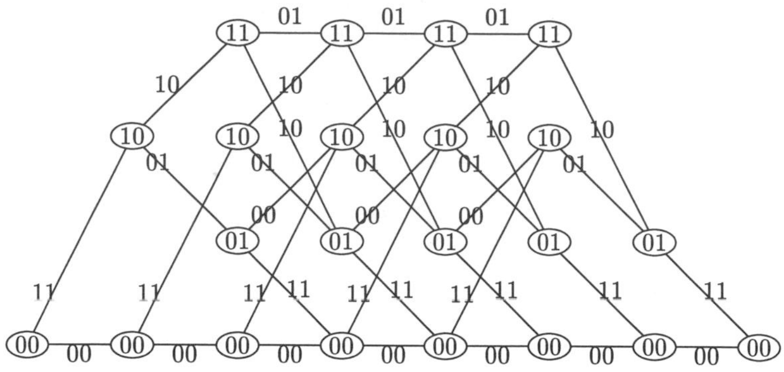

Figure 9.4. The terminated trellis diagram for the $(14, 5)$ linear block code obtained by zero-tail termination of the $(2, 1, 2)$ convolutional code with $x = 5$ of Example 9.3.

It follows from Chapter 6 and Section 9.2 that the trellis diagram for the block code generated by the matrix G of (9.15) consists of $x + m$ sections of n bits each and terminates at the all-zero state. Each path in this trellis diagram corresponds to an information sequence

$$\boldsymbol{a} = (\boldsymbol{a}_0, \boldsymbol{a}_1, \ldots, \boldsymbol{a}_{x-1}, \boldsymbol{a}_x, \ldots, \boldsymbol{a}_{x+m-1}),$$

with $\boldsymbol{a}_x = \cdots = \boldsymbol{a}_{x+m-1} = \mathbf{0}$. Since the termination is achieved by appending a tail of mk zeroes to each transmitted information sequence, the above termination technique is referred to as **zero tail termination**.

An (n, k) cyclic code may be regarded as a $(1, 1, n - k)$ convolutional code terminated for every k information bits.

Example 9.3 Consider the rate-1/2 convolutional code given in Example 9.1. Suppose $x = 5$. Then zero tail termination results in a $(14, 5)$ block code with the following generator matrix,

$$G = \begin{bmatrix} 1 & 1 & 0 & 1 & 1 & 1 & & & & & & & & \\ & & 1 & 1 & 0 & 1 & 1 & 1 & & & & & & \\ & & & & 1 & 1 & 0 & 1 & 1 & 1 & & & & \\ & & & & & & 1 & 1 & 0 & 1 & 1 & 1 & & \\ & & & & & & & & 1 & 1 & 0 & 1 & 1 & 1 \end{bmatrix}$$

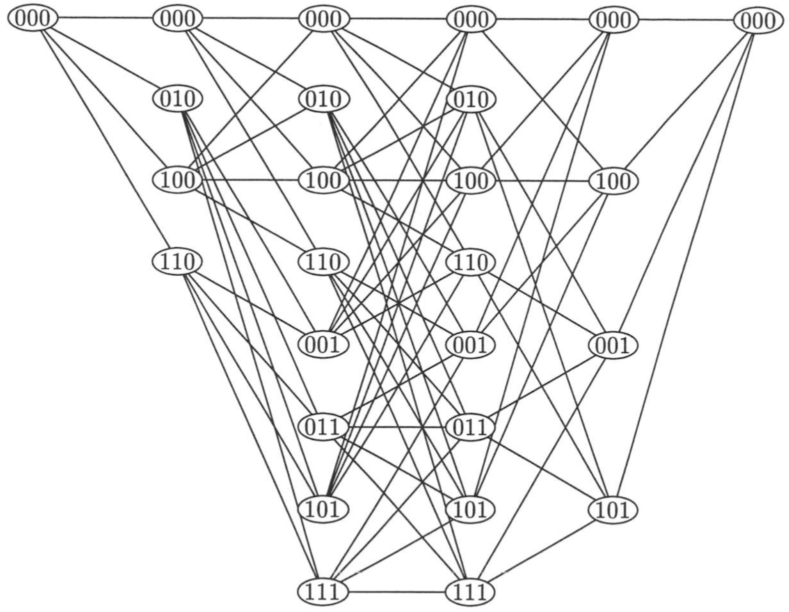

Figure 9.5. Trellis diagram of the $(15, 7)$ block code obtained by zero-tail termination of the $(3, 2, 2)$ code of Example 9.4.

The terminated trellis diagram for this code is shown in Figure 9.4.

$\triangle\triangle$

In the case $p = km - M > 0$, there exists at least one row $\boldsymbol{g}^{(l)}$, $1 \leq l \leq k$, in the generating pattern $[G_0 G_1 G_2 \cdots G_m]$ terminating with q zero n-bit blocks, where $1 \leq q \leq p$. It follows that each information bit associated with row $\boldsymbol{g}^{(l)}$ remains in the encoder memory for only $m - q$ time units. In that case, row $\boldsymbol{g}^{(l)}$ properly shifted can appear q more times at the bottom of the matrix G given by (9.15) without modifying the final state of the trellis diagram. By adding p such rows to G in (9.15), we obtain an $(n(x+m), kx+p)$ linear block code.

Example 9.4 Consider the $(3, 2, 2)$ convolutional code with generating pattern

$$\begin{bmatrix} 1 & 0 & 1 & 0 & 1 & 1 & 0 & 0 & 0 \\ 0 & 1 & 1 & 0 & 0 & 1 & 1 & 0 & 1 \end{bmatrix}.$$

For this code, $m_1 = 1$, $m_2 = 2$, $m = 2$, and $M = 3$. The first row in the above generating pattern terminates with one 3-bit zero block. Therefore, $p = 1$. For $x = 3$, zero tail termination of this convolutional code after the transmission of 7 information bits results in a $(15, 7)$ linear block code with generator matrix

$$G = \begin{bmatrix} 1 & 0 & 1 & 0 & 1 & 1 & 0 & 0 & 0 & & & & & & \\ 0 & 1 & 1 & 0 & 0 & 1 & 1 & 0 & 1 & & & & & & \\ & & & 1 & 0 & 1 & 0 & 1 & 1 & 0 & 0 & 0 & & & \\ & & & 0 & 1 & 1 & 0 & 0 & 1 & 1 & 0 & 1 & & & \\ & & & & & & 1 & 0 & 1 & 0 & 1 & 1 & 0 & 0 & 0 \\ & & & & & & 0 & 1 & 1 & 0 & 0 & 1 & 1 & 0 & 1 \\ & & & & & & & & & 1 & 0 & 1 & 0 & 1 & 1 \end{bmatrix}$$

The corresponding trellis diagram is depicted in Figure 9.5. The labeling of this figure is obtained based on Section 9.2.

△△

Zero tail termination of a rate-k/n convolutional code always results in a loss of code rate from $R = k/n$ to $R^* = (kx + p)/(n(x + m)) = (k/n)(x + p/k)/(x + m)$. If x is small and m is large, the loss in rate is significant. For example, consider the rate-2/3 code given in Example 9.4. Suppose $x = 10$. Then there is a 12.5 % rate loss.

Termination of a convolutional code without any rate loss can be achieved by a technique known as **tail biting**. This is simply done by taking $x + m$ consecutive n-bit sections from the fully developed trellis diagram of a rate-k/n convolutional code C. The truncated trellis, denoted T^*, has 2^M states at either end. Let $\sigma^{(i)}$ denote a state in the state space $\Sigma(C)$ of C with $1 \leq i \leq 2^M$. Let $L(\sigma^{(i)}, \sigma^{(i)})$ denote the set of paths in T^* connecting $\sigma^{(i)}$ at the left end of T^* to the same state $\sigma^{(i)}$ at the right end of T^*. Define

$$L^* \triangleq \left\{ L(\sigma^{(i)}, \sigma^{(i)}) : 1 \leq i \leq 2^M \right\}. \tag{9.16}$$

Let C^* denote the block code of length $n(x + m)$ whose codewords correspond to the label sequences of the paths in L^*. Let $\sigma^{(1)}$ denote the all-zero state in

$\Sigma(C)$. Then $L(\sigma^{(1)}, \sigma^{(1)})$ simply gives the $(n(x+m), kx)$ linear block code C_1^* obtained by the zero tail termination (or generated by the matrix of (9.15)). For $1 \leq i \leq 2^M$, the block code C_i^* that corresponds to $L(\sigma^{(i)}, \sigma^{(i)})$ is simply a coset of C_1^* in the partition C^*/C_1^*. Therefore,

$$C^* = \bigcup_{i=1}^{2^M} C_i^*. \tag{9.17}$$

The code C^* is an $(n(x+m), k(x+m))$ linear block code generated by the following matrix,

$$G^* = \begin{bmatrix}
G_m & & & & & G_0 & G_1 & G_2 & \cdots & G_{m-1} \\
G_{m-1} & G_m & & & & & G_0 & G_1 & G_2 & \cdots \\
 & & \ddots & & & & & & \ddots & \\
G_2 & \cdots & G_{m-1} & G_m & & & & & G_0 & G_1 \\
G_1 & G_2 & \cdots & G_{m-1} & G_m & & & & & G_0 \\
- & - & - & - & - & - & - & - & - & - \\
G_0 & G_1 & G_2 & \cdots & G_{m-1} & G_m & & & & \\
 & G_0 & G_1 & G_2 & \cdots & G_{m-1} & G_m & & & \\
 & & G_0 & G_1 & G_2 & \cdots & G_{m-1} & G_m & & \\
 & & & \ddots & & & & \ddots & & \\
 & & & & G_0 & G_1 & G_2 & \cdots & G_{m-1} & G_m
\end{bmatrix}. \tag{9.18}$$

Note that the submatrix below the dashed line is identical to the matrix of (9.15), which generates C_1^*. The submatrix above the dashed line generates the coset representatives of the partition C^*/C_1^*. The non-zero components of each coset representative are confined to only the first m blocks and the last m blocks. Consequently, all the cosets in the partition C^*/C_1^* are identical in the middle $x - m$ blocks. This implies that all the trellises for the cosets in C^*/C_1^* are identical in the middle $x - m$ sections. In decoding C^* with a trellis-based decoding algorithm, such as the Viterbi algorithm, all the trellises for C_i^* with $1 \leq i \leq 2^M$ can be processed in parallel.

The overall trellis for C^* consists of 2^M initial states and 2^M final states, where the initial and final states are the same state. In encoding, the initial state is determined by the first m information blocks.

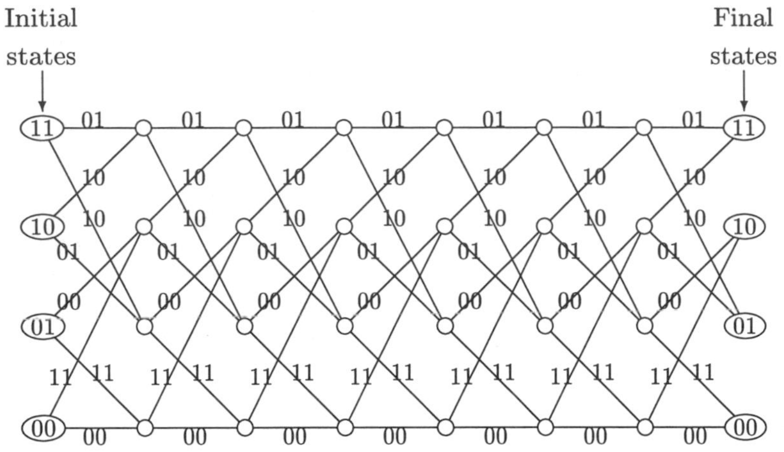

Figure 9.6. Trellis diagram for the $(14, 7)$ linear block code obtained from the rate-1/2 $(2, 1, 2)$ convolutional code with tail biting and $x = 5$.

Example 9.5 Again consider the $(2, 1, 2)$ convolutional code given in Example 9.1. Let $x = 5$. Using the tail biting termination technique, we obtain a $(14, 7)$ linear block code C^* with generator matrix,

$$G^* = \left[\begin{array}{cccccccccccccc}
1 & 1 & & & & & & & & & 1 & 1 & 0 & 1 \\
0 & 1 & 1 & 1 & & & & & & & & & 1 & 1 \\
\hline
1 & 1 & 0 & 1 & 1 & 1 & & & & & & & & \\
 & & 1 & 1 & 0 & 1 & 1 & 1 & & & & & & \\
 & & & & 1 & 1 & 0 & 1 & 1 & 1 & & & & \\
 & & & & & & 1 & 1 & 0 & 1 & 1 & 1 & & \\
 & & & & & & & & 1 & 1 & 0 & 1 & 1 & 1
\end{array}\right].$$

The trellis diagram for C^* is shown in Figure 9.6.

△△

Tail biting termination maintains the code rate of the original convolutional code. However it may reduce the minimum free distance of the original convolutional code and hence, degrades the error performance. In general, the amount of reduction in the minimum free distance depends on the number of information blocks x.

9.5 RM CODES VIEWED AS TERMINATED CONVOLUTIONAL CODES

In this section, we further explore the parallel and regular structures associated with the iterative squaring construction of RM codes described in Section 8.1 and show that these codes can be represented as the union of cosets of terminated time-invariant convolutional codes, as presented in Section 9.4. This representation allows us to use conventional methods devised for convolutional codes to design the encoder of an RM code, construct its trellis diagram, label the states and branches of this trellis diagram, and finally decode it.

In the following, we simplify the notation introduced in Chapters 2 and 8, so that $G_{\mathrm{RM}}(r,m) \triangleq G_{r,m}$ and $\Delta_{\mathrm{RM}}(r/r-1,m) \triangleq \Delta_{r,m}$. Also, by convention, we assume that $G_{-1,m} = [0]$, which implies that $\Delta_{0,m} = G_{0,m}$, and define $\Delta_{r,m} = [0]$ for $r > m$. With these definitions, G refers to the generator matrix of an RM code, Δ refers to coset selectors, and P refers to a generating pattern. This general guideline is used to describe the decompositions presented in the following. In general, P defines a convolutional code C and Δ describes the number of cosets of C. We associate with each level of the l-level iterative squaring construction a different convolutional code. For each value of l, each convolutional code is defined by a generating pattern of different length 2^ν, referred to as the **l-level generating pattern of length 2^ν**.

For the l-level squaring construction, the l-level generating pattern of length one is defined as

$$P^1_{r,m}(l) \triangleq G_{r-l,m-l}, \tag{9.19}$$

which generates a trivial convolutional code of zero memory order. Its associated trellis diagram contains a single state and the $2^{K_{r-l,m-l}}$ parallel branches compose the unique transition. These parallel branches define the $\mathrm{RM}_{r-l,m-l}$ code. With these definitions, (8.19) (the one-level squaring construction) can be expressed as

$$G_{r,m} = \begin{bmatrix} \Delta_{r,m-1} & \Delta_{r,m-1} \\ P^1_{r,m}(1) & 0 \\ 0 & P^1_{r,m}(1) \end{bmatrix}. \tag{9.20}$$

Similarly, based on (8.11), (8.12) and (8.23), we obtain for the two-level squaring construction

$$G_{r,m} = \begin{bmatrix} \Delta_{r,m-2} & \Delta_{r,m-2} & \Delta_{r,m-2} & \Delta_{r,m-2} \\ \Delta_{r-1,m-2} & \Delta_{r-1,m-2} & \Delta_{r-1,m-2} & \Delta_{r-1,m-2} \\ 0 & \Delta_{r-1,m-2} & 0 & \Delta_{r-1,m-2} \\ 0 & 0 & \Delta_{r-1,m-2} & \Delta_{r-1,m-2} \\ P^1_{r,m}(2) & 0 & 0 & 0 \\ 0 & P^1_{r,m}(2) & 0 & 0 \\ 0 & 0 & P^1_{r,m}(2) & 0 \\ 0 & 0 & 0 & P^1_{r,m}(2) \end{bmatrix}. \tag{9.21}$$

This simple construction can be repeated iteratively. For each coset of the l-level coset decomposition $\text{RM}_{r,m-l}/\text{RM}_{r-1,m-l}/\cdots/\text{RM}_{r-l,m-l}$ corresponding to the l-level squaring construction, we obtain 2^l independent sections, each section representing the $\text{RM}_{r-l,m-l}$ code. The advantage of this approach is the independence of the 2^l sections, which can be exploited to speed up the decoding process. However, the number of cosets increases rapidly.

For the l-level squaring construction, the l-level generating pattern of length two is defined as

$$P^2_{r,m}(l) \triangleq [P^{20}_{r,m}(l) \quad P^{21}_{r,m}(l)], \tag{9.22}$$

where

$$P^{20}_{r,m}(l) = \begin{bmatrix} G_{r-l,m-l} \\ \Delta_{r-l+1,m-l} \end{bmatrix}, \tag{9.23}$$

$$P^{21}_{r,m}(l) = \begin{bmatrix} 0 \\ \Delta_{r-l+1,m-l} \end{bmatrix}, \tag{9.24}$$

for $l \leq r$. For $l > r$, we simply define $P^{20}_{r,m}(l) = P^{21}_{r,m}(l) = G_{0,m-l}$. The generating pattern $P^2_{r,m}(l)$ represents a rate-$K_{r-l+1,m-l}/2^{m-l}$ unit-memory convolutional code whose encoder contains $K_{r-l+1,m-l} - K_{r-l,m-l}$ input shift registers, each connected to a different input. Due to the repetition of the matrix $\Delta_{r-l+1,m-l}$ in $P^2_{r,m}(l)$, the convolutional encoding can also be viewed as a differential encoding of $K_{r-l+1,m-l} - K_{r-l,m-l}$ bits of the information sequence. This fact implies that $P^2_{r,m}(l)$ describes a catastrophic encoder [30, 68].

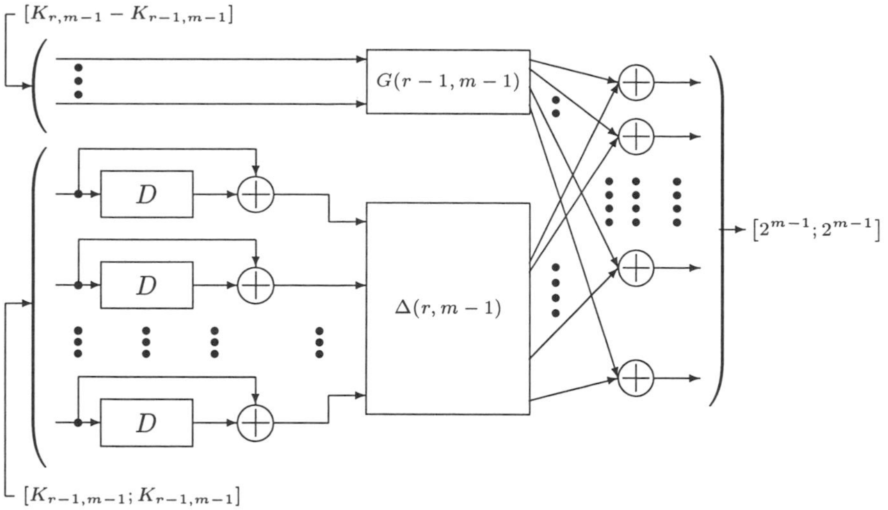

Figure 9.7. Convolutional code encoder for one-level squaring construction.

It follows from (9.22), (9.23), and (9.24) that for the one-level generating pattern of length two, we can rewrite (9.20) as

$$G_{r,m} = \begin{bmatrix} P_{r,m}^{20}(1) & P_{r,m}^{21}(1) \\ 0 & G_{r-1,m-1} \end{bmatrix}. \tag{9.25}$$

The corresponding encoder is represented in Figure 9.7. In this figure, the notation $[a; b]$ represents the number of activated inputs for two consecutive information blocks in each part of the encoder (with and without shift registers). Therefore, for the encoder part considered, a information bits are encoded at the first time instant, and b information bits are encoded at the next time instant. In Figure 9.7, we observe that after the encoding of two consecutive information blocks, all shift registers in the encoder are set back to zero. It follows that the trellis diagram associated with (9.25) has two sections corresponding to the two encoding stages of Figure 9.7.

Similarly, for the two-level squaring construction, after elementary row operations, (9.21) becomes (based on (8.23) and the two-level generating pattern of length two),

$$G_{r,m} = \begin{bmatrix} \Delta_{r,m-2} & \Delta_{r,m-2} & \Delta_{r,m-2} & \Delta_{r,m-2} \\ P^{20}_{r,m}(2) & P^{21}_{r,m}(2) & 0 & 0 \\ 0 & P^{20}_{r,m}(2) & P^{21}_{r,m}(2) & 0 \\ 0 & 0 & P^{20}_{r,m}(2) & P^{21}_{r,m}(2) \\ 0 & 0 & 0 & G_{r-2,m-2} \end{bmatrix}. \tag{9.26}$$

Based on Section 9.4, (9.26) shows that the $\mathrm{RM}_{r,m}$ code can be viewed as the union of $2^{K_{r,m-2}-K_{r-1,m-2}}$ cosets of a zero tail terminated convolutional code whose generator matrix is

$$G = \begin{bmatrix} P^{20}_{r,m}(2) & P^{21}_{r,m}(2) & 0 & 0 \\ 0 & P^{20}_{r,m}(2) & P^{21}_{r,m}(2) & 0 \\ 0 & 0 & P^{20}_{r,m}(2) & P^{21}_{r,m}(2) \\ 0 & 0 & 0 & G_{r-2,m-2} \end{bmatrix}. \tag{9.27}$$

For each coset, the associated encoder is structurally identical to the encoder shown in Figure 9.7. Again, this construction can be repeated iteratively. Generalizing (9.26) to the l-level squaring construction, we obtain, for each coset, a 2^l-section trellis diagram corresponding to a terminated convolutional code with the l-level generating pattern of length two $P^2_{r,m}(l)$ for the first $2^l - 1$ sections. Since the trellis terminates after the 2^l-th section, only the parallel branches corresponding to the row space of $G_{r-l,m-l}$ remain for this last section.

Example 9.6 Consider the $\mathrm{RM}_{1,3} = (8,4,4)$ RM code. For the two-level squaring construction of this code, we obtain in (9.26) the coset selector

$$\Delta_{1,1} = [0 \quad 1], \tag{9.28}$$

and the two matrices corresponding to the two-level generating pattern of length two,

$$P^{20}_{1,3}(2) = P^{21}_{1,3}(2) = [1 \quad 1]. \tag{9.29}$$

For this code, (9.26) becomes

$$G = \begin{bmatrix} 0 & 1 & 0 & 1 & 0 & 1 & 0 & 1 \\ 1 & 1 & 1 & 1 & & & & \\ & & 1 & 1 & 1 & 1 & & \\ & & & & 1 & 1 & 1 & 1 \end{bmatrix}. \tag{9.30}$$

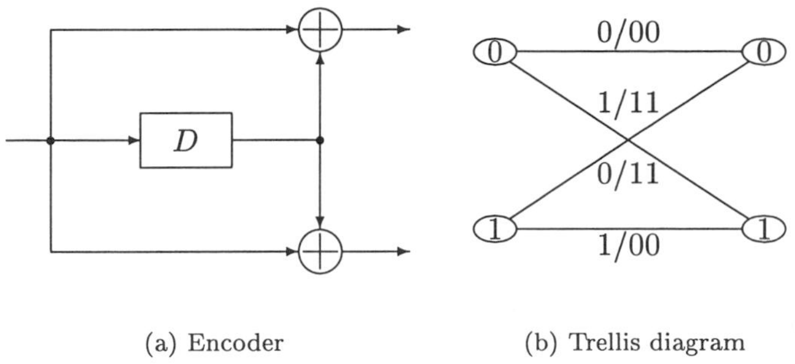

Figure 9.8. Encoder and trellis diagram for the $(2,1,1)$ convolutional code for the $(8,4,4)$ RM code.

Since $G_{-1,1} = [0]$, we have $P_{1,3}^{20}(2) = G_{0,1}$, so that only the lower part with shift registers remains in the encoder shown in Figure 9.7. The corresponding rate-1/2 encoder and trellis diagram are shown in Figure 9.8. We can easily verify that the encoder of the embedded convolutional code is catastrophic. However, terminating the trellis breaks the catastrophic behavior. For this code, no additional bits are encoded during the termination stage, or equivalently, the trellis diagram contains no parallel branches. For $m \geq 3$, this example can be generalized in a straightforward way to the two-level squaring construction of any $\mathrm{RM}_{1,m}$ code which is associated with a rate-$1/2^{m-2}$ convolutional code [30].

Example 9.7 Consider the $\mathrm{RM}_{2,4} = (16,11,4)$ RM code. For the two-level squaring construction of this code, the matrices in (9.22), (9.23), and (9.24) are the coset selector

$$\Delta_{2,2} = [0 \quad 0 \quad 0 \quad 1] \tag{9.31}$$

and the two matrices

$$P_{2,4}^{20}(2) = \begin{bmatrix} 1 & 1 & 1 & 1 \\ 0 & 0 & 1 & 1 \\ 0 & 1 & 0 & 1 \end{bmatrix}, \tag{9.32}$$

$$P_{2,4}^{21}(2) = \begin{bmatrix} 0 & 0 & 0 & 0 \\ 0 & 0 & 1 & 1 \\ 0 & 1 & 0 & 1 \end{bmatrix}. \tag{9.33}$$

From these matrices, the generator matrix of the (16,11,4) RM code corresponding to (9.26) is

$$G = \begin{bmatrix} 0 & 0 & 0 & 1 & 0 & 0 & 0 & 1 & 0 & 0 & 0 & 1 & 0 & 0 & 0 & 1 \\ 1 & 1 & 1 & 1 & 0 & 0 & 0 & 0 & & & & & & & & \\ 0 & 0 & 1 & 1 & 0 & 0 & 1 & 1 & & & & & & & & \\ 0 & 1 & 0 & 1 & 0 & 1 & 0 & 1 & & & & & & & & \\ & & & & 1 & 1 & 1 & 1 & 0 & 0 & 0 & 0 & & & & \\ & & & & 0 & 0 & 1 & 1 & 0 & 0 & 1 & 1 & & & & \\ & & & & 0 & 1 & 0 & 1 & 0 & 1 & 0 & 1 & & & & \\ & & & & & & & & 1 & 1 & 1 & 1 & 0 & 0 & 0 & 0 \\ & & & & & & & & 0 & 0 & 1 & 1 & 0 & 0 & 1 & 1 \\ & & & & & & & & 0 & 1 & 0 & 1 & 0 & 1 & 0 & 1 \\ & & & & & & & & & & & & 1 & 1 & 1 & 1 \end{bmatrix}. \tag{9.34}$$

The corresponding rate-3/4 convolutional code is generated by the shifts of the generating pattern

$$\begin{bmatrix} 1 & 1 & 1 & 1 & 0 & 0 & 0 & 0 \\ 0 & 0 & 1 & 1 & 0 & 0 & 1 & 1 \\ 0 & 1 & 0 & 1 & 0 & 1 & 0 & 1 \end{bmatrix}. \tag{9.35}$$

The encoder and trellis diagram of this convolutional code are depicted in Figure 9.9. Since the encoder contains only two memory elements, the trellis diagram has four states and each transition is represented by two parallel branches. As in the previous example, this convolutional code encoder is catastrophic but the catastrophic behavior stops at the fourth stage if encoding is realized based on (9.26). At the last stage of encoding, $K_{0,1} = 1$ bit is encoded. Again, for $m \geq 4$, this example can be generalized in a straightforward way to the two-level squaring construction of any $\text{RM}_{2,m}$ code which is associated with a rate-$(m-1)/2^{m-2}$ convolutional code [30].

$\triangle\triangle$

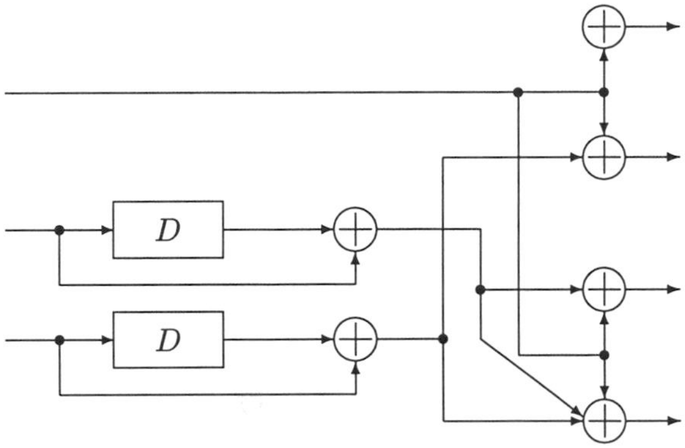

(a) Encoder

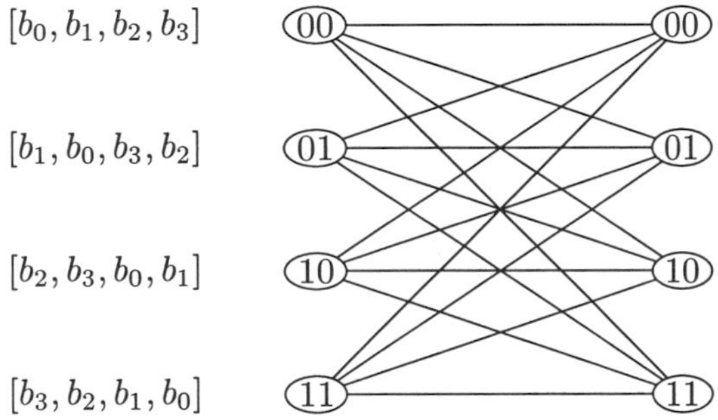

(b) Trellis diagram with composite branches $b_0 = \{0000, 1111\}$, $b_1 = \{0101, 1010\}$, $b_2 = \{0011, 1100\}$, $b_3 = \{0110, 1001\}$.

Figure 9.9. Encoder and trellis diagram for the $(4, 3, 2)$ convolutional code for the $(16, 11, 4)$ RM code.

For the l-level squaring construction, the l-level generating pattern of length four is defined as

$$P^4_{r,m}(l) \triangleq [P^{40}_{r,m}(l) \quad P^{41}_{r,m}(l) \quad P^{42}_{r,m}(l) \quad P^{43}_{r,m}(l)], \tag{9.36}$$

where

$$P^{40}_{r,m}(l) = \begin{bmatrix} G_{r-l,m-l} \\ \Delta_{r-l+1,m-l} \\ \Delta_{r-l+2,m-l} \end{bmatrix} = \begin{bmatrix} P^{20}_{r,m}(l) \\ \Delta_{r-l+2,m-l} \end{bmatrix}, \tag{9.37}$$

$$P^{41}_{r,m}(l) = \begin{bmatrix} 0 \\ \Delta_{r-l+1,m-l} \\ \Delta_{r-l+2,m-l} \end{bmatrix} = \begin{bmatrix} P^{21}_{r,m}(l) \\ \Delta_{r-l+2,m-l} \end{bmatrix}, \tag{9.38}$$

$$P^{42}_{r,m}(l) = P^{43}_{r,m}(l) = \begin{bmatrix} 0 \\ 0 \\ \Delta_{r-l+2,m-l} \end{bmatrix}, \tag{9.39}$$

for $l \leq r$. If $l > r$, the conventions introduced previously are applied to the definitions of these matrices. Equation (9.36) defines a rate-$K_{r-l+2,m-l}/2^{m-l}$ convolutional code. This convolutional code also corresponds to a differential encoding and its encoder is catastrophic.

Using these definitions, (9.26) can be rewritten as

$$G_{r,m} = \begin{bmatrix} P^{40}_{r,m}(2) & P^{41}_{r,m}(2) & P^{42}_{r,m}(2) & P^{43}_{r,m}(2) \\ 0 & P^{20}_{r,m}(2) & P^{21}_{r,m}(2) & 0 \\ 0 & 0 & P^{20}_{r,m}(2) & P^{21}_{r,m}(2) \\ 0 & 0 & 0 & G_{r-2,m-2} \end{bmatrix}, \tag{9.40}$$

which corresponds to the zero tail termination of the convolutional code associated with (9.36) for the case $x = 1$ in Section 9.4.

Based on the 3-level squaring construction of $G_{r,m}$ illustrated in Example 8.2, the matrix $G_{r,m}$ can be put into the following form:

$$\begin{bmatrix}
\Delta_{r,m-3} & \Delta_{r,m-3} & \Delta_{r,m-3} & \Delta_{r,m-3} & \Delta_{r,m-3} & \Delta_{r,m-3} & \Delta_{r,m-3} & \Delta_{r,m-3} \\
0 & \Delta_{r-1,m-3} & \Delta_{r-1,m-3} & 0 & 0 & \Delta_{r-1,m-3} & \Delta_{r-1,m-3} & 0 \\
P^{40}_{r,m}(3) & P^{41}_{r,m}(3) & P^{42}_{r,m}(3) & P^{43}_{r,m}(3) & 0 & 0 & 0 & 0 \\
0 & P^{20}_{r,m}(3) & P^{21}_{r,m}(3) & 0 & 0 & 0 & 0 & 0 \\
0 & 0 & P^{40}_{r,m}(3) & P^{41}_{r,m}(3) & P^{42}_{r,m}(3) & P^{43}_{r,m}(3) & 0 & 0 \\
0 & 0 & 0 & P^{20}_{r,m}(3) & P^{21}_{r,m}(3) & 0 & 0 & 0 \\
0 & 0 & 0 & 0 & P^{40}_{r,m}(3) & P^{41}_{r,m}(3) & P^{42}_{r,m}(3) & P^{43}_{r,m}(3) \\
0 & 0 & 0 & 0 & 0 & P^{20}_{r,m}(3) & P^{21}_{r,m}(3) & 0 \\
0 & 0 & 0 & 0 & 0 & 0 & P^{20}_{r,m}(3) & P^{21}_{r,m}(3) \\
0 & 0 & 0 & 0 & 0 & 0 & 0 & G_{r-3,m-3}
\end{bmatrix}. \tag{9.41}$$

We see that (9.41) does not correspond exactly to the conventional form given in (9.7), since the inputs connected to the shift registers of length three are not represented in each shift of the generating pattern. Equivalently, the code represented by (9.41) can be obtained from the convolutional code corresponding to (9.36) by simply periodically feeding zeroes to the inputs connected to the shift registers of length three, with a period of two. As a result, trellis states represented by non-zero values in these positions do not belong to the trellis diagram associated with (9.41). This trellis diagram can be constructed from the trellis diagram associated with the convolutional code corresponding to (9.36) by simply discarding these states, as well as all the branches connected to these states. This state reduction reduces $\rho_{\max}(C)$ defined in Section 7.1.

By following the recursive decomposition presented in Section 8.1, generalization to the l-level generating pattern of length 2^ν for $\nu \geq 3$ follows the same lines. Also, as discussed in Section 9.4, any generating pattern generates an infinite family of $(n(x+m), kx+p)$ block codes obtained by terminating the encoding of the associated convolutional code after $kx+p$ information bits. Similarly an infinite family of block codes can be associated with the l-level generating pattern of length 2^ν. For example, based on (9.26), the two-level generating pattern of length two obtained from the $\mathrm{RM}_{r,m}$ code defines a family of codes with

$$\begin{aligned}
N &= 2^{m-2}x, \\
K &= (x-1)K_{r-1,m-2} + \lfloor x/4 \rfloor (K_{r,m-2} - K_{r-1,m-2}) \\
&\quad + K_{r-2,m-2},
\end{aligned}$$

$$d_H = 2^{m-r}.$$

To guarantee the same minimum distance as for (9.26), the coset selector inputs are activated independently every four blocks. For the $\mathrm{RM}_{2,5}$ code, we obtain a family of $(8x, 4(x-1)+3\lfloor x/4 \rfloor + 1, 8)$ codes. This family contains the $(8,1,8)$, $(16,5,8)$ and $(32,16,8)$ RM codes. Also, for $x = 8$, the corresponding $(64,35,8)$ code is a subcode of the $(64,42,8)$ RM code.

10 THE VITERBI AND DIFFERENTIAL TRELLIS DECODING ALGORITHMS

Decoding algorithms based on the trellis representation of a code (block or convolutional) drastically reduce decoding complexity. The best known and most commonly used trellis-based decoding algorithm is the Viterbi algorithm [23, 79, 105]. It is a maximum likelihood decoding algorithm. Convolutional codes with the Viterbi decoding have been widely used for error control in digital communications over the last two decades. This chapter is concerned with the application of the Viterbi decoding algorithm to linear block codes. First, the Viterbi algorithm is presented. Then, optimum sectionalization of a trellis to minimize the computational complexity of a Viterbi decoder is discussed and an algorithm is presented. Some design issues for IC (integrated circuit) implementation of a Viterbi decoder are considered and discussed. Finally, a new decoding algorithm based on the principle of **compare-select-add** is presented. This new algorithm can be applied to both block and convolutional codes and is more efficient than the conventional Viterbi algorithm based on the **add-compare-select** principle. This algorithm is particularly efficient for rate-$1/n$ antipodal convolutional codes and their high-rate punctured codes. It reduces computational complexity by one-third compared with the Viterbi algorithm.

10.1 THE VITERBI DECODING ALGORITHM

The Viterbi algorithm is based on the simple idea that among the paths merging into a state in the code trellis, only the most probable path needs to be saved for future processing and all the other paths can be eliminated without affecting decoding optimality. This elimination of the less probable paths from further consideration drastically reduces decoding complexity. The path being saved is called the **survivor**. Therefore, there is a survivor at each state in the trellis at every level. The survivors at each level of the code trellis are extended to the next level through the composite branches between the two levels. The paths that merge into a state at the next level are then compared and the most probable path is selected as the survivor. This process continues until the end of the trellis is reached. At the end of the trellis, there is only one state, the final state σ_f, and there is only one survivor, which is the most likely codeword. Decoding is then completed.

Viterbi decoding of a linear block code based on a sectionalized trellis diagram $T(\{h_0, h_1, \ldots, h_L\})$, with section boundary locations $0 = h_0 < h_1 < \cdots < h_L = N$, is carried out serially, section by section, from the initial state σ_0 to the final state σ_f. Suppose the decoder has processed j trellis sections up to time-h_j. There are $|\Sigma_{h_j}(C)|$ **survivors**, one for each state in $\Sigma_{h_j}(C)$. These survivors together with their **path** (or **state**) **metrics** are stored in memory. To process the $(j+1)$-th section, the decoder executes the following steps:

(1) Each survivor is extended through the composite branches diverging from it to the next state level at time-h_{j+1}.

(2) For each composite branch into a state in $\Sigma_{h_{j+1}}(C)$, find the single branch with the largest (correlation) metric. The metric computed is the branch metric.

(3) Replace each composite branch by the branch with the largest metric.

(4) **Add** the metric of a branch to the metric of the survivor from which the branch diverges. For each state σ in $\Sigma_{h_{j+1}}(C)$, **compare** the metrics of the paths converging into it and **select** the path with the largest path metric as the survivor terminating at state σ. This step is called the **add-compare-select (ACS)** procedure in the Viterbi algorithm.

The decoder executes the above steps repeatedly, section by section, until it reaches the final state σ_f. At this point, there is one and only one survivor, which is the decoded codeword and the most likely codeword. The information bits corresponding to this decoded codeword are then delivered to the user. The decoding window is simply the code length.

Using the above decoding algorithm, the total number of operations (additions and comparisons) can be computed easily. This number can be reduced significantly if sectionalization of a trellis is done properly [60]. This will be discussed in the next section.

10.2 OPTIMUM SECTIONALIZATION OF A CODE TRELLIS: LAFOURCADE-VARDY ALGORITHM

In decoding a block code with the Viterbi algorithm, the total number of computations depends on the sectionalization of the trellis diagram for the code. A sectionalization of a code trellis for a code C that gives the smallest total number of computations is called an optimum sectionalization for C. An optimum sectionalization is not necessarily unique. In the following, an algorithm for finding an optimum sectionalization is presented. This algorithm was devised by Lafourcade and Vardy [60].

The Lafourcade-Vardy (LV) algorithm is based on the following simple fact:

(F) For any integers x and y with $0 \leq x < y \leq N$, a section from time-x to time-y in any sectionalized trellis $T(U)$ with $x, y \in U$ and $x+1, x+2, \ldots, y-1 \notin U$ is identical.

Let $\varphi(x, y)$ denote the number of computations required in steps (1) to (4) of the Viterbi algorithm to process the trellis section from time-x to time-y in any sectionalized code trellis $T(U)$ with $x, y \in U$ and $x+1, x+2, \ldots y-1 \notin U$. It follows from the above simple fact (F) that $\varphi(x, y)$ is determined only by x and y. Let $\varphi_{\min}(x, y)$ denote the smallest number of computations of steps (1)

Table 10.1. Optimum sectionalizations.

Code	Optimum Sectionalization U	Complexity $\varphi_{\min}(0,N)$	Complexity N-section
$RM_{1,6}$	$\{0,4,8,12,16,24,32,40,48,$ $52,56,60,63,64\}$	806	2,825
$RM_{2,6}$	$\{0,8,16,32,48,56,61,63,64\}$	101,786	425,209
$RM_{3,6}$	$\{0,8,16,24,32,40,48,56,64\}$	538,799	773,881

to (4) to process the trellis section(s) from time-x to time-y in any sectionalized code trellis $T(U)$ with $x, y \in U$. The value, $\varphi_{\min}(0,N)$, gives the total number of computations of the Viterbi algorithm for the code trellis with an optimum sectionalization. Then, it follows from (F) and the definitions of $\varphi(x,y)$ and $\varphi_{\min}(x,y)$ that

$$\varphi_{\min}(0,y) = \begin{cases} \min\left\{\varphi(0,y), \min_{0<x<y}\{\varphi_{\min}(0,x)+\varphi(x,y)\}\right\}, & \text{for } 1 < y \leq N, \\ \varphi(0,1), & \text{for } y = 1. \end{cases} \tag{10.1}$$

We can compute $\varphi_{\min}(0,y)$ for every y with $0 < y \leq N$ efficiently in the following way: The values of $\varphi(x,y)$ for $0 \leq x < y \leq N$ are computed using the structure of the trellis section from time-x to time-y. First, the value of $\varphi_{\min}(0,1)$ is computed. For an integer y with $1 < y \leq N$, $\varphi_{\min}(0,y)$ can be computed from $\varphi_{\min}(0,x)$ and $\varphi(x,y)$ with $0 < x < y$. By storing the information when the minimum value occurs in the right-hand side of (10.1), an optimum sectionalization is found from the computation of $\varphi_{\min}(0,N)$.

Example 10.1 Table 10.1 gives the optimum sectionalizations for three RM codes of length 64 using the LV algorithm.

△△

10.3 SOME DESIGN ISSUES FOR IC IMPLEMENTATION OF VITERBI DECODERS FOR LINEAR BLOCK CODES

Theoretically, any linear block code can be decoded by applying the Viterbi algorithm to a trellis for the code. However, practical limitations preclude the application of this algorithm to many good codes with long block lengths. The main reasons are the increases in state complexity, state connectivity, and branch complexity of the trellises for good block codes as the length of the codes increases. Much of the research on maximum likelihood decoding of linear block codes with the Viterbi algorithm over a code trellis has focussed on the minimization of the number of computations required for decoding a received sequence. If the actual decoding is intended to be performed using a stored program approach (a software implementation) that executes the operations needed to decode a received sequence sequentially, then this approach will lead to the fastest decoding speed. However, if an IC (hardware) implementation is intended, then many other factors besides the number of decoding computations must be considered. We must consider the factors that affect the circuit requirements, wire-routing within an IC chip, chip size, circuit utilization, power consumption, ACS computation speed, and other implementation issues. As a result, an alternate approach that is more suitable for IC implementation is desired.

For IC implementation of a Viterbi decoder for a linear block code, besides the state and branch complexities, other important trellis structural properties that should be included in the design considerations are state connectivity, the parallel structure, regularity, and symmetry. Proper use of these structural properties may result in a simpler decoding circuit and a higher decoding speed.

Optimum sectionalization in terms of minimizing the computational complexity, in general, results in a non-uniformly sectionalized trellis diagram. In a Viterbi decoder, quantities such as the branch labels, survivor path metrics, and survivor path labels generally reside in word registers, which are basically an ordered sequence of bit registers. The same hardware is used to process all trellis sections. If a register must store a particular variable, such as a branch metric or a state metric, it must be designed to accommodate the largest value of the variable over all trellis sections. Since the section lengths for a non-uniformly sectionalized trellis vary from one section to another, the registers involved must be designed based on the longest section. This may increase the relative complexity of an IC Viterbi decoder. Therefore, for IC implementation of a Viterbi decoder, a uniformly sectionalized trellis is more desirable.

Although a minimal trellis reduces the state and branch complexities, the states are densely connected. For long codes, this dense connection between the states causes serious wire-routing (interconnection) problems within an IC chip for hardware implementation of a Viterbi decoder and requires a large area of the chip (or a multilayer chip) to accommodate the decoding circuit. Furthermore, interconnections increase internal communications between various parts of the decoding circuit, which slow down the decoding speed and increase power consumption. Let $\rho_{\max}(C)$ be the maximum state space dimension of a minimal trellis for a code C. Then the number of registers required to store the survivor paths and their metrics must be $2^{\rho_{\max}(C)}$. If a separate ACS circuit is required for processing each state at each trellis level, then $2^{\rho_{\max}(C)}$ ACS circuits are needed. If the differences between $\rho_{\max}(C)$ and the state space dimensions at many section boundary locations are large, then many of the registers and ACS circuits are not used during the decoding process. This results in poor hardware utilization efficiency. All the above problems may be solved or partially solved by using a non-minimal trellis with a proper parallel decomposition, as discussed in Chapter 7. Regularity among the trellis sections also helps to overcome the above problems and reduces decoding complexity. Symmetry structure, such as mirror symmetry, allows bidirectional decoding, which speeds up the decoding process. Therefore, for hardware implementation of a Viterbi decoder for a linear block code, a non-minimal trellis may result in a simpler and faster decoding circuit with a higher hardware utilization efficiency. In design, both minimal and non-minimal trellises should be considered and the one that results in a simpler circuit and a higher decoding speed should be used.

In the following, we examine some key factors that affect the decoding complexity and speed of a Viterbi decoder based on a minimal or non-minimal trellis. The non-minimal trellis structure presented in Section 7.1 reduces internal communications and allows independent parallel processing of the subtrellises, while decreasing the complexity of an IC Viterbi decoder. It has significant advantages over the minimal trellis for IC implementation of a Viterbi decoder.

10.3.1 Hardware Utilization Efficiency and Effective Computational Complexity

Consider an IC Viterbi decoder based on an L-section trellis for a linear (N, K) block code C with section boundary locations at $h_0 = 0, h_1, h_2, \ldots, h_L = N$. While many VLSI structures have been described for a Viterbi decoder [10, 38, 100], the most widely implemented structure is based on the ACS-array architecture, wherein each abstract state in the trellis manifests itself as a physical ACS circuit on the IC, and the same ACS circuits are repeatedly used for all levels in the trellis. The ACS circuits can be labeled ACS-l for $1 \leq l \leq 2^{\rho_{L,\max}(C)}$, where $\rho_{L,\max}(C)$ is the maximum state space dimension of the L-section minimal trellis for C. We assume that $\rho_{L,\max}(C)$ is fixed no matter whether a minimal trellis or a non-minimal trellis is used in the decoder design.

The ACS circuits work as follows. At time-0, the metrics of the ACS circuits corresponding to the originating states of each parallel subtrellis are initialized to 0. At time-h_1, the ACS-l corresponding to state $\sigma^{(l)} \in \Sigma_{h_1}(C)$ at the end of section-1 of the trellis, for $1 \leq l \leq |\Sigma_{h_1}(C)|$, has the metric of state $\sigma^{(l)}$. The index of the surviving branch into state $\sigma^{(l)}$ is stored in ACS-l. Continuing in this way, at time-h_i, for $1 \leq i \leq L$, ACS-l corresponding to state $\sigma^{(l)} \in \Sigma_{h_i}(C)$ will have the metric for $\sigma^{(l)}$ and a sequence of i survivor branch indices corresponding to the most likely path from the initial state to $\sigma^{(l)}$.

Whenever the decoder is processing the trellis at a level at which the size of the state space is smaller than $2^{\rho_{L,\max}(C)}$, a number of ACS circuits will be idle. If the number of inactive ACS circuits is large and occurs often during the decoding process, the hardware utilization efficiency becomes poor. For example, consider the minimal 8-section trellis for the $(64, 42)$ RM code, $\mathrm{RM}_{3,6}$. This trellis has a state space dimension profile $(0, 7, 10, 13, 10, 13, 10, 7, 0)$ with $\rho_{8,\max}(C) = 13$. For a Viterbi decoder designed based on this trellis, at time-h_1 and -h_7, there are $2^{13} - 2^7 = 8,064$ inactive ACS circuits. At time-h_2, -h_4 and -h_6, there are $2^{13} - 2^{10} = 7,168$ inactive ACS circuits. Only at time-h_3 and -h_5, all the ACS circuits are active. We see that the hardware utilization efficiency is very poor for a Viterbi decoder for the $\mathrm{RM}_{3,6}$ code based on the minimal 8-section trellis using the ACS-array architecture.

Hardware utilization efficiency can be improved by a proper parallel decomposition of a minimal trellis into parallel isomorphic subtrellises. The decomposition results in a non-minimal trellis with the same maximum state space dimension $\rho_{L,\max}(C)$. Therefore, the number of ACS circuits in the ACS-array is still the same, but the number of active ACS circuits is increased at many, if not all, section boundary locations. We illustrate this with an example. Using the method presented in Section 7.1, the minimal 8-section trellis for the $(64, 42)$ RM code, $\mathrm{RM}_{3,6}$, can be decomposed into a non-minimal 8-section trellis with 128 parallel isomorphic subtrellises, each having a state space dimension profile $(0, 6, 6, 6, 3, 6, 6, 6, 0)$. Therefore, the state space dimension profile for the overall trellis is $(0, 13, 13, 13, 10, 13, 13, 13, 0)$. We see that the maximum state space dimension is still $\rho_{8,\max}(C) = 13$. However, for a decoder based on this non-minimal trellis, all 8,192 ACS circuits are active all the time, except at time-h_4. This greatly improves the hardware utilization efficiency.

For a trellis (minimal or non-minimal) that consists of parallel subtrellises, all the subtrellis decoders operate independently in parallel without communication between them. From the standpoint of speed, the effective computational complexity of decoding a received sequence is defined as the computational complexity of a single parallel subtrellis (viz. the minimal trellis for a subcode C') plus the cost of the final comparison among the survivors presented by each of the subtrellis decoders. The time required for final comparison is generally small relative to the time required for processing a subtrellis and this comparison can be pipelined. Since all the subtrellises are processed in parallel, the speed of decoding is therefore limited only by the time required to process one subtrellis. If a minimal trellis does not have enough parallel structure and decoding speed is critical, parallel decomposition can be used to reduce the effective computational complexity and thus to gain speed.

10.3.2 Complexity of the ACS Circuit

The converging branch dimension profile (CBDP), $(\delta_1, \delta_2, \ldots, \delta_L)$, defined in Section 6.2 also affects decoding speed and complexity. Each component δ_i is the base-2 logarithm of the number of composite branches converging into a state at a particular level of the trellis. The number 2^{δ_i} is called a **radix number** in the IC literature. At level-i of the trellis, each ACS circuit has to perform δ_i stages of a tree-type two-way comparison to find the best incoming branch. Hence, a reduction in the values of the components in the CBDP of a trellis will improve decoding speed and reduce the complexity of each ACS circuit. If the radix numbers in a minimal trellis are too large, then parallel decomposition can be used to achieve smaller radix numbers, and hence to reduce the complexity of each ACS circuit and to increase the decoding speed.

10.3.3 Traceback Complexity

Even though the branch and state metrics are computed and updated in the Viterbi decoder at every level of the trellis, the best (or most likely) path through the trellis must be determined. The process of determining the best path is called **traceback** in the literature.

Recall that the number of parallel branches in a composite branch in the i-th section of an L-section trellis for C with section boundary locations in $\{h_0, h_1, \ldots, h_L\}$ is

$$\left|C^{\mathrm{tr}}_{h_{i-1},h_i}\right| = 2^{k(C_{h_{i-1},h_i})}.$$

For a Viterbi decoder based on the minimal trellis, the ACS-l corresponding to state $\sigma^{(l)} \in \Sigma_{h_i}(C)$ for $1 \leq i \leq L$ must store $\delta_i(C) + k(C_{h_{i-1},h_i})$ bits in order to identify which of the $2^{\delta_i(C)}$ composite branches converging into state $\sigma^{(l)}$ is chosen and which of the $2^{k(C_{h_{i-1},h_i})}$ parallel branches survives. Therefore, each ACS-l needs to store

$$\sum_{i=1}^{L}(\delta_i(C) + k(C_{h_{i-1},h_i})) = K \tag{10.2}$$

bits in order to identify the sequence of survived incoming branches and to determine the decoded path. If this number is too large, parallel decomposition can be used to reduce it. Consider a non-minimal trellis with 2^q parallel subtrellises obtained by parallel decomposition of the minimal L-section trellis for C based on a subcode C'. If a Viterbi decoder is designed based on this non-minimal trellis, then the number of bits that must be stored for each ACS-l is

$$\sum_{i=1}^{L}(\delta_i(C') + k(C'_{h_{i-1},h_i})) = \dim(C'). \tag{10.3}$$

Since $\dim(C') = K - q \leq K$, the ACS circuits based on this non-minimal trellis design require less storage than for the design based on the minimal trellis. The total savings in storage in all the $2^{\rho_{L,\max}(C)}$ ACS circuits are

$$2^{\rho_{L,\max}(C)}(K - \dim(C')). \tag{10.4}$$

This is a significant savings.

10.3.4 ACS-Connectivity

The hardware implementation of a Viterbi decoder is severely affected by the physical placement of the ACS circuits and the need to route information between them. The routing complexity should be minimized in a Viterbi decoder IC design in order to reduce the size of the IC chip.

The basic operations performed by an ACS circuit are: addition of the branch metrics of the incoming branches to the state metrics of the corresponding originating states, comparison of the resulting sums to find the best one, selection of the surviving sum as the new state metric and the corresponding surviving branch label. The ACS-array architecture is usually dominated by the area required by the interconnections to transfer the state metrics. For a state $\sigma^{(l)} \in \Sigma_{h_i}(C)$ with $1 \leq l \leq |\Sigma_{h_i}(C)|$ and $0 \leq i < L$, let $Q_i(\sigma^{(l)})$ denote the set of states in $\Sigma_{h_{i+1}}(C)$ that are adjacent to $\sigma^{(l)}$. For $l > |\Sigma_{h_i}(C)|$, $Q_i(\sigma^{(l)}) = \emptyset$. Then in the ACS-array implementation of a Viterbi decoder, paths to transfer the state metrics exist between ACS-l and all the ACS circuits that correspond to the states in

$$Q_0(\sigma^{(l)}) \cup Q_1(\sigma^{(l)}) \cup \cdots \cup Q_{(L-1)}(\sigma^{(l)}). \tag{10.5}$$

The above set, denoted $\tilde{Q}^{(l)}$, defines the connectivity of ACS-l in the ACS-array corresponding to state $\sigma^{(l)}$. We call $|\tilde{Q}^{(l)}|$ and $\tilde{q}^{(l)} \triangleq \log_2 |\tilde{Q}^{(l)}|$ the connectivity and connectivity dimension of the ACS-l, respectively. The connectivities of ACS circuits determine the areas on an IC chip needed for wiring [10, 38]. This area should be kept as small as possible.

The ACS-connectivity can be reduced by using a non-minimal trellis with a proper number of parallel isomorphic subtrellises. With such a trellis, the ACS circuits can be divided into blocks such that the ACS circuits corresponding to states in a single subtrellis form a block. A particular ACS circuit only needs to transfer its metric to a subset of ACS circuits within its own block. This greatly reduces the ACS-connectivity and hence the hardware complexity and the wiring area on the IC chip.

10.3.5 Branch Complexity

The decoding speed of a Viterbi decoder depends on the total number of branches in the trellis to be processed and how fast they are being processed. If the processing load is shared by many ACS circuits at any time instant, then each ACS circuit will carry a small amount of processing load. This will speed up the decoding process. Therefore, a more meaningful measure of branch complexity is the number of branches to be processed by an ACS circuit [73, 101].

As pointed out earlier in this section, the number of active ACS circuits can be increased by parallel decomposition of a minimal trellis. However, parallel decomposition, in general, results in an increase in the number of composite branches in a trellis section. If the rate of increase of active ACS circuits is larger than the increase rate of composite branches, then the number of branches to be processed by each ACS circuit will decrease. The processing load of an ACS circuit at time-h_i is determined by the number of composite branches diverging from its corresponding state in $\Sigma_{h_i}(C)$. Therefore the total number of branches to be processed by an ACS circuit is determined by the diverging branch dimension profile (DBDP) of the trellis being used in the design.

Consider the minimal 8-section trellis of the $(64, 42)$ RM code, $\text{RM}_{3,6}$. The state space dimension profile of this code is $(0, 7, 10, 13, 10, 13, 10, 7, 0)$ and its DBDP is $(7, 6, 6, 3, 6, 3, 3, 0)$. Consider section-2 of the trellis. The number of composite branches in this section is 2^{13}. However, the number of active ACS circuits corresponding to the states of the trellis at the end of section-1 is $2^7 = 128$. Since each state has 64 composite branches diverging from it, each active ACS circuit must process 64 composite branches. Now consider the parallel decomposition of this minimal trellis into 128 parallel isomorphic subtrellises. The resultant non-minimal trellis has a state space dimension profile $(0, 13, 13, 13, 10, 13, 13, 13, 0)$ and each subtrellis has a state space dimension profile $(0, 6, 6, 6, 3, 6, 6, 6, 0)$. The DBDP of this non-minimal trellis is $(13, 3, 3, 3, 6, 3, 3, 0)$. All the components of this DBDP, except for the first one, are smaller than (or equal to) the corresponding components of the DBDP for the minimal 8-section trellis of this code. Consider section-2 of this non-minimal trellis. The total number of composite branches is now 2^{16}, a large increase from 2^{13} for the minimal trellis. However, all the 2^{13} ACS circuits at time-h_1 are active and they share the processing load. Each ACS circuit processes only 8 composite branches, compared to 64 for the minimal trellis. It is the same at the other time instants, except at time-h_4, where each active ACS circuit needs to process 64 branches, the same as for the minimal trellis. Therefore, the number of operations performed per ACS circuit is smaller for a Viterbi decoder designed based on the above non-minimal trellis. Reducing the diverging branch profile also results in a reduction of ACS-connectivity and hence a reduction in implementation complexity and wiring area on an IC chip.

Based on the above analysis and discussions, we may conclude that in designing a hardware Viterbi decoder for a specific linear block code, if the minimal trellis for the code is not desirable, then a non-minimal trellis with proper structural properties should be considered.

10.4 DIFFERENTIAL TRELLIS DECODING

The Viterbi algorithm was first devised for decoding convolutional codes. This decoding algorithm is based on the simple principle of add-compare-select (ACS) to process the code trellis and eliminate the less probable paths at each trellis level. This simple ACS principle has been used for implementing Viterbi decoders over the last two decades. However, a trellis-based decoding algorithm for convolutional codes can be devised based on a different processing principle, namely **compare-select-add (CSA)**. This decoding algorithm is devised based on a specific partition of a trellis section and the CSA processing principle. It is more efficient than the conventional Viterbi decoding algorithm. This decoding algorithm is called the **differential trellis decoding (DTD)** algorithm [32].

Consider a rate-$1/n$ $(n, 1, m)$ convolutional code of memory order m. The encoder of this code has one input and n outputs. Let $\boldsymbol{a} = (a_0, a_1, \ldots, a_i, \ldots)$ be the input information sequence. The n corresponding output code sequences are

$$\boldsymbol{u}^{(1)} = (u_0^{(1)}, u_1^{(1)}, \ldots, u_l^{(1)}, \ldots),$$

$$\begin{aligned} \boldsymbol{u}^{(2)} &= (u_0^{(2)}, u_1^{(2)}, \ldots, u_l^{(2)}, \ldots), \\ &\vdots \\ \boldsymbol{u}^{(n)} &= (u_0^{(n)}, u_1^{(n)}, \ldots, u_l^{(n)}, \ldots). \end{aligned}$$

At time-i, the input to the encoder is a_i and the output of the encoder is a block of n code bits $(u_i^{(1)}, u_i^{(2)}, \ldots, u_i^{(n)})$. The trellis for this code consists of 2^m states with two branches entering and leaving each state at any time (or level) greater than m.

A rate-$1/n$ $(n, 1, m)$ convolutional code is said to be **antipodal** if, in the generator matrix of (9.7), $G_0 = G_m = [11\ldots1]$. Most of the best rate-$1/n$ convolutional codes are antipodal. For an antipodal convolutional code, the two branches entering (or leaving) a state in its code trellis are **one's complement** to each other, i.e., if one branch is labeled with $(u_i^{(1)}, u_i^{(2)}, \ldots, u_i^{(n)})$, then the other branch is labeled with $(1 \oplus u_i^{(1)}, 1 \oplus u_i^{(2)}, \ldots, 1 \oplus u_i^{(n)})$, where $\oplus$ denotes the modulo-2 addition.

At time-i, the state of the encoder is defined by and labeled with the information bits $(a_{i-1}, a_{i-2}, \ldots, a_{i-m})$, stored in the input shift register. Consider the trellis section from time-i to time-$(i+1)$ for $i \geq m$. This section can be partitioned into 2^{m-1} two-state **fully connected subtrellises** with the following structural properties: (1) the two states at time-i are labeled with

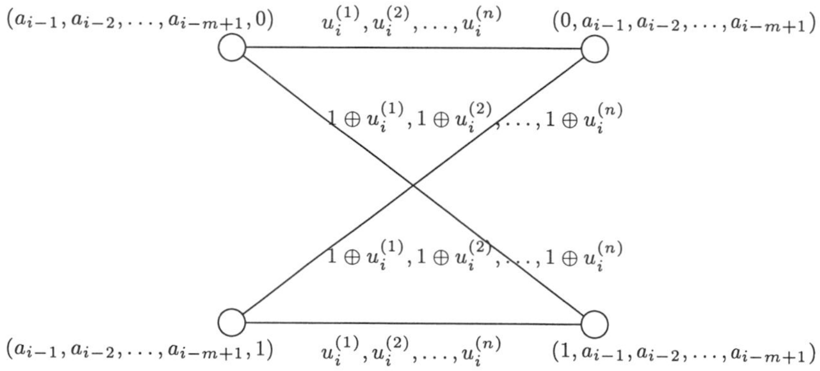

Figure 10.1. The structure of a 2-state subtrellis.

$(a_{i-1}, a_{i-2}, \ldots, a_{i-m+1}, 0)$ and $(a_{i-1}, a_{i-2}, \ldots, a_{i-m+1}, 1)$, respectively; (2) the two states at time-$(i+1)$ are labeled with $(0, a_{i-1}, a_{i-2}, \ldots, a_{i-m+1})$ and $(1, a_{i-1}, a_{i-2}, \ldots, a_{i-m+1})$, respectively; (3) the branches connecting the state $(a_{i-1}, a_{i-2}, \ldots, a_{i-m+1}, 0)$ to the states $(0, a_{i-1}, a_{i-2}, \ldots, a_{i-m+1})$ and $(1, a_{i-1}, a_{i-2}, \ldots, a_{i-m+1})$ are labeled with the code blocks $(u_i^{(1)}, u_i^{(2)}, \ldots, u_i^{(n)})$ and $(1 \oplus u_i^{(1)}, 1 \oplus u_i^{(2)}, \ldots, 1 \oplus u_i^{(n)})$, respectively; and (4) the branches connecting state $(a_{i-1}, a_{i-2}, \ldots, a_{i-m+1}, 1)$ to the states $(0, a_{i-1}, a_{i-2}, \ldots, a_{i-m+1})$ and $(1, a_{i-1}, a_{i-2}, \ldots, a_{i-m+1})$ are labeled with the code blocks $(1 \oplus u_i^{(1)}, 1 \oplus u_i^{(2)}, \ldots, 1 \oplus u_i^{(n)})$ and $(u_i^{(1)}, u_i^{(2)}, \ldots, u_i^{(n)})$, respectively. The structure of such a two-state subtrellis is depicted in Figure 10.1. These 2^{m-1} fully connected subtrellises are commonly called "butterflies".

Based on the above state grouping and trellis partitioning between time-i and time-$(i+1)$, each subtrellis can be labeled by an $(m-1)$-tuple $\boldsymbol{\alpha} = (a_{i-1}, a_{i-2}, \ldots, a_{i-m+1})$. In each subtrellis-$\boldsymbol{\alpha}$, the states at time-$i$ and the states at time-$(i+1)$ are represented by $(\boldsymbol{\alpha}, a_{i-m})$ and $(a_i, \boldsymbol{\alpha})$, respectively, with $a_{i-m}, a_i \in \{0, 1\}$.

The decoding algorithm to be presented in the following is based on the above trellis partition. Assume that BPSK is used for transmission and each BPSK signal has unit energy. A code sequence is mapped into a bipolar signal sequence for transmission. The i-th code block $(u_i^{(1)}, u_i^{(2)}, \ldots, u_i^{(n)})$ is mapped into the following bipolar sequence:

$$(2u_i^{(1)} - 1, 2u_i^{(2)} - 1, \ldots, 2u_i^{(n)} - 1). \tag{10.6}$$

Suppose correlation is used as the decoding metric. Let $\boldsymbol{r}_i = (r_i^{(1)}, r_i^{(2)}, \ldots, r_i^{(n)})$ be the received block in the interval between time-i and time-$(i+1)$. It follows from properties (3), (4) of a butterfly subtrellis given above, that the four branch metrics between time-i and time-$(i+1)$ in subtrellis-$\boldsymbol{\alpha}$ take two opposite values $\pm N_{i+1}^{\boldsymbol{\alpha}}$, with

$$N_{i+1}^{\boldsymbol{\alpha}} \triangleq \sum_{j=1}^{n} (2u_i^{(j)} - 1) \cdot r_i^{(j)}. \tag{10.7}$$

Let $M_i(\boldsymbol{\alpha}, 0)$ and $M_i(\boldsymbol{\alpha}, 1)$ denote the cumulative correlation metrics that have survived at time-i for states $(\boldsymbol{\alpha}, 0)$ and $(\boldsymbol{\alpha}, 1)$, respectively. Define

$$\Delta_{i+1}^{\boldsymbol{\alpha}}(0,1) \triangleq M_i(\boldsymbol{\alpha}, 0) - M_i(\boldsymbol{\alpha}, 1) \tag{10.8}$$

as the difference between these two metrics computed at time-$(i+1)$. Then at time-$(i+1)$, the difference between the cumulative metric candidates corresponding to transitions from states $(\boldsymbol{\alpha}, 0)$ and $(\boldsymbol{\alpha}, 1)$ to state $(a_i, \boldsymbol{\alpha})$ is given by

$$D_{i+1}^{\boldsymbol{\alpha}}(a_i) = \Delta_{i+1}^{\boldsymbol{\alpha}}(0,1) - 2(2a_i - 1)N_{i+1}^{\boldsymbol{\alpha}} \tag{10.9}$$

for $a_i \in \{0, 1\}$.

Note that MLD maximizes the correlation metric. Hence, from (10.9), we conclude that at state $(a_i, \boldsymbol{\alpha})$ of subtrellis-$\boldsymbol{\alpha}$, we select the branch diverging from state $(\boldsymbol{\alpha}, 0)$ if $\Delta_{i+1}^{\boldsymbol{\alpha}}(0,1) > 2(2a_i - 1)N_{i+1}^{\boldsymbol{\alpha}}$, and the branch diverging from state $(\boldsymbol{\alpha}, 1)$ otherwise. Therefore, this decision can be made by first determining

$$|M_{\Delta,N}| = \max\{|\Delta_{i+1}^{\boldsymbol{\alpha}}(0,1)|, |2N_{i+1}^{\boldsymbol{\alpha}}|\}, \tag{10.10}$$

and then checking the sign of the value $M_{\Delta,N}$ corresponding to this maximum, denoted $\mathrm{sgn}(M_{\Delta,N})$. Based on the comparison result given in (10.10) and $\mathrm{sgn}(M_{\Delta,N})$, the selection of the surviving branches into states $(a_i, \boldsymbol{\alpha})$ with $a_i \in \{0, 1\}$ is made. All the four selections of surviving branches are shown in Figure 10.2. The selection rules are given below:

(1) If $|\Delta_{i+1}^{\boldsymbol{\alpha}}(0,1)| > |2N_{i+1}^{\boldsymbol{\alpha}}|$ and $\Delta_{i+1}^{\boldsymbol{\alpha}}(0,1) > 0$, the two branches diverging from state $(\boldsymbol{\alpha}, 0)$ into states $(0, \boldsymbol{\alpha})$ and $(1, \boldsymbol{\alpha})$ are selected as the surviving branches.

(2) If $|\Delta_{i+1}^{\boldsymbol{\alpha}}(0,1)| > |2N_{i+1}^{\boldsymbol{\alpha}}|$ and $\Delta_{i+1}^{\boldsymbol{\alpha}}(0,1) \leq 0$, the two branches diverging from state $(\boldsymbol{\alpha}, 1)$ into states $(0, \boldsymbol{\alpha})$ and $(1, \boldsymbol{\alpha})$ are selected as the surviving branches.

(3) If $|\Delta_{i+1}^{\boldsymbol{\alpha}}(0,1)| \leq |2N_{i+1}^{\boldsymbol{\alpha}}|$ and $2N_{i+1}^{\boldsymbol{\alpha}} > 0$, the branch diverging from state $(\boldsymbol{\alpha}, 0)$ into state $(0, \boldsymbol{\alpha})$ and the branch diverging from state $(\boldsymbol{\alpha}, 1)$ into state $(1, \boldsymbol{\alpha})$ are selected as the surviving branches.

(4) If $|\Delta_{i+1}^{\boldsymbol{\alpha}}(0,1)| \leq |2N_{i+1}^{\boldsymbol{\alpha}}|$ and $2N_{i+1}^{\boldsymbol{\alpha}} \leq 0$, the branch diverging from state $(\boldsymbol{\alpha}, 0)$ into state $(1, \boldsymbol{\alpha})$ and the branch diverging from state $(\boldsymbol{\alpha}, 1)$ into state $(0, \boldsymbol{\alpha})$ are selected as the surviving branches.

For each subtrellis-$\boldsymbol{\alpha}$, the decoding process from time-i to time-$(i+1)$ can be carried out as follows:

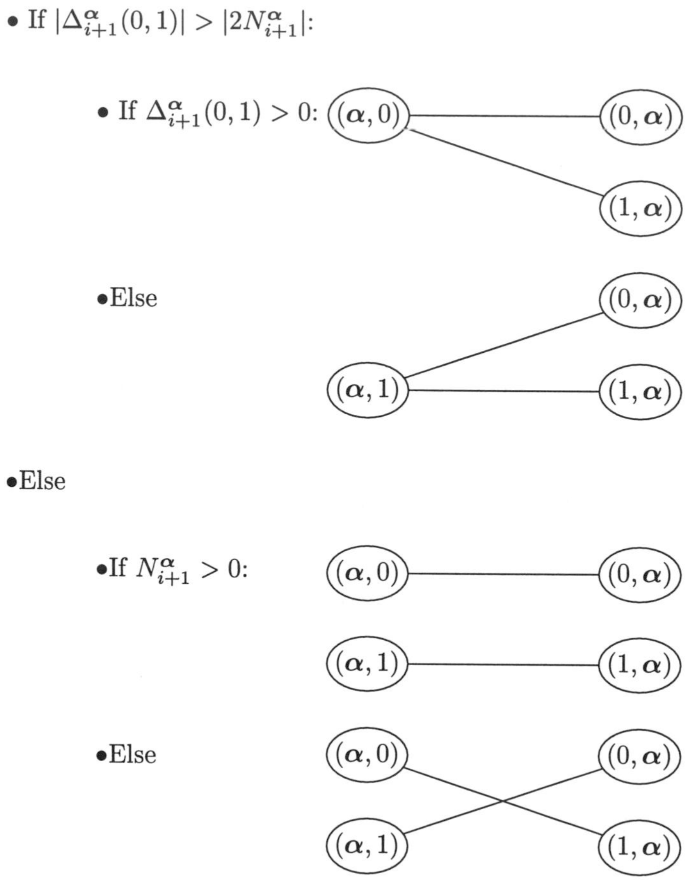

Figure 10.2. Branch selections for subtrellis-$\boldsymbol{\alpha}$.

Step-1 Compute the four possible branch metrics $\pm N_{i+1}^{\boldsymbol{\alpha}}$ (preprocessing) and scale them by 2.

Step-2 From Step-1, identify $2N_{i+1}^{\boldsymbol{\alpha}}$ and compute the metric difference $\Delta_{i+1}^{\boldsymbol{\alpha}}(0,1)$.

Step-3 Compare $|\Delta_{i+1}^{\boldsymbol{\alpha}}(0,1)|$ with $|2N_{i+1}^{\boldsymbol{\alpha}}|$.

Step-4 Based on the comparison result of Step-3, determine either $\text{sgn}(\Delta_{i+1}^{\boldsymbol{\alpha}}(0,1))$ or $\text{sgn}(N_{i+1}^{\boldsymbol{\alpha}})$, and select the surviving branches based on the selection rule shown in in Figure 10.2.

Step-5 For each state at time-$(i+1)$, update the new survivor metric based on Step-4.

The above decoding algorithm is called the **differential CSA-algorithm**. The metric computations in Step-1, which are also performed by the conventional Viterbi algorithm, can be preprocessed since at most 2^{n-1} values must be computed. Also, if the branch from state $(\boldsymbol{\alpha}, a_{i-m})$ to state $(a_i, \boldsymbol{\alpha})$ survives, the surviving metric at state $(a_i, \boldsymbol{\alpha})$ in Step-5 can be computed as follows:

$$M_{i+1}(a_i, \boldsymbol{\alpha}) = M_i(\boldsymbol{\alpha}, a_{i-m}) + (2a_{i-m} - 1)(2a_i - 1)N_{i+1}^{\boldsymbol{\alpha}}. \tag{10.11}$$

Note that the scaling by 2 of the preprocessed values $\pm N_{i+1}^{\boldsymbol{\alpha}}$ at Step-1 and the sign checks at Step-4 are elementary binary operations (scaling is done by shifting the register once). The real number operations are performed at Step-2, Step-3, and Step-5. There are 2^{m-1} subtractions at Step-2, 2^{m-1} comparisons at Step-3 and 2^m additions at Step-5. Therefore **a total of** $2\cdot 2^m$ **real number operations** is required to process a trellis section. However, after Step-1, the conventional Viterbi algorithm requires 2^{m+1} additions to evaluate the cumulative metrics for 2^m states and 2^m comparisons to determine the 2^m survivors. This results in **a total of** $3\cdot 2^m$ **real number operations** to process a trellis section. As a result, the the differential CSA-algorithm requires about 1/3 less real number operations than the conventional Viterbi algorithm for rate-$1/n$ antipodal convolutional codes as well as high-rate punctured codes obtained from them.

Example 10.2 Consider the (2,1,6) convolutional code with generating pattern

$$[1\ \ 1\ \ 0\ \ 1\ \ 1\ \ 1\ \ 1\ \ 1\ \ 0\ \ 0\ \ 1\ \ 0\ \ 1\ \ 1],$$

which is the most commonly used convolutional code. This code is antipodal. Its trellis consists of 64 states and can be decomposed into 32 fully connected 2-state subtrellises as shown in Figure 10.1. Since $n = 2$, at time-i, there are four possible branch metrics of the form $\pm(r_{i,1} \pm r_{i,2})$, which are computed with two real additions, and then scaled by 2. For this code, at time-i, the Viterbi algorithm computes 128 cumulative metric candidates and then performs 64 comparisons, so a total of 192 real value operations is required. The differential CSA-algorithm first computes 32 metric differences at Step-2, and then performs 32 comparisons at Step-3. Finally, based on the 32 sign checks of Step-4, 64 surviving cumulative metrics are updated at Step-5. As a result, only 128 real value additions are executed. Therefore, 64 real value operations are saved by the differential CSA-algorithm at the expense of 32 sign checks and 2 scalings by 2.

In practical applications, high-rate convolutional codes are often constructed from a low-rate $(n, 1, m)$ convolutional code by puncturing. The trellis for the punctured code has the same structure and state complexity as that of the original rate-$1/n$ convolutional code, except that the lengths of its sections vary periodically. As a result, the decoder for the rate-$1/n$ convolutional code can be used for decoding the punctured code. If the base rate-$1/n$ convolutional code is antipodal, then any punctured code constructed from it is also antipodal. Each trellis section for the punctured code can be partitioned into 2^{m-1} butterfly subtrellises in exactly the same manner as described above. The two branches leaving (or entering) a state in a butterfly subtrellis are one's complement of each other. Consequently, the differential CSA-algorithm can be used for decoding the punctured code. All the rate-$k/(k+1)$ punctured convolutional codes presented in [16] are time-varying antipodal codes. Also, this construction can be generalized to the case where k rate-$1/n$ base convolutional codes rather than only one are periodically selected, with period k. Again, if the resulting time-varying punctured code is antipodal, then the differential CSA-algorithm can be used.

The application of the differential CSA-algorithm to rate-k/n convolutional codes with $k > 1$ also allows $1/3$ real value computation saving after proper pairing of the states in the code trellis [32]. The differential CSA-algorithm can also be applied efficiently to trellis decoding of block codes. For example, trellis decoding based on the 4-section trellis diagram for the $(16, 5)$ RM code requires 59 real value operations for the differential CSA-algorithm and 95 real value operations for the conventional Viterbi algorithm.

11 A RECURSIVE MAXIMUM LIKELIHOOD DECODING

The Viterbi algorithm is indeed a very simple and efficient method of implementing the maximum likelihood decoding. However, if we take advantage of the structural properties in a trellis section, other efficient trellis-based decoding algorithms can be devised. Recently, an efficient trellis-based **recursive maximum likelihood decoding** (**RMLD**) algorithm for linear block codes has been proposed [37]. This algorithm is more efficient than the conventional Viterbi algorithm in both computation and hardware requirements. Most importantly, the implementation of this algorithm does not require the construction of the entire code trellis, only some special one-section trellises of relatively small state and branch complexities are needed for constructing path (or branch) metric tables recursively. At the end, there is only one table which contains only the most likely codeword and its metric for a given received sequence $\boldsymbol{r} = (r_1, r_2, \ldots, r_N)$. This algorithm basically uses the **divide and conquer strategy**. Furthermore, it allows **parallel/pipeline** processing of received sequences to speed up decoding.

11.1 BASIC CONCEPTS

Consider a binary (N, K) linear block code C. Suppose a codeword is transmitted and $\boldsymbol{r} = (r_1, r_2, \ldots, r_N)$ is the received vector at the output of the matched filter of the receiver.

Let T be the minimal trellis diagram for C. Consider the trellis section from time-x to time-y. As shown in Section 6.2, a composite branch between two adjacent states in this trellis section is a coset in $p_{x,y}(C)/C^{\rm tr}_{x,y}$, and a composite branch may appear many times as shown in (6.13). Using this fact, we can reduce the decoding complexity by just processing the distinct composite branches in each trellis section. To achieve this, we form a table for the metrics of composite branches, which for each coset D in $p_{x,y}(C)/C^{\rm tr}_{x,y}$, stores the largest metric D, denoted $m(D)$, and the label for the branch with the largest metric, denoted $l(D)$. This table is called the **composite branch metric table**, denoted $\mathrm{CBT}_{x,y}$, for the trellis section between time-x and time-y. Since the set of cosets $p_{0,N}(C)/C^{\rm tr}_{0,N} = C/C$ consists of C only, the table $\mathrm{CBT}_{0,N}$ contains only the codeword in C that has the largest metric. This is the most likely codeword. The RMLD algorithm is simply an algorithm to construct a composite branch metric table **recursively** from tables for trellis sections of shorter lengths to reduce computational complexity. When the table $\mathrm{CBT}_{0,N}$ is constructed, the decoding is completed and $\mathrm{CBT}_{0,N}$ contains the decoded codeword.

A straightforward method to construct the table $\mathrm{CBT}_{x,y}$ is to compute the metrics of all the vectors in the punctured code $p_{x,y}(C)$, and then find the vector with the largest metric for every coset in $p_{x,y}(C)/C^{\rm tr}_{x,y}$ by comparing the metrics of vectors in the coset. This method is efficient only when $y - x$ is small and should only be used at the **bottom** (or the **beginning**) of the recursive construction procedure. When $y - x$ is large, $\mathrm{CBT}_{x,y}$ is constructed from $\mathrm{CBT}_{x,z}$ and $\mathrm{CBT}_{z,y}$ for a properly chosen integer z with $x < z < y$.

Therefore, the key part of the RMLD algorithm is to construct the metric table $\mathrm{CBT}_{x,y}$ from tables $\mathrm{CBT}_{x,z}$ and $\mathrm{CBT}_{z,y}$. First we must show that this can be done. For two adjacent states, σ_x and σ_y, with $\sigma_x \in \Sigma_x(C)$ and $\sigma_y \in \Sigma_y(C)$, let

$$\Sigma_z(\sigma_x, \sigma_y) = \{\sigma_z^{(1)}, \sigma_z^{(2)}, \ldots, \sigma_z^{(\mu)}\} \tag{11.1}$$

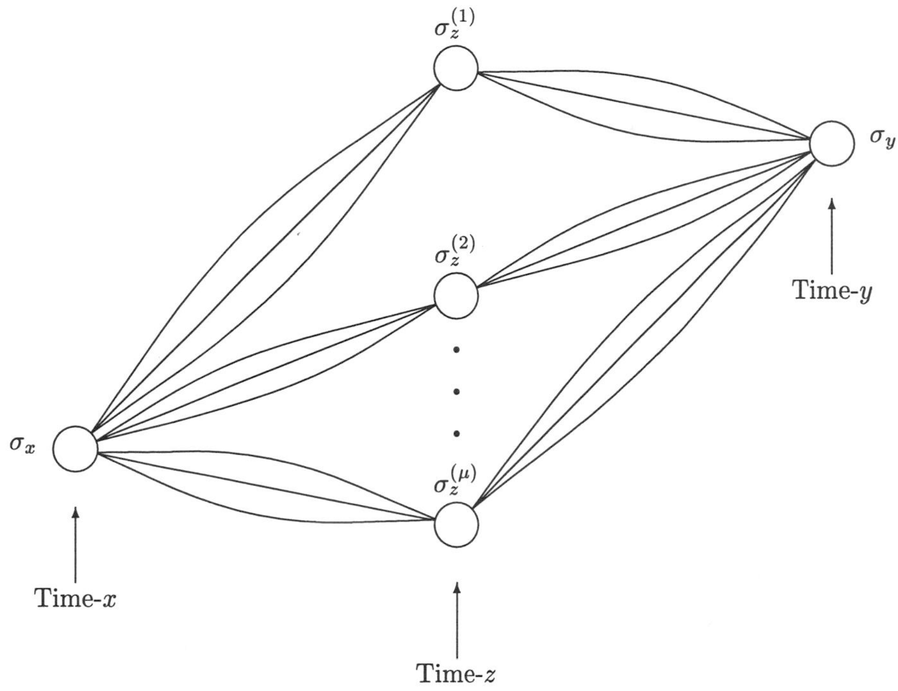

Figure 11.1. Connection between two states.

denote the subset of states in $\Sigma_z(C)$ through which the paths in $L(\sigma_x, \sigma_y)$ connect σ_x to σ_y as shown in Figure 11.1. Then

$$L(\sigma_x, \sigma_y) = \bigcup_{\sigma_z^{(i)} \in \Sigma_z(\sigma_x, \sigma_y)} L(\sigma_x, \sigma_z^{(i)}) \circ L(\sigma_z^{(i)}, \sigma_y). \tag{11.2}$$

It follows from (11.2) and the definitions of metric and label of a coset (or a composite branch) that we have

$$m(L(\sigma_x, \sigma_y)) \triangleq \max_{\sigma_z^{(i)} \in \Sigma_z(\sigma_x, \sigma_y)} \{m(L(\sigma_x, \sigma_z^{(i)})) + m(L(\sigma_z^{(i)}, \sigma_y))\} \tag{11.3}$$

and

$$l(L(\sigma_x, \sigma_y)) = l(L(\sigma_x, \sigma_z^{(i_{\max})})) \circ l(L(\sigma_z^{(i_{\max})}, \sigma_y)), \tag{11.4}$$

where $i_{\max}$ is the index for which the sum in (11.3) takes its maximum.

Note that the metrics, $m(L(\sigma_x, \sigma_z^{(i)}))$ and $m(L(\sigma_z^{(i)}, \sigma_x))$, and labels, $l(L(\sigma_x, \sigma_z^{(i)}))$ and $l(L(\sigma_z^{(i)}, \sigma_y))$, are stored in the composite branch metric tables $\mathrm{CBT}_{x,y}$ and $\mathrm{CBT}_{z,y}$. The state set $\Sigma_z(\sigma_x, \sigma_y)$ can be determined from the code trellis. Therefore, (11.3) and (11.4) show that the composite branch metric table $\mathrm{CBT}_{x,y}$ can be constructed from $\mathrm{CBT}_{x,z}$ and $\mathrm{CBT}_{z,y}$.

Based on the structural properties of a sectionalized trellis, we can readily show that

$$\mu = |\Sigma_z(\sigma_x, \sigma_y)| = 2^{k(C_{x,y})-k(C_{x,z})-k(C_{z,y})}. \tag{11.5}$$

This says that if we compute the metric $m(L(\sigma_x, \sigma_y))$ from (11.3) using tables, $\mathrm{CBT}_{x,z}$ and $\mathrm{CBT}_{z,y}$, we need to perform $|\Sigma_z(\sigma_x, \sigma_y)|$ additions and $|\Sigma_z(\sigma_x, \sigma_y)| - 1$ comparisons. However, if we compute the metric $m(L(\sigma_x, \sigma_y))$ directly from the parallel branches in $L(\sigma_x, \sigma_y)$, we need to compute $\left|C_{x,y}^{\mathrm{tr}}\right| = 2^{k(C_{x,y})}$ branch metrics and perform $2^{k(C_{x,y})} - 1$ comparisons. For large $y - x$, $\left|C_{x,y}^{\mathrm{tr}}\right|$ is much larger than $|\Sigma_z(\sigma_x, \sigma_y)|$ and hence constructing the metric table $\mathrm{CBT}_{x,y}$ from tables, $\mathrm{CBT}_{x,z}$ and $\mathrm{CBT}_{z,y}$, requires much less additions and comparisons than the direct construction of $\mathrm{CBT}_{x,y}$ from vectors in $p_{x,y}(C)$ and cosets in $p_{x,y}(C)/C_{x,y}^{\mathrm{tr}}$. Therefore, recursive construction of composite branch metric tables for trellis sections of longer lengths from tables for trellis sections of shorter lengths reduces decoding computational complexity.

11.2 THE GENERAL ALGORITHM

Now we describe the general framework of the RMLD algorithm for constructing the composite branch metric table $\mathrm{CBT}_{x,y}$ for decoding a received sequence $\boldsymbol{r}$. We denote this algorithm with $\mathbf{RMLD}(x, y)$. This algorithm uses two procedures, denoted $\mathbf{MakeCBT}(x, y)$ and $\mathbf{CombCBT}(x, y; z)$, which are defined as follows:

- MakeCBT(x, y): construct the table $\mathrm{CBT}_{x,y}$ directly as described later.
- CombCBT$(x, y; z)$: Given tables $\mathrm{CBT}_{x,z}$ and $\mathrm{CBT}_{z,y}$ as inputs, where $x < z < y$, combine these tables to form $\mathrm{CBT}_{x,y}$ as shown in (11.3) and (11.4).

The procedure CombCBT$(x, y; z)$ can be expressed as

$$\mathrm{CombCBT}(\mathrm{RMLD}(x, z), \mathrm{RMLD}(z, y))$$

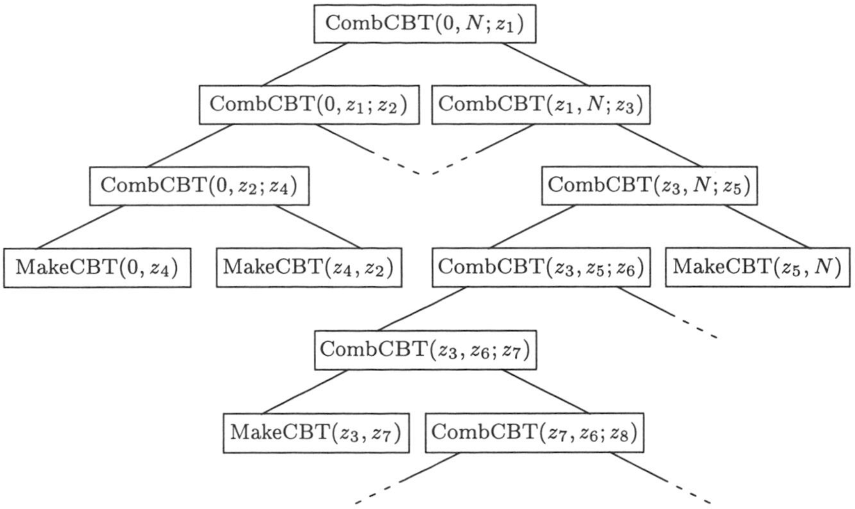

Figure 11.2. Illustration of the recursion process of the RMLD algorithm.

to show its recursive nature.

[Algorithm RMLD(x, y)]
Construct $\mathrm{CBT}_{x,y}$ using the least complex of the following two options:

(1) Execute MakeCBT(x, y), or

(2) Execute CombCBT(RMLD(x, z), RMLD(z, y)), where z with $x < z < y$ is selected to minimize computational complexity.

Decoding is accomplished by executing RMLD$(0, N)$. The recursion process is depicted in Figure 11.2. We see that the RMLD algorithm allows parallel/pipeline processing of received words. This speeds up the decoding process.

The MakeCBT(x, y) procedure is efficient only when $y-x$ is small and should only be used at the **bottom** (or the **beginning**) of the recursive construction procedure. When $y-x$ is large, $\mathrm{CBT}_{x,y}$ is constructed from $\mathrm{CBT}_{x,z}$ and $\mathrm{CBT}_{z,y}$ for a properly chosen z with $x < z < y$. At the bottom of the recursion process, $y - x$ is small and the computation done by the MakeCBT procedure during the entire decoding process is also small. Therefore, the major computation is carried out by the CombCBT procedure. Hence, the CombCBT procedure is the major procedure in the RMLD algorithm and should be devised to reduce either the total number of computations for software implementation or the circuit requirement and chip size for IC implementation.

In a soft-decision decoding algorithm, addition and comparison operations for metrics are considered as the basic operations. An addition operation and a comparison operation are in general assumed to have equal weight (or cost).

Let $\psi_M(x,y)$ and $\psi_C(x,y;z)$ denote the number of basic operations required to execute the procedure MakeCBT(x,y) and the procedure CombCBT$(x,y;z)$, respectively. The values of $\psi_M(x,y)$ and $\psi_C(x,y;z)$ depend on the implementation of the RMLD algorithm. Assume that the formulas for $\psi_M(x,y)$ and $\psi_C(x,y;z)$ are given. To minimize the overall decoding complexity of the RMLD algorithm, sectionalization of a trellis (choices of z) must be done properly. A sectionalization which gives the smallest overall decoding complexity for given $\psi_M(x,y)$ and $\psi_C(x,y;z)$ is called the **optimum sectionalization** for the code.

Let $\psi_{\min}(x,y)$ denote the smallest number of operations required to construct the table CBT$_{x,y}$. Then it follows from the algorithm RMLD(x,y) given above that

$$\psi_{\min}(x,y) \triangleq \begin{cases} \psi_M(x,y), & \text{if } x+1=y, \\ \min\left\{\psi_M(x,y), \min_{x<z<y}\{\psi_{R,\min}(x,y;z)\}\right\}, & \text{otherwise,} \end{cases} \tag{11.6}$$

where

$$\psi_{R,\min}(x,y;z) \triangleq \psi_{\min}(x,z) + \psi_{\min}(z,y) + \psi_C(x,y;z). \tag{11.7}$$

The total number of operations required to decode a received word is given by $\psi_{\min}(0,N)$.

By using (11.6) and (11.7) together with formulas for $\psi_M(x,y)$ and $\psi_C(x,y;z)$, we can compute $\psi_{\min}(x,y)$ for every (x,y) with $0 \le x < y \le N$ efficiently in the following way: The values of $\psi_{\min}(x,x+1)$ for $0 \le x < N$ are computed using the given formula for $\psi_M(x,y)$. For an integer i with $0 \le x < x+i \le N$, $\psi_{\min}(x,x+i)$ can be computed from $\psi_{\min}(x',y')$ with $y'-x' < i$ and the given formulas for $\psi_M(x,y)$ and $\psi_C(x,y;z)$. By keeping track of the values of z selected in the above procedure, it is easy to find an optimum sectionalization.

If $\psi_M(x,y)$ and $\psi_C(x,y;z)$ are independent of the received sequence $\boldsymbol{r}$ for any $0 \le x < y \le N$ and $x < z < y$, then the optimum sectionalization can be fixed.

11.3 DIRECT METHODS FOR CONSTRUCTING COMPOSITE BRANCH METRIC TABLES

For two integers x and y such that $0 \le x < y \le N$, a straightforward way to construct the composite branch metric table $\text{CBT}_{x,y}$ directly is to compute the metrics of all the vectors in the punctured code $p_{x,y}(C)$ independently, and then find the vector (branch) with the largest metric for every coset in $p_{x,y}(C)/C^{\text{tr}}_{x,y}$ by comparing the metrics of vectors in the coset. Each surviving vector and its metric are stored in the table $\text{CBT}_{x,y}$. Let **MakeCBT-I**(x,y) denote this procedure.

The number of addition-equivalent operations required to construct the table $\text{CBT}_{x,y}$ by executing MakeCBT-I(x,y), denoted $\psi_M^{(I)}(x,y)$, is given as follows:

$$\psi_M^{(I)}(x,y) = (y-x-1)2^{k(p_{x,y}(C))} + 2^{k(p_{x,y}(C))-k(C_{x,y})}(2^{k(C_{x,y})}-1). \quad (11.8)$$

The first term is the number of additions to compute all the metrics for the vectors in $p_{x,y}(C)$, and the second term is the number of comparisons for finding the vectors with the largest metrics by comparing the metrics of vectors in each coset in $p_{x,y}(C)/C^{\text{tr}}_{x,y}$.

A more efficient method for constructing the table $\text{CBT}_{x,y}$ is to compute the metrics of the 2^{y-x} branch labels following the order of the Gray code as proposed in [60, 102]. Let **MakeCBT-G**(x,y) denote this procedure, where G stands for Gray code. Assume that the bit metric satisfies the following condition: $M(r,0) = -M(r,1)$, where r is a received symbol. This condition holds for the AWGN channel with BPSK transmission and $M(r,1) \triangleq r$. We also assume that the all-one vector of length $y-x$, denoted $\mathbf{1}_{y-x}$, is in $p_{x,y}(C)$ for any x and y with $0 \le x < y \le N$. In this case, the metrics of 2^{y-x-1} labels are computed first in the order of the Gray code, and then the remaining metrics are computed by negating the first 2^{y-x-1} metrics. If $\mathbf{1}_{y-x} \in C^{\text{tr}}_{x,y}$, for any vector in a coset of $C^{\text{tr}}_{x,y}$, the complementary vector is in the same coset. In this case, we can simply discard the branches with negative metrics [60, 102] for finding the largest metric in each coset.

Let $\psi_M^{(G)}(x,y)$ denote the values of $\psi_M(x,y)$ using the MakeCBT-G(x,y). Assume that the negation is costless. Then,

$$\psi_M^{(G)}(x,y) = \begin{cases} 2^{y-x-1} + y - x - 2 + 2^{k(p_{x,y}(C))-k(C_{x,y})}(2^{k(C_{x,y})-1} - 1), \\ \qquad \text{if } \mathbf{1}_{y-x} \in C^{\text{tr}}_{x,y}, \\ 2^{y-x-1} + y - x - 2 + 2^{k(p_{x,y}(C))-k(C_{x,y})}(2^{k(C_{x,y})} - 1), \\ \qquad \text{otherwise.} \end{cases} \tag{11.9}$$

For small $y-x$, the dimension of $p_{x,y}(C)$ is close to $y-x$. The computational complexities of both MakeCBT-I and MakeCBT-G procedures are small for small $y-x$. The RMLD algorithm with the MakeCBT-G procedure requires slightly less computational complexity than that with the MakeCBT-I procedure; however, the MakeCBT-I procedure is simpler for IC implementation. Using the MakeCBT-G procedure, the metrics of the first 2^{y-x-1} labels in $p_{x,y}(C)$ must be computed serially, however with the MakeCBT-I procedure, the metrics for all the labels in $p_{x,y}(C)$ can be computed independently in parallel.

11.4 THE COMBCBT PROCEDURE

The CombCBT$(x,y;z)$ procedure simply performs the computation of (11.3) and finds the label of (11.4). It is important to note that in the construction of the metric table CBT$_{x,y}$, we do not need to compute the metric

$$m(\sigma_x,\sigma_y) \triangleq m(L(\sigma_x,\sigma_y))$$

for every adjacent state pair (σ_x,σ_y). We only need to compute $m(\sigma_x,\sigma_y)$ for those adjacent state pairs for which the paths between each state pair form a distinct coset in $p_{x,y}(C)/C^{\text{tr}}_{x,y}$. Therefore, we only compute the metrics for $2^{k(p_{x,y}(C))-k(C_{x,y})}$ distinct adjacent state pairs between time-x and time-y. This is the key to reduce computational complexity.

In principle we can construct the metric table CBT$_{x,y}$ using the section of the code trellis T from time-x to time-y as follows:

(i) For each coset $D \in p_{x,y}(C)/C^{\text{tr}}_{x,y}$, identify a state pair (σ_x,σ_y) such that $L(\sigma_x,\sigma_y) = D$;

(ii) Determine the state set $\Sigma_z(\sigma_x,\sigma_y)$; and

(iii) Compute the metric $m(\sigma_x, \sigma_y)$ and the label $l(\sigma_x, \sigma_y)$ from (11.3) and (11.4), respectively.

However for long codes, it is a big effort to construct the large trellis section from time-x to time-y and execute the above steps (i) to (iii). The total number of composite branches in the trellis section between time-x and time-y can be very large and the number of distinct composite branches is only a small fraction. Examining this trellis section can be very time consuming and effort wasting. Consequently, implementation will be complex and costly.

To overcome the complexity problem and facilitate the computation of (11.3), we construct a much simpler special two-section trellis for the punctured code $p_{x,y}(C)$ with section boundary locations in $\{x, z, y\}$ and multiple "final" states at time-y, one for each coset in $p_{x,y}(C)/C^{\text{tr}}_{x,y}$. This special two-section trellis contains only the needed information for constructing the metric table $\text{CBT}_{x,y}$ from $\text{CBT}_{x,z}$ and $\text{CBT}_{z,y}$ (no redundancy). For a coset $D_y \in p_{x,y}(C)/C^{\text{tr}}_{x,y}$, define

$$S_z(D_y) \triangleq \{D_z \in p_{x,z}(C)/C^{\text{tr}}_{x,z} : D_z \subseteq p_{x,z}(D_y)\}, \tag{11.10}$$

where $p_{x,z}(D_y)$ is the truncation of the coset D_y from time-x to time-z. For each $D_z \in S_z(D_y)$, there is exactly one coset in $p_{z,y}(C)/C^{\text{tr}}_{z,y}$, denoted $\text{adj}(D_z, D_y)$, such that $D_z \circ \text{adj}(D_z, D_y) \subseteq D_y$ (see Figure 11.1). Then,

$$D_y = \bigcup_{D_z \in S_z(D_y)} D_z \circ \text{adj}(D_z, D_y). \tag{11.11}$$

From (11.11) we see that the metric of D_y can be computed from metrics of cosets in $p_{x,z}(C)/C^{\text{tr}}_{x,z}$ and cosets in $p_{z,y}(C)/C^{\text{tr}}_{z,y}$ (or from tables $\text{CBT}_{x,z}$ and $\text{CBT}_{z,y}$) once the set $S_z(D_y)$ and $\text{adj}(D_z, D_y)$ for each $D_z \in S_z(D_y)$ are identified. The special two-section trellis to be constructed is simply to display the relationship given by (11.11) and identify the set $S_z(D_y)$ for each coset $D_y \in p_{x,y}(C)/C^{\text{tr}}_{x,y}$.

Let Σ_z and Σ_y denote the state spaces of the special two-section trellis for $p_{x,y}(C)$ at time-z and time-y, respectively. To achieve the purpose as described above, the special two-section trellis for $p_{x,y}(C)$ must have the following structural properties:

(1) There is an initial state, denoted $\sigma_{x,0}$ at time-x.

(2) There is a one-to-one correspondence between the states in the state space Σ_z and the cosets in $p_{x,z}(C)/C^{\text{tr}}_{x,z}$. Let D_z denote a coset in $p_{x,z}(C)/C^{\text{tr}}_{x,z}$ and $\sigma(D_z)$ denote its corresponding state at time-z. Then the composite branch label between $\sigma_{x,0}$ and $\sigma(D_z)$ is $L(\sigma_{x,0}, \sigma(D_z)) = D_z$.

(3) There is a one-to-one correspondence between the states in the state space Σ_y and the cosets $p_{x,y}(C)/C^{\text{tr}}_{x,y}$. Let D_y denote a coset in $p_{x,y}(C)/C^{\text{tr}}_{x,y}$ and $\sigma(D_y)$ denote its corresponding state at time-y. For any state $\sigma(D_y) \in \Sigma_y$, $L(\sigma(D_z), \sigma(D_y)) = \text{adj}(D_z, D_y)$ if $D_z \in S_z(D_y)$. Otherwise, $L(\sigma(D_z), \sigma(D_y)) = \emptyset$.

From the structural properties of the above special two-section trellis, we see that: (1) For every state $\sigma(D_z) \in \Sigma_z$, its (state) metric $m(D_z)$ is given in the table $\text{CBT}_{x,z}$; and (2) For each composite branch between a state $\sigma(D_z)$ at time-z and an adjacent state $\sigma(D_y)$ at time-y, its composite branch metric, $m(\sigma(D_z), \sigma(D_y))$, is given in the table $\text{CBT}_{z,y}$.

It follows from (11.11) and the structural properties of the above special two-section trellis for $p_{x,y}(C)$ that for each coset $D_y \in p_{x,y}(C)/C^{\text{tr}}_{x,y}$, the metric $m(D_y)$ is given by

$$m(D_y) = \max_{D_z \in S_z(D_y)} \{m(D_z) + m(\sigma(D_z), \sigma(D_y))\}, \tag{11.12}$$

where the set of states correspond to $S_z(D_y)$ and the state pairs $(\sigma(D_z), \sigma(D_y))$ can be easily identified from the special two-section trellis. Eq.(11.12) is simply equivalent to (11.3). Therefore, $\text{CombCBT}(x, y; z)$ will be designed to compute the metrics for the table $\text{CBT}_{x,y}$ based on (11.12) using the special two-section trellis. In general, this special trellis is much simpler than the section of the entire code trellis T from time-x to time-y except for the cases where $x = 0$ or $y = N$, and is much easier to construct. As a result, the construction of the metric table $\text{CBT}_{x,y}$ is much simpler.

The construction of the above special two-section trellis for $p_{x,y}(C)$ is done as follows: Choose a basis $\{\boldsymbol{v}_1, \boldsymbol{v}_2, \ldots, \boldsymbol{v}_{k(p_{x,y}(C))}\}$ of $p_{x,y}(C)$ such that the first $k(C^{\text{tr}}_{x,y}) = k(C_{x,y})$ vectors form a basis of $C^{\text{tr}}_{x,y}$. Define

$$n_{x,y} \triangleq y - x + k(p_{x,y}(C)) - k(C_{x,y}). \tag{11.13}$$

Let $G(x,y)$ be the following $k(p_{x,y}(C)) \times n_{x,y}$ matrix:

$$G(x,y) = \left[\begin{array}{cc} \boldsymbol{v}_1 & \\ \vdots & \boldsymbol{0} \\ \boldsymbol{v}_{k(C_{x,y})} & \\ & ----- \\ \boldsymbol{v}_{k(C_{x,y})+1} & \\ \vdots & \boldsymbol{I} \\ \boldsymbol{v}_{k(p_{x,y}(C))} & \end{array} \right],$$

where $\boldsymbol{0}$ denotes the $k(C_{x,y}) \times (k(p_{x,y}(C)) - k(C_{x,y}))$ all-zero matrix, and $\boldsymbol{I}$ denotes the identity matrix of dimension $(k(p_{x,y}(C)) - k(C_{x,y}))$. Let $C(x,y)$ be the binary linear code of length $n_{x,y}$ generated by $G(x,y)$. Construct a 3-section trellis diagram $T(\{x,z,y,x+n_{x,y}\})$ for $C(x,y)$ with section boundaries at times x, z, y and $x+n_{x,y}$ as shown in Figure 11.3. Then the first two sections of $T(\{x,z,y,x+n_{x,y}\})$ give the desired special two-section trellis for computing (11.12).

In fact from (11.12) and the properties of the special two-section trellis for $p_{x,y}(C)$, we only need the second section of $T(\{x,z,y,x+n_{x,y}\})$ to construct the table $\text{CBT}_{x,y}$. For convenience, we denote this special one-section trellis with $T_x(z,y)$. Table $\text{CBT}_{x,z}$ gives the state metrics of $T_x(z,y)$ at time-z and Table $\text{CBT}_{z,y}$ gives the composite branch metrics of $T_x(z,y)$ between time-z and time-y. Therefore, the implementation of the RMLD algorithm does not require the construction of the code trellis T for the entire code C, it only requires the construction of the special one-section trellises, one for each recursion step. Each of these special one-section trellises has the minimum (state and branch) complexity for constructing a composite branch metric table using the CombCBT procedure. This reduces decoding complexity considerably.

Example 11.1 Consider the RM code $C = \text{RM}_{2,4}$ given in Example 6.3. Let $x = 4$, $y = 12$, and $z = 8$. Then, $p_{x,y}(C) = \text{RM}_{2,3}$, $C^{\text{tr}}_{x,y} = \text{RM}_{1,3}$, $C^{\text{tr}}_{x,z} =$

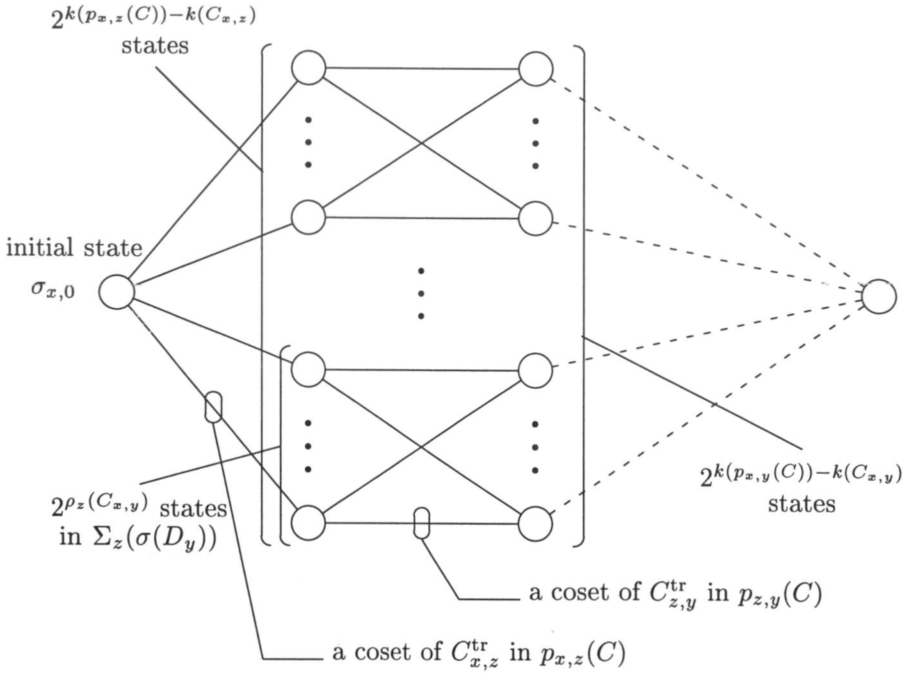

Figure 11.3. Structure of the trellis diagram $T(\{x, z, y, x + n_{x,y}\})$.

$C^{\rm tr}_{z,y} = \mathrm{RM}_{0,2}$, and $n_{x,y} = 11$. $C(x, y)$ is the (11,9) code generated by

$$G(x,y) = \begin{bmatrix} 1 & 1 & 1 & 1 & 1 & 1 & 1 & 1 & 0 & 0 & 0 \\ 0 & 1 & 0 & 1 & 0 & 1 & 0 & 1 & 0 & 0 & 0 \\ 0 & 0 & 1 & 1 & 0 & 0 & 1 & 1 & 0 & 0 & 0 \\ 0 & 0 & 0 & 0 & 1 & 1 & 1 & 1 & 0 & 0 & 0 \\ & & & & & & & & & & \\ 0 & 0 & 0 & 1 & 0 & 0 & 0 & 1 & 1 & 0 & 0 \\ 0 & 0 & 0 & 0 & 0 & 1 & 0 & 1 & 0 & 1 & 0 \\ 0 & 0 & 0 & 0 & 0 & 0 & 1 & 1 & 0 & 0 & 1 \end{bmatrix}.$$

It can be put in trellis oriented form by simple row operations. The one-section minimal trellis diagram, $T_4(8, 12)$, consists of two 4-state parallel components. One of the components is depicted in Figure 11.4. The other can be obtained

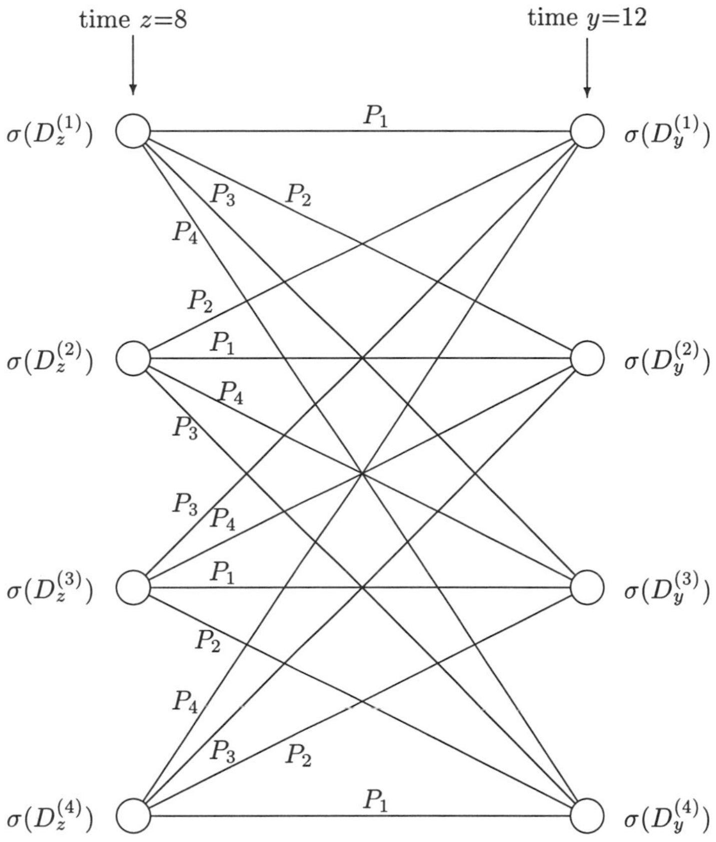

$P_1 \triangleq \{0000, 1111\}, P_2 \triangleq \{0011, 1100\}$
$P_3 \triangleq \{0101, 1010\}, P_4 \triangleq \{0110, 1001\}$

Figure 11.4. A parallel component of $T_4(8, 12)$ for the $\text{RM}_{2,4}$ code.

by adding $(0, 0, 0, 1)$ to each branch label.

$\triangle\triangle$

From (11.12), we see that the computation of the composite branch metric $m(D_y)$ depends on the size of the set $S_z(D_y)$. Since for a coset $D_y \in$

$p_{x,y}(C)/C^{\text{tr}}_{x,y}$, the truncation $p_{x,z}(D_y)$ is a union of cosets in $p_{x,z}(C)/C^{\text{tr}}_{x,z}$, it follows from property (3) of the special two-section trellis that for every state $\sigma(D_y) \in \Sigma_y$, the number of composite branches merging into the state $\sigma(D_y)$ is

$$\begin{aligned} |S_z(D_y)| &= \frac{|D_y|}{|D_z| \cdot |L(\sigma(D_z), \sigma(D_y))|} = \frac{\left|C^{\text{tr}}_{x,y}\right|}{\left|C^{\text{tr}}_{x,z}\right| \cdot \left|C^{\text{tr}}_{z,y}\right|} \\ &= 2^{k(C_{x,y})-k(C_{x,z})-k(C_{z,y})}, \end{aligned} \tag{11.14}$$

which is exactly the same as (11.5). From (11.14), we can readily determine the number of computation operations required to compute $m(D_y)$.

Next we need to devise efficient methods to solve (11.12) using the one-section trellis $T_x(x, y)$ so that either the computational complexity of the CombCBT procedure is reduced or the circuit requirement and chip size of IC implementation of the CombCBT procedure are reduced. Two methods for solving (11.12) will be presented in the next two sections and they result in two specific CombCBT procedures, named the **CombCBT-V** and the **CombCBT-U** procedures.

11.5 COMBCBT-V$(X, Y; Z)$ PROCEDURE

A straightforward procedure to solve (11.12) based on the one-section trellis $T_x(z, y)$ is to apply the conventional add-compare-select (**ACS**) procedure that is used in the conventional Viterbi algorithm. For each coset D_y in $p_{x,y}(C)/C^{\text{tr}}_{x,y}$, the metric sum, $m(D_z) + m(\sigma(D_z), \sigma(D_y))$, is computed for every state $\sigma(D_z)$ with $D_z \in S_z(D_y)$, and $m(D_y)$ is found by comparing all the computed metric sums. This procedure is called the CombCBT-V$(x, y; z)$ procedure, where V stands for Viterbi algorithm.

Since the Viterbi algorithm is applied to a one-section trellis diagram to construct a composite branch metric table from two smaller tables, the IC implementation of the CombCBT-V procedure is quite simple and straightforward.

Let $\psi^{(V)}_C(x, y; z)$ denote the value of $\psi_C(x, y; z)$ for the CombCBT-V$(x, y; z)$ procedure. Note that the number of states at time-y in $T_x(z, y)$ is $2^{k(p_{x,y}(C))-k(C_{x,y})}$, and for each state $\sigma(D_y)$ at time-y, the number of states $\sigma(D_z)$ at time-z in $T_x(z, y)$ which are adjacent to $\sigma(D_y)$ is given by (11.14).

Let

$$\rho_z(C_{x,y}) = k(C_{x,y}) - k(C_{x,z}) - k(C_{z,y}). \tag{11.15}$$

Then, the total number of additions and that of comparisons executed by the CombCBT-V$(x, y; z)$ procedure are

$$2^{k(p_{x,y}(C))-k(C_{x,y})+\rho_z(C_{x,y})} \quad \text{and} \quad 2^{k(p_{x,y}(C))-k(C_{x,y})}(2^{\rho_z(C_{x,y})} - 1),$$

respectively. Consequently, the computational complexity of the CombCBT-V$(x, y; z)$ procedure is given by

$$\psi_C^{(V)}(x, y; z) = 2^{k(p_{x,y}(C))-k(C_{x,y})}(2^{\rho_z(C_{x,y})+1} - 1). \tag{11.16}$$

11.6 RMLD-(I,V) AND RMLD-(G,V) ALGORITHMS

Combining the CombCBT-V procedure with either the MakeCBT-I procedure or the MakeCBT-G procedure, we obtain two specific RMLD algorithms, denoted **RMLD-(I,V)** and **RMLD-(G,V)**. From (11.6), (11.7), (11.8) and (11.16), we can compute the total number of addition-equivalent operations required by the RMLD-(I,V) algorithm for decoding a received word. The computational complexity of the RMLD-(G,V) algorithm can be computed from (11.6), (11.7), (11.9) and (11.16).

For either the RMLD-(I,V) algorithm or the RMLD-(G,V) algorithm, we need to know for what value of $y - x$ that the CombCBT-V procedure should be executed to construct the table $\text{CBT}_{x,y}$. This is answered by the following two theorems. We simply state the theorems here without the proofs which can be found in [111].

Theorem 11.1 Consider a binary linear code C of length N such that the minimum Hamming distances of C and its dual code are both greater than one.

(i) If $y - x > 2$, then for any z with $x < z < y$, the CombCBT-V$(x, y; z)$ procedure requires less computation to form the metric table $\text{CBT}_{x,y}$ than the MakeCBT-I(x, y) procedure. If $y - x = 2$, the complexities of CombCBT-V$(x, y; x + 1)$ and that of MakeCBT-I(x, y) are the same.

(ii) If $y-x > 2$, $k(p_{x,y}(C)) = y-x$ and $C^{\text{tr}}_{x,y} = \{\mathbf{0}_{y-x}, (1,*,\ldots,*,1)\}$, where $\mathbf{0}_{y-x}$ denotes the all-zero vector of length $y-x$ and $(1,*,\ldots,*,1)$ denotes a vector of length $y-x$ such that the first and the last components are 1, then the right-hand side of (11.6) takes its minimum for both $z = \lfloor (x+y)/2 \rfloor$ and $z = \lceil (x+y)/2 \rceil$. △△

Theorem 11.1 simply says that for $y - x > 2$, procedure CombCBT-V$(x, y; z)$ should be used to construct the metric table CBT$_{x,y}$ in the RMLD-(I,V) algorithm.

Theorem 11.2 Consider a binary linear code C of length N such that the minimum Hamming distance of C and its dual code are both greater than one. For the RMLD-(G,V) algorithm,

(i) If $k(p_{x,y}(C)) = y - x$ and $C^{\text{tr}}_{x,y} = \{\mathbf{0}\}$ or $\{\mathbf{0}_{y-x}, (1,*,\ldots,*,1)\}$, then the MakeCBT-G$(x, y)$ procedure requires less computation than the CBT$(x, y; z)$ procedure for any z with $x < z < y$ to form the metric table CBT$_{x,y}$ for $y - x > 2$. When $y - x = 2$, they are the same.

(ii) If the conditions of (i) do not hold, then the CombCBT-V$(x, y; z)$ procedure with some z is more efficient than the MakeCBT-G(x, y) for constructing the metric table CBT$_{x,y}$ for $y - x > 2$. Moreover, if $k(p_{x,y}(C)) < y - x$ and $C^{\text{tr}}_{x,y} = \{\mathbf{0}\}$ or $\{\mathbf{0}_{y-x}, (1,*,\ldots,*,1)\}$, then the right-hand side of (11.6) takes its minimum for both $z = \lfloor (x+y)/2 \rfloor$ and $z = \lceil (x+y)/2 \rceil$. △△

Since the Viterbi algorithm is applied to a one-section trellis diagram to construct a composite branch metric table from two smaller tables, the IC implementations of both the RMLD-(I,V) and RMLD-(G,V) algorithms are quite simple and straightforward. For high speed decoders, the MakeCBT-I procedure is more suitable than the MakeCBT-G procedure, since branch metrics can be computed in parallel. As shown in Theorem 11.1, in the optimum sectionalization, the value of $y - x$ for the MakeCBT-I procedure to be executed can be kept equal to 2, but this is not necessarily the case for the MakeCBT-G procedure (see Theorem 11.2). Hence, IC implementation of the MakeCBT-I procedure is easier. Furthermore, with the MakeCBT-G procedure, the metrics must be computed serially.

11.7 COMBCBT-U$(X, Y; Z)$ PROCEDURE

This procedure is based on the decomposition of the one-section trellis $T_x(z, y)$ into simple uniform subtrellises as described in Section 6.4. The one-section trellis $T_x(z, y)$ may consist of parallel isomorphic components. These parallel components can be partitioned into groups of the same size in such a way that: (1) two parallel components in the same group are identical up to path labeling; and (2) two parallel components in two different groups do not have any path label in common [44]. Each group consists of 2^λ identical parallel components, where λ can be computed from (6.36) with $C(x, y)$ as the code.

Furthermore, each parallel component of $T_x(z, y)$ can be decomposed into subtrellises with simple uniform structures as shown in Figure 11.5 by applying Theorem 3 of [44] (also see Section 6.3) to the code $C(x, y)$ that was used for constructing the one-section trellis $T_x(z, y)$. Consider a parallel component Λ. The state spaces at the two ends of the parallel component can be partitioned into blocks of the same size 2^ν, called **left U-blocks** and **right U-blocks**, respectively, where ν can be computed from (6.41) with C replaced by $C(x, y)$.

A pair of a left U-block and a right U-block is called a U-block pair, and each U-block pair (B_z, B_y) has the following **uniform structure**, denoted U: For any two states σ_y and σ'_y in B_y,

$$\{L(\sigma_z, \sigma_y) : \sigma_z \in B_z\} = \{L(\sigma_z, \sigma'_y) : \sigma_z \in B_z\}. \tag{11.17}$$

The above property simply says that for a U-block pair (B_z, B_y), the set of composite branches from the states in the left U-block B_z to any state in the right U-block B_y is the same. This property can be used in solving (11.12) to reduce the computational complexity. The label set of composite branches defined by (11.17) is called the **composite branch label set** of the U-block pair (B_z, B_y). Two different U-block pairs have mutually disjoint composite branch label sets.

The CombCBT-U$(x, y; z)$ procedure is devised based on the uniform structure of a U-block pair. In contrast to the CombCBT-V$(x, y; z)$ which solves (11.12) independently for every state $\sigma(D_y)$ with $D_y \in p_{x,y}(C)/C^{\mathrm{tr}}_{x,y}$, CombCBT-U$(x, y; z)$ solves (11.12) simultaneously for each U-block pair (B_z, B_y) of a parallel component of $T_x(z, y)$ by taking into account of the uniform

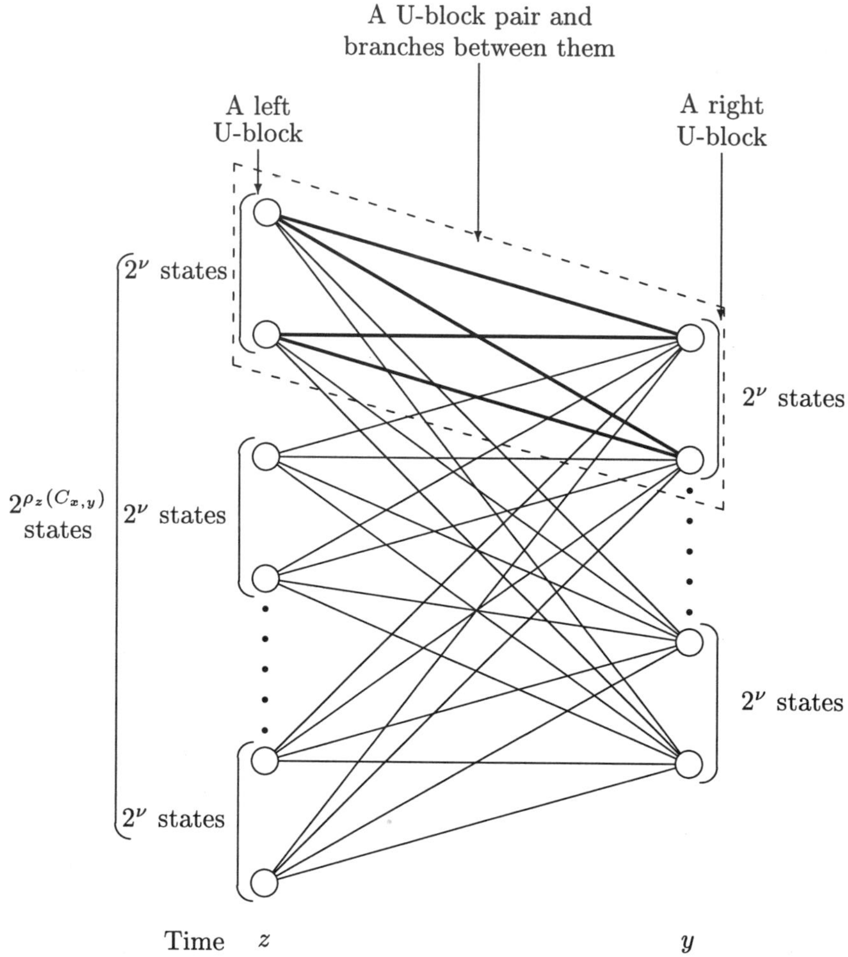

Figure 11.5. The left U-blocks and right U-blocks of a parallel component in $T_x(z, y)$.

property U given by (11.17). For a parallel component Λ of $T_x(z, y)$, let LU(Λ) denote the set of left U-blocks in Λ. Based on (11.17), (11.12) can be put in the following form: For a state $\sigma(D_y)$ in a right U-block B_y,

$$m_{B_z}(D_y) \triangleq \max_{D_z \in \{D_z : \sigma(D_z) \in B_z\}} \{m(D_z) + m(\sigma(D_z), \sigma(D_y))\}, \text{ for } B_z \in \mathrm{LU}(\Lambda), \tag{11.18}$$

$$m(D_y) = \max_{B_z \in LU(\Lambda)} m_{B_z}(D_y). \tag{11.19}$$

Equations (11.18) and (11.19) show that (11.12) can be solved simultaneously for each U-block pair. This allows parallel processing to speed up the computation. In fact the computations of (11.18) and (11.19) can be carried out for all the parallel components of $T_x(z, y)$ in parallel.

For easy understanding, an example is used to explain how to solve (11.18) for each U-block pair (B_z, B_y).

Example 11.2 Again consider the RM code $C = \mathrm{RM}_{2,4}$. As shown in Example 11.1, the one-section trellis diagram $T_x(z, y)$ with $x = 4$, $y = 12$ and $z = 8$ consists of two four-state parallel components. From (6.36) and (6.41), we find that $\lambda = 0$ and $\nu = 2$. Therefore, the two parallel components are not identical, and each consists of only one left U-block and one right U-block. As shown in Figure 11.4, the four end states of one parallel component at time-8, denoted $\sigma(D_z^{(1)})$, $\sigma(D_z^{(2)})$, $\sigma(D_z^{(3)})$, $\sigma(D_z^{(4)})$, form a single left U-block, and the 4 end states at time-12, denoted $\sigma(D_y^{(1)})$, $\sigma(D_y^{(2)})$, $\sigma(D_y^{(3)})$, $\sigma(D_y^{(4)})$, form a single right U-block. There are four different composite branch labels between them, denoted

$$P_1 \triangleq \{0000, 1111\}, \quad P_2 \triangleq \{0011, 1100\},$$
$$P_3 \triangleq \{0101, 1010\}, \quad P_4 \triangleq \{0110, 1001\}.$$

The set of the composite branch labels merging into any state $\sigma(D_y^{(j)})$ at time-12 is $\{P_1, P_2, P_3, P_4\}$.

From (11.18) and (11.19), the largest metric, denoted $m(D_y^{(j)})$, for the coset $D_y^{(j)}$ with $1 \le j \le 4$ is given by

$$m(D_y^{(j)}) = \max_{1 \le i \le 4} \{m(D_z^{(i)}) + m(P_{b(i,j)})\}, \tag{11.20}$$

where $b(i,j)$ is the unique integer such that $L(\sigma(D_z^{(i)}), \sigma(D_y^{(j)})) = P_{b(i,j)}$.

To compute the metrics $m(D_y^{(j)})$ for $1 \leq j \leq 4$, form the following set of metric sums:

$$M \triangleq \{m(D_z^{(i)}) + m(P_b) : 1 \leq i \leq 4, 1 \leq b \leq 4\}.$$

Each sum in M is associated with a path in the trellis of Figure 11.4. Clearly the largest sum in M corresponds to the survivor path for the associated state $\sigma(D_y^{(q)})$ at time-y., i.e., $m(D_y^{(q)})$ is equal to the largest sum in M. Thus this value for the coset $D_y^{(q)}$ is entered in Table CBT$_{x,y}$. This can be proceeded by examining the second, third, ... largest sum in M. If the j-th largest sum M_j corresponds to state $\sigma(D_y^{(q')})$, and CBT$_{x,y}$ contains no entry for the coset $D_y^{(q')}$, then M_j is entered in CBT$_{x,y}$. This process continues until CBT$_{x,y}$ contains entries for each $D_y^{(j)}$.

Similarly, the metrics of cosets that correspond to the four states at time-12 in the other parallel component in the one-section trellis $T_x(z,y)$ can be computed. This completes the construction of table CBT$_{x,y}$.

We can find the j-th largest sum of M more efficiently by pre-sorting

$$\{m(D_z^{(i)}) : 1 \leq i \leq 4\} \quad \text{and} \quad \{m(P_b) : 1 \leq b \leq 4\}.$$

△△

In general, for a U-block pair (B_z, B_y) with $B_z = \{\sigma(D_z^{(1)}), \sigma(D_z^{(2)}), \ldots, \sigma(D_z^{(2^\nu)})\}$, $B_y = \{\sigma(D_y^{(1)}), \sigma(D_y^{(2)}), \ldots, \sigma(D_y^{(2^\nu)})\}$, and the composite branch label set of (B_z, B_y), $\{P_1, P_2, \ldots, P_{2^\nu}\}$, (11.18) is solved in the following way:

(S1) Sort $m(D_z^{(1)})$, $m(D_z^{(2)})$, ..., $m(D_z^{(2^\nu)})$ in the decreasing order.

(S2) Sort $m(P_1)$, $m(P_2)$, ..., $m(P_{2^\nu})$ in the decreasing order.

(S3) Form $M \triangleq \{m(D_z^{(i)}) + m(P_b) : 1 \leq i \leq 2^\nu, 1 \leq b \leq 2^\nu\}$. Determine $m_{B_z}(D_y^{(j)})$ with $1 \leq j \leq 2^\nu$ as described in Example 11.2 by using the following partial ordering on M:

$$m(D_z^{(i)}) + m(P_b) \geq m(D_z^{(i')}) + m(P_{b'}), \quad \text{if } m(D_z^{(i)}) \geq m(D_z^{(i')}) \text{ and } m(P_b) \geq m(P_{b'}).$$

Clearly the above procedure for a U-block pair can be executed for all the U-block pairs in all the distinct parallel components in the one-section trellis $T_x(z, y)$ simultaneously.

Note that the CombCBT-U$(x, y; z)$ procedure is identical to the CombCBT-V$(x, y; z)$ procedure only for the case of $\nu = 0$ (the trivial case in which each left U-block and right U-block consist of a single state).

Let $\psi_C^{(U)}(x, y; z)$ denote the number of addition-equivalent operations of CombCBT-U$(x, y; z)$. The computational complexity for solving (11.18) depends on the received sequence. In the following, an upper bound on $\psi_C^{(U)}(x, y; z)$ for the worst case is given, which is independent of the received sequence. Without derivation, the bound is given below [37]:

$$\begin{aligned}\psi_C^{(U)}(x, y; z) \quad \leq \quad & (2^{k(p_{x,z}(C))-k(C_{x,z})-\nu} + 2^{k(p_{x,y}(C))-k(C_{x,y})-\nu})\eta(2^\nu) \\ & +2^{k(p_{x,y}(C))-k(C_{x,y})}((1+\frac{1}{2^\nu})2^{\rho_z(C_{x,y})} - 1) \end{aligned} \tag{11.21}$$

where

$$\eta(2^\nu) = \begin{cases} \nu, & \text{for } \nu = 0, 1, \\ 2(\nu-1)(2^\nu - 1) - 1, & \text{otherwise.} \end{cases} \tag{11.22}$$

Let $\bar{\psi}_C^{(U)}(x, y; z)$ denote the upper bound given by the right-hand side of (11.21). It can be shown that [37]

$$\bar{\psi}_C^{(U)}(x, y; z) \leq \psi_C^{(V)}(x, y; z). \tag{11.23}$$

The inequality of (11.23) holds for $\nu \geq 2$. As ν becomes large, the ratio

$$\bar{\psi}_C^{(U)}(x, y; z)/\psi_C^{(V)}(x, y; z)$$

decreases rapidly. This says that the CombCBT-U procedure is more efficient than the CombCBT-V procedure computation-wise.

11.8 RMLD-(G,U) ALGORITHM

The CombCBT-U procedure can be combined with either the MakeCBT-I procedure or the MakeCBT-G procedure to form specific RMLD algorithms. Since the MakeCBT-G procedure requires less computational complexity than the MakeCBT-I procedure. The MakeCBT-G procedure and CombCBT-U procedure are combined to form an RMLD algorithm, called the RMLD-(G,U) algorithm.

Table 11.1. Numbers of addition-equivalent operations with various maximum likelihood decoding algorithms for some RM and extended BCH codes of length 64.

Code, (Basis)	64-section	RMLD-(I,V)	RMLD-(G,V)	RMLD-(G,U) upper bound $\psi_{\min}^{(G,U)}$	Lafourcade & Vardy [60]
$\mathrm{RM}_{2,6}(64,22)$	425,209	78,209	77,896	66,824	101,786
$\mathrm{RM}_{3,6}(64,42)$	773,881	326,017	323,759	210,671	538,799
$\mathrm{RM}_{4,6}(64,57)$	7,529	5,281	4,999	4,087	6,507
EBCH(64,10), (C)	20,073	3,201	3,108	3,108	4,074
EBCH(64,16), (B)	764,153	120,193	119,880	96,840	148,566
EBCH(64,18), (B)	2,865,401	468,040	468,040	372,808	509,120
EBCH(64,24), (B)	1,327,353	271,745	271,432	171,823	316,608
EBCH(64,30), (C)	35,028,985	16,091,009	16,056,668	9,408,567	16,598,063
EBCH(64,36), (C)	18,710,521	9,995,617	9,961,580	7,684,276	12,829,263
EBCH(64,39), (C)	38,436,857	24,741,161	24,707,149	19,841,161	30,982,731
EBCH(64,45), (C)	1,082,105	893,489	891,695	665,713	891,819
EBCH(64,51), (A)	418,553	312,721	312,382	257,300	393,528

From (11.6), (11.7), (11.9) and (11.21), we can compute an upper bound, denoted $\psi_{\min}^{(G,U)}$, on the worst-case computational complexity of the RMLD-(G,U) algorithm for decoding a received word.

11.9 COMPARISONS

Among the three specific RMLD algorithms, RMLD-(I,V), RMLD-(G,V) and RMLD-(G,U), the RMLD-(G,U) algorithm is the most efficient one computation-wise, while the RMLD-(I,V) algorithm is the simplest for IC implementation.

In the following, the three specific RMLD algorithms are applied to some well known codes of length 64 to show their effectiveness in terms of computational complexity.

Let EBCH$(64,k)$ denote the extended code obtained from the binary primitive $(63,k)$ BCH code by adding an overall parity bit. The computational complexities of decoding the $\mathrm{RM}_{r,6}$ codes with $2 \leq r \leq 4$ and the permuted EBCH$(64,k)$ codes with $10 \leq k \leq 51$ are computed based on certain symbol position permutations and given in Table 11.1. Hereafter, $\mathrm{RM}_{r,m}$ is denoted $\mathrm{RM}_{r,m}(2^m, \sum_{i=0}^{r}\binom{m}{i})$ to show the number of information bits explicitly.

To reduce the state complexities of the trellis diagrams for EBCH$(64,k)$ codes, the order of symbol positions must be permuted. For the RM codes, the natural symbol ordering is optimal for the state complexity [45]. For the EBCH codes, consider the following permutations π [46]: Let α be a primitive element of GF(2^6) and $\{\beta_1,\ldots,\beta_6\}$ a basis of GF(2^6) over GF(2). For a positive integer i less than 2^6, let α^{i-1} be expressed as $\alpha^{i-1} = \sum_{j=1}^{6} b_{i,j}\beta_j$, with $b_{i,j} \in$ GF(2). For $i = 0$, let $b_{0,j} \triangleq 0$ for $1 \le j \le 6$. Then π is the following permutation on $\{1,2,\ldots,2^6\}$, $\pi(i) \triangleq 1 + \sum_{j=1}^{6} b_{i-1,j}2^{6-j}$, for $1 \le i \le 2^6$. Consider the following three bases for codes of length 64: (1) Basis A is the polynomial basis, $\{1,\alpha,\alpha^2,\ldots,\alpha^5\}$; (2) Basis B is $\{1,\alpha,\alpha^2,\alpha^{21},\alpha^{22},\alpha^{23}\}$, which is obtained by combining a basis of GF(2^6) over GF(2^2), $\{1,\alpha,\alpha^2\}$, and a basis of GF(2^2) over GF(2), $\{1,\alpha^{21}\}$; (3) Basis C is $\{1,\alpha,\alpha^9,\alpha^{10},\alpha^{18},\alpha^{19}\}$, which is obtained by combining a basis of GF(2^6) over GF(2^3), $\{1,\alpha\}$, and a basis of GF(2^3) over GF(2), $\{1,\alpha^9,\alpha^{18}\}$.

Table 11.1 gives the total numbers of addition-equivalent operations required by the three specific RMLD algorithms, RMLD-(I,V), RMLD-(G,V) and RMLD-(G,U), for decoding the above codes. For the RMLD-(G,U) algorithm, only the values of the upper bound $\psi_{\min}^{(G,U)}$ on the worst-case computational complexity are given. For comparison purpose, the numbers of addition-equivalent operations required for decoding the above codes with the Viterbi decoding algorithm based on optimum sectionalization presented in Section 10.2 (Lafourcade and Vardy algorithm) are also included. The column labeled 64-section gives the numbers of operations required in the conventional Viterbi decoding based on the bit-level 64-section minimal trellis diagram.

For all EBCH codes, other than the EBCH$(64,51)$ code, the symbol permutations indicated in Table 11.1 give the smallest optimum values for each column among the three symbol permutations given above. For EBCH(64,51), Basis B gives the smallest number of operations required in the conventional Viterbi decoding based on the 64-section trellis diagram among the three permutations, but Basis A gives the smallest values for the other columns. This shows that a good bit ordering for the N-section trellis diagram is not always good for the proposed RMLD decoding procedures. The last column in Table 11.1 shows the numbers of addition-equivalent operations given by Lafourcade and Vardy [60].

Table 11.1 shows that the RMLD-(G,U) algorithm is the most efficient trellis-based decoding algorithm even in terms of the worst case computational complexity, and the difference between the computational complexities of the RMLD-(I,V) and RMLD-(G,V) algorithms is very small. All three RMLD algorithms are more efficient than the Viterbi decoding algorithm based on optimum sectionalization [60], except only for the RMLD-(I,V) algorithm for the EBCH$(64,45)$ code. For each algorithm, the number of basic operations executed by the MakeCBT procedure is relatively small compared with that executed by the CombCBT procedure. Consider the EBCH$(64,45)$ code. Decoding this code with the RMLD-(I,V) algorithm, the number of basic operations executed by the MakeCBT-I procedure is 108 out of a total of 893,489 basic operations. Using the RMLD-(G,U) algorithm, the MakeCBT-G procedure executes 818 basic operations out of a total $665,713$ basic operations.

Let $< x, y >$ denote the MakeCBT-G(x, y) operation, and let $\cdot$ denote the CombCBT-U operation. The optimum trellis sectionalizations for RM$_{3,6}(64, 42)$ and EBCH$(64, 24)$ with Basis B for the complexity measure $\bar{\psi}_{\min}^{(G,U)}$ of the RMLD-(G,U) algorithm are identical, and represented as

$$((< 0,8 > \cdot < 8,16 >) \cdot (< 16,24 > \cdot < 24,32 >)) \cdot \\ ((< 32,40 > \cdot < 40,48 >) \cdot (< 48,56 > \cdot < 56,64 >)).$$

The optimum trellis sectionalization for RM$_{2,6}(64, 22)$ for $\bar{\psi}_{\min}^{(G,U)}$ is

$$((((((< 0,8 > \cdot < 8,16 >) \cdot (< 16,24 > \cdot < 24,32 >)) \\ \cdot (< 32,40 > \cdot < 40,48 >)) \cdot < 48,56 >) \cdot < 56,61 >) \\ \cdot < 61,63 >) \cdot < 63,64 >,$$

and that for EBCH$(64, 30)$ with Basis C is

$$(< 0,16 > \cdot < 16,32 >) \cdot (< 32,48 > \cdot < 48,64 >).$$

The optimum trellis sectionalization for a code using the algorithm RMLD is generally not unique. The above optimum trellis sectionalizations are chosen in the following manner: If $\psi_{\min}(x, y) = \psi_M^{(G)}(x, y)$, then execute the

Table 11.2. Average numbers of addition-equivalent operations using the RMLD-(G,U) algorithm for the $\mathrm{RM}_{3,6}(64,42)$ and EBCH$(64,24)$ codes.

Code	Algorithm	Measure	SNR	Value
$\mathrm{RM}_{3,6}(64,42)$	RMLD-(G,U)	Upper bound $\bar{\psi}_{\min}^{(G,U)}$ on the worst case complexity		$210,671$
		The average number of operations $\psi_{\min}^{(G,U)}$	1dB	$66,722$
			4dB	$66,016$
			7dB	$63,573$
			10dB	$61,724$
EBCH$(64,24)$, Basis B	RMLD-(G,U)	Upper bound $\bar{\psi}_{\min}^{(G,U)}$ on the worst case complexity		$171,823$
		The average number of operations $\psi_{\min}^{(G,U)}$	1dB	$70,676$
			4dB	$70,420$
			7dB	$69,325$
			10dB	$68,158$

MakeCBT-G(x,y) procedure. Otherwise, the CombCBT-U$(x,y;z)$ procedure is executed for an integer z such that $\psi_{\min}(x,y) = \psi_{\min}(x,z) + \psi_{\min}(z,y) + \psi_C^{(U)}(x,y;z)$ and $|z-(x+y)/2|$ are the smallest.

Using the trellis sectionalizations which are optimum with respect to the measure $\bar{\psi}_{\min}^{(G,U)}$, we evaluate the average values of $\psi_{\min}(0,N)$, denoted $\psi_{\min}^{(G,U)}$, for the RMLD-(G,U) algorithm which are given in Table 11.2. It is assumed that BPSK modulation is used on an AWGN channel. The average values at the SNRs per information bit, 1, 4, 7 and 10 (dB), are listed in the rows labeled $\psi_{\min}^{(G,U)}$ for $\mathrm{RM}_{3,6}(64,42)$ and EBCH$(64,24)$. We see that these values vary only slightly and are much smaller than the worst-case upper bound $\bar{\psi}_{\min}^{(G,U)}$.

12 AN ITERATIVE DECODING ALGORITHM FOR LINEAR BLOCK CODES BASED ON A LOW-WEIGHT TRELLIS SEARCH

For long linear block codes, maximum likelihood decoding based on full code trellises would be very hard to implement if not impossible. In this case, we may wish to trade error performance for the reduction in decoding complexity. Suboptimum soft-decision decoding of a linear block code based on a low-weight subtrellis can be devised to provide an effective trade-off between error performance and decoding complexity. This chapter presents such a suboptimal decoding algorithm for linear block codes. This decoding algorithm is iterative in nature and based on an optimality test. It has the following important features: (1) a simple method to generate a sequence of candidate codewords, one at a time, for test; (2) a sufficient condition for testing a candidate codeword for optimality; and (3) a low-weight subtrellis search for finding the most likely (ML) codeword.

12.1 GENERAL CONCEPTS

A simple **low-cost** decoder, such as an algebraic decoder, is used to generate a sequence of **candidate codewords** iteratively one at a time using a set of **test error patterns** based on the reliability information of the received symbols. When a candidate is generated, it is tested based on an **optimality condition**. If it satisfies the optimality condition, then it is the **most likely** (ML) codeword and decoding stops. If it fails the optimality test, a **search** for the ML codeword is conducted in a **region** which contains the ML codeword. The search region is determined by the current candidate codeword (or codewords) and the reliability of the received symbols. The search is conducted through a **purged trellis diagram** for the given code using a trellis-based decoding algorithm. If the search fails to find the ML codeword, a new candidate is generated using a new test error pattern (or any simple method), and the optimality test and search are renewed. The process of testing and searching continues until either the ML codeword is found or all the test error patterns are exhausted (or a stopping criterion is met) and the decoding process is terminated.

The key elements in this decoding algorithm are:

(1) Generation of candidate codewords,

(2) Optimality test,

(3) A search criterion,

(4) A low-weight trellis diagram,

(5) A search algorithm, and

(6) A stopping criterion.

12.2 OPTIMALITY CONDITIONS

Suppose C is used for error control over the **AWGN** channel using **BPSK** signaling. Let $\boldsymbol{c} = (c_1, c_2, \ldots, c_N)$ be the transmitted codeword. For BPSK transmission, $\boldsymbol{c}$ is mapped into a bipolar sequence $\boldsymbol{x} = (x_1, x_2, \ldots, x_N)$ with $x_i = (2c_i - 1) \in \{\pm 1\}$ for $1 \leq i \leq N$. Suppose $\boldsymbol{x}$ is transmitted and $\boldsymbol{r} = (r_1, r_2, \ldots, r_N)$ is received at the output of the matched filter of the receiver. Let

$\boldsymbol{z} = (z_1, z_2, \ldots, z_N)$ be the binary **hard-decision** received sequence obtained from $\boldsymbol{r}$ using the **hard-decision function** given by

$$z_i = \begin{cases} 1 & \text{for } r_i > 0 \\ 0 & \text{for } r_i \leq 0 \end{cases} \tag{12.1}$$

for $1 \leq i \leq N$. We use $|r_i|$ as the **reliability measure** of the received symbol r_i since this value is proportional to the log-likelihood ratio associated with the symbol hard-decision, the larger the magnitude the greater its reliability.

For any binary N-tuple $\boldsymbol{u} = (u_1, u_2, \ldots, u_N) \in \{0,1\}^N$, the **correlation** between $\boldsymbol{u}$ and the received sequence $\boldsymbol{r}$ is given by

$$M(\boldsymbol{u}, \boldsymbol{r}) \triangleq \sum_{i=1}^{N} r_i \cdot (2u_i - 1). \tag{12.2}$$

It follows from (12.1) and (12.2) that

$$M(\boldsymbol{z}, \boldsymbol{r}) = \sum_{i=1}^{N} |r_i| \geq 0, \tag{12.3}$$

and for any $\boldsymbol{u} \in \{0,1\}^N$,

$$M(\boldsymbol{z}, \boldsymbol{r}) \geq M(\boldsymbol{u}, \boldsymbol{r}).$$

Define the following index sets:

$$D_0(\boldsymbol{u}) \triangleq \{i : u_i = z_i \text{ and } 1 \leq i \leq N\}, \tag{12.4}$$

$$\begin{aligned} D_1(\boldsymbol{u}) &\triangleq \{i : u_i \neq z_i \text{ and } 1 \leq i \leq N\} \\ &= \{1, 2, \ldots, N\} \backslash D_0(\boldsymbol{u}). \end{aligned} \tag{12.5}$$

Let

$$n(\boldsymbol{u}) \triangleq |D_1(\boldsymbol{u})|. \tag{12.6}$$

Consider

$$\begin{aligned} M(\boldsymbol{u}, \boldsymbol{r}) &= \sum_{i=1}^{N} r_i(2u_i - 1) \\ &= \sum_{i \in D_0(\boldsymbol{u})} r_i(2u_i - 1) + \sum_{i \in D_1(\boldsymbol{u})} r_i(2u_i - 1) \end{aligned}$$

$$
\begin{aligned}
&= \sum_{i \in D_0(\boldsymbol{u})} r_i(2z_i - 1) - \sum_{i \in D_1(\boldsymbol{u})} r_i(2z_i - 1) \\
&= \sum_{i=1}^{N} r_i(2z_i - 1) - 2 \sum_{i \in D_1(\boldsymbol{u})} r_i(2z_i - 1) \\
&= M(\boldsymbol{z}, \boldsymbol{r}) - 2 \sum_{i \in D_1(\boldsymbol{u})} |r_i| . \qquad (12.7)
\end{aligned}
$$

Let

$$L(\boldsymbol{u}, \boldsymbol{r}) \triangleq \sum_{i \in D_1(\boldsymbol{u})} |r_i| . \qquad (12.8)$$

Then, $M(\boldsymbol{u}, \boldsymbol{r})$ can be expressed in terms of $M(\boldsymbol{z}, \boldsymbol{r})$ and $L(\boldsymbol{u}, \boldsymbol{r})$ as follows:

$$M(\boldsymbol{u}, \boldsymbol{r}) = M(\boldsymbol{z}, \boldsymbol{r}) - 2L(\boldsymbol{u}, \boldsymbol{r}). \qquad (12.9)$$

$L(\boldsymbol{u}, \boldsymbol{r})$ is called the **correlation discrepancy** of $\boldsymbol{u}$.

From (12.9), the MLD can be stated in terms of the correlation discrepancies of codewords as follows: The decoder computes the correlation discrepancy $L(\boldsymbol{c}, \boldsymbol{r})$ for each codeword $\boldsymbol{c} \in C$, and decode $\boldsymbol{r}$ into the codeword $\boldsymbol{c}_{\text{opt}}$ for which

$$L(\boldsymbol{c}_{\text{opt}}, \boldsymbol{r}) = \min_{\boldsymbol{c} \in C} L(\boldsymbol{c}, \boldsymbol{r}). \qquad (12.10)$$

From (12.10), we see that if there exists a codeword $\boldsymbol{c}^*$ for which

$$L(\boldsymbol{c}^*, \boldsymbol{r}) \leq \alpha(\boldsymbol{c}^*, \boldsymbol{r}) \triangleq \min_{\boldsymbol{c} \in C, \boldsymbol{c} \neq \boldsymbol{c}^*} L(\boldsymbol{c}, \boldsymbol{r}),$$

then $\boldsymbol{c}^* = \boldsymbol{c}_{\text{opt}}$. It is not possible to determine $\alpha(\boldsymbol{c}^*, \boldsymbol{r})$ without evaluating $L(\boldsymbol{c}, \boldsymbol{r})$ for all $\boldsymbol{c} \in C$. However, if it is possible to determine a **tight lower bound** on $\alpha(\boldsymbol{c}^*, \boldsymbol{r})$, then we have a **sufficient condition** for testing the optimality of a candidate codeword.

Suppose $\boldsymbol{c}$ is a candidate codeword for testing. The index set $D_0(\boldsymbol{c})$ consists of $N - n(\boldsymbol{c})$ indices. Order the indices in $D_0(\boldsymbol{c})$ as follows:

$$D_0(\boldsymbol{c}) = \{k_1, k_2, \ldots, k_{N-n(\boldsymbol{c})}\} \qquad (12.11)$$

such that for $1 \leq i < j \leq N - n(\boldsymbol{c})$,

$$|r_{k_i}| < |r_{k_j}|. \qquad (12.12)$$

Let $D_0^{(j)}(\boldsymbol{c})$ denote the set of first j indices in the ordered set $D_0(\boldsymbol{c})$, i.e.,

$$D_0^{(j)}(\boldsymbol{c}) \triangleq \{k_1, k_2, \ldots, k_j\}. \tag{12.13}$$

For $j \leq 0$, $D_0^{(j)}(\boldsymbol{c}) \triangleq \emptyset$ and for $j \geq N - n(\boldsymbol{c})$, $D_0^{(j)}(\boldsymbol{c}) \triangleq D_0(\boldsymbol{c})$.

Let $W = \{0, w_1, w_2, \ldots, w_m\}$ be the weight profile of code C and w_k be the k-th smallest non-zero weight in W. Define

$$s_k \triangleq w_k - n(\boldsymbol{c}), \tag{12.14}$$

$$G(\boldsymbol{c}, w_k) \triangleq \sum_{i \in D_0^{(s_k)}(\boldsymbol{c})} |r_i|, \tag{12.15}$$

and

$$R(\boldsymbol{c}, w_k) \triangleq \{\boldsymbol{c}' \in C : d(\boldsymbol{c}', \boldsymbol{c}) < w_k\}, \tag{12.16}$$

where $d(\boldsymbol{c}', \boldsymbol{c})$ denotes the Hamming distance between $\boldsymbol{c}'$ and $\boldsymbol{c}$.

Theorem 12.1 For a codeword $\boldsymbol{c}$ in C and a nonzero weight $w_k \in W$, if

$$L(\boldsymbol{c}, \boldsymbol{r}) \leq G(\boldsymbol{c}, w_k), \tag{12.17}$$

then the optimal solution $\boldsymbol{c}_{\text{opt}}$ is in the region $R(\boldsymbol{c}, w_k)$ [75].

Proof: Let $\boldsymbol{c}'$ be a codeword in $C \backslash R(\boldsymbol{c}, w_k)$, i.e.,

$$d(\boldsymbol{c}', \boldsymbol{c}) \geq w_k. \tag{12.18}$$

We want to show that $L(\boldsymbol{c}, \boldsymbol{r}) \leq L(\boldsymbol{c}', \boldsymbol{r})$. Let n_{01} and n_{10} be defined as

$$n_{01} \triangleq |D_0(\boldsymbol{c}) \cap D_1(\boldsymbol{c}')|, \tag{12.19}$$

$$n_{10} \triangleq |D_1(\boldsymbol{c}) \cap D_0(\boldsymbol{c}')|. \tag{12.20}$$

Since

$$d(\boldsymbol{c}', \boldsymbol{c}) = n_{01} + n_{10} \geq w_k, \tag{12.21}$$

we have

$$\begin{aligned} n_{01} &\geq w_k - n_{10} \\ &\geq w_k - |D_1(\boldsymbol{c})| \\ &= w_k - n(\boldsymbol{c}). \end{aligned} \tag{12.22}$$

From (12.19) and (12.22), we find that

$$\begin{aligned} |D_1(\boldsymbol{c}')| &\geq |D_0(\boldsymbol{c}) \cap D_1(\boldsymbol{c}')| \\ &\geq w_k - n(\boldsymbol{c}). \end{aligned} \tag{12.23}$$

It follows from (12.19), (12.22) and (12.23) that

$$\begin{aligned} L(\boldsymbol{c}', \boldsymbol{r}) &= \sum_{i \in D_1(\boldsymbol{c}')} |r_i| \\ &\geq \sum_{i \in D_0^{(w_k - n(\boldsymbol{c}))}(\boldsymbol{c})} |r_i| \\ &= G(\boldsymbol{c}; w_k) \\ &\geq L(\boldsymbol{c}, \boldsymbol{r}). \end{aligned} \tag{12.24}$$

Eq.(12.24) implies that the most likely codeword $\boldsymbol{c}_{\text{opt}}$ must be in the Region $R(\boldsymbol{c}, w_k)$.

△△

Given a codeword $\boldsymbol{c}$, Theorem 12.1 simply defines a region in which $\boldsymbol{c}_{\text{opt}}$ can be found. It says that $\boldsymbol{c}_{\text{opt}}$ is among those codewords in C that are at distance w_{k-1} or less from the codeword $\boldsymbol{c}$, i.e.,

$$d(\boldsymbol{c}, \boldsymbol{c}_{\text{opt}}) \leq w_{k-1}. \tag{12.25}$$

If w_{k-1} is small, we can make a search in the region $R(\boldsymbol{c}, w_k)$ to find $\boldsymbol{c}_{\text{opt}}$. If w_{k-1} is too big, then it is better to generate another candidate codeword $\boldsymbol{c}'$ for testing and hopefully the search region $R(\boldsymbol{c}', w_k)$ is small.

Two special cases are particularly important:

(1) If $k = 1$, the codeword $\boldsymbol{c}$ is the optimal MLD codeword $\boldsymbol{c}_{\text{opt}}$.

(2) If $k = 2$, the optimal MLD codeword $\boldsymbol{c}_{\text{opt}}$ is either $\boldsymbol{c}$ or a **nearest neighbor** of $\boldsymbol{c}$.

Corollary 12.1 Let $\boldsymbol{c} \in C$.

(1) If $L(\boldsymbol{c}, \mathbf{r}) \leq G(\boldsymbol{c}, w_1)$, then $\boldsymbol{c} = \boldsymbol{c}_{\text{opt}}$.

(2) If $L(\boldsymbol{c}, \mathbf{r}) > G(\boldsymbol{c}, w_1)$ but $L(\boldsymbol{c}, \mathbf{r}) \leq G(\boldsymbol{c}, w_2)$, then $\boldsymbol{c}_{\text{opt}}$ is at a distance not greater than the minimum distance $d_H = w_1$ from $\boldsymbol{c}$. △△

The first part of Corollary 12.1 provides a sufficient condition for **optimality** of a codeword. The second part of Corollary 12.1 gives the condition that $\boldsymbol{c}_{\text{opt}}$ is either a **nearest neighbor** of a tested codeword or the tested codeword itself. We call $G(\boldsymbol{c}, w_1)$ and $G(\boldsymbol{c}, w_2)$ the **optimality** and **nearest neighbor test thresholds**, respectively. They will be used in an iterative decoding algorithm for testing. The sufficient condition on optimality given in Corollary 12.1 was first derived by Taipale and Pursley [94].

12.3 GENERATION OF CANDIDATE CODEWORDS AND TEST ERROR PATTERNS

The iterative decoding algorithm to be presented depends on the generation of a sequence of candidate codewords with a simple low-cost decoder. There are a number of ways of generating these candidate codewords. The simplest way is to use a set of probable test error patterns to modify the hard-decision received vector $\boldsymbol{z}$ and then decode each modified received sequence with an algebraic decoder. The test error patterns are generated in the decreasing likelihood order, one at a time. The most probable test error pattern is generated first and the least probable one is generated last. When a test error pattern $\boldsymbol{e}$ is generated, the sum $\boldsymbol{e} + \boldsymbol{z}$ is formed. Then the algebraic decoder decodes the modified received vector $\boldsymbol{e} + \boldsymbol{z}$ into a candidate codeword $\boldsymbol{c}$ for optimality test based on the two sufficient conditions given in Corollary 12.1.

Let p be a positive integer not greater than N. Let Q_p denote the set of the p **least reliable positions** of the received sequence $\boldsymbol{r}$. Let E denote the set of 2^p binary error patterns of length N with errors confined to the positions in Q_p. The set E forms the basic set of test error patterns. The error patterns in E are more likely to occur than the other error patterns. In the Chase decoding algorithm-II [14], $p = \lfloor d_H/2 \rfloor$ is chosen and E consists of $2^{\lfloor d_H/2 \rfloor}$ test error patterns where d_H is the minimum distance of the code to be decoded and $\lfloor d_H/2 \rfloor$ denotes the largest integer equal to or less than $d_H/2$. Using this set of test error patterns, Chase proved that his decoding algorithm achieves asymptotically optimum error performance. In the iterative decoding algorithm to be presented in the next section, the same basic set of test error patterns will be used for generating candidate codewords.

For a test error pattern $\boldsymbol{e} \in E$, let dec($\boldsymbol{e}$) denote the decoded codeword of the algebraic decoder with $\boldsymbol{e} + \boldsymbol{z}$ as the input vector. In case of a decoding failure (it may occur in a bounded distance-t decoder), let dec($\boldsymbol{e}$) $\triangleq (*)$ (undefined). In this case, the next test error pattern $\boldsymbol{e}' \in E$ is generated for decoding. A test error $\boldsymbol{e} \in E$, is said to be **decodable** if dec($\boldsymbol{e}$) $\neq (*)$. Two decodable error patterns, $\boldsymbol{e}$ and $\boldsymbol{e}'$ in E are said to be **equivalent** if dec($\boldsymbol{e}$) = dec($\boldsymbol{e}'$) $\neq (*)$. Let $\boldsymbol{e}$ be a decodable error pattern in E and let $Q(\boldsymbol{e})$ denote the set of all test error patterns in E which are equivalent to $\boldsymbol{e}$. $Q(\boldsymbol{e})$ is called the **equivalence class** containing $\boldsymbol{e}$, and a test error pattern in $Q(\boldsymbol{e})$ is chosen as the class representative. Since all the test error patterns in an equivalence class generate the same candidate codeword, only the class representative should be used. How to partition E into equivalence classes and generate equivalence class representatives affects the efficiency of any decoding algorithm that utilizes test patterns.

Let E_{rep} denote the set of all representatives of the equivalence classes of E. Then every decodable error pattern in E_{rep} generates a distinct candidate codeword for testing. To construct E_{rep} for a given received sequence $\boldsymbol{r}$, preprocessing is needed before decoding. This preprocessing of E is effective only if it is simpler than an algebraic decoding operation. An effective procedure for generating test error patterns in E_{rep} is presented in [75].

12.4 AN ITERATIVE DECODING ALGORITHM

This decoding algorithm is iterative in nature and devised based the reliability measures of the received symbols. It consists of the following key steps:

(1) Generate a candidate codeword $\boldsymbol{c}$ by using a test error pattern in E_{rep}.

(2) Perform the optimality test or the nearest neighbor test for each generated candidate codeword $\boldsymbol{c}$.

(3) If the optimality test fails but the nearest neighbor test succeeds, a search in the region $R(\boldsymbol{c}, w_2)$ is initiated. The search is conducted through the minimum-weight subtrellis, $T_{\min}(\boldsymbol{c})$, centered around $\boldsymbol{c}$ using a trellis-based decoding algorithm, say Viterbi or RMLD algorithms.

(4) If both optimality and nearest neighbor tests fail, a new test error pattern in E_{rep} is generated for the next decoding iteration.

Suppose the optimal MLD codeword has not been found at the end of the $(j-1)$-th decoding iteration. Then the j-th decoding iteration is initiated. Let $\boldsymbol{c}_{\text{best}}$ and $L(\boldsymbol{c}_{\text{best}}, \boldsymbol{r})$ denote the **best** codeword and its correlation discrepancy that have been found so far and are stored in a buffer memory. The j-th decoding iteration consists of the following steps:

Step 1: Fetch $\boldsymbol{e}_j$ from E_{rep} and decode $\boldsymbol{e}_j + \boldsymbol{z}$ into a codeword $\boldsymbol{c} \in C$. If the decoding succeeds, go to Step 2. Otherwise, go to Step 1.

Step 2: If $L(\boldsymbol{c}, \boldsymbol{r}) \leq G(\boldsymbol{c}, w_1)$, $\boldsymbol{c}_{\text{opt}} = \boldsymbol{c}$ and stop the decoding process. Otherwise, go to Step 3.

Step 3: If $L(\boldsymbol{c}, \boldsymbol{r}) \leq G(\boldsymbol{c}, w_2)$, search $T_{\text{min}}(\boldsymbol{c})$ to find $\boldsymbol{c}_{\text{opt}}$ and stop the decoding process. Otherwise, go to Step 4.

Step 4: If $L(\boldsymbol{c}, \boldsymbol{r}) < L(\boldsymbol{c}_{\text{best}}, \boldsymbol{r})$, replace $\boldsymbol{c}_{\text{best}}$ by $\boldsymbol{c}$ and $L(\boldsymbol{c}_{\text{best}}, \boldsymbol{r})$ by $L(\boldsymbol{c}, \boldsymbol{r})$. Otherwise, go to Step 5.

Step 5: If $j < |E_{\text{rep}}|$, go to Step 1. Otherwise search $T_{\text{min}}(\boldsymbol{c}_{\text{best}})$ and output the codeword with the **least** correlation discrepancy. Stop.

The decoding process is depicted by the flow diagram shown in Figure 12.1. The only case for which the decoded codeword may not be optimal is the output from the search of $T_{\text{min}}(\boldsymbol{c}_{\text{best}})$. It is important to point out that when a received vector causes the decoding algorithm to perform $2^{\lfloor d_H/2 \rfloor}$ iterations without satisfying the sufficient conditions for optimality, optimum decoding is not guaranteed. Most of the decoding errors occur in this situation. The main cause of this situation is that the number of errors caused by the channel in the most reliable $N - \lfloor d_H/2 \rfloor$ positions is larger than $\lfloor (d_H - 1)/2 \rfloor$.

12.5 COMPUTATIONAL COMPLEXITY

We assume that the algebraic decoding complexity is small compared with the computational complexity required to process the minimum-weight trellis. The computational complexity is measured only in terms of real operations,

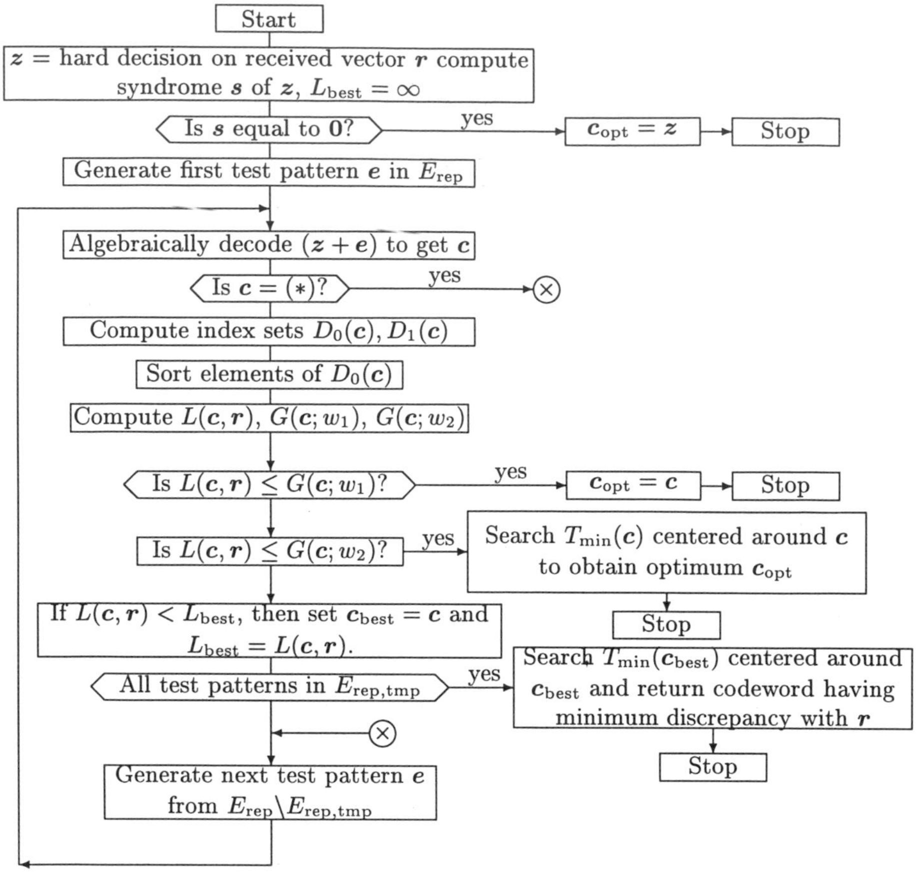

Figure 12.1. Flowchart of the Iterative Decoding Algorithm with minimum weight trellis search, where $E_{\text{rep,tmp}}$ denotes the set of those representative test error patterns that have been generated.

(real additions and comparisons). This number is a variable depending on the SNR. Let $C_{\max}$ denote the **worst case maximum number** of real operations required at any SNR. Then $C_{\max}$ can be computed by analyzing the flowchart of Figure 12.1.

Let A_0 denote the fixed number of real operations required in sorting the components of the received vector in increasing order of reliability. Let A_{loop} denote the number of real operations required in:

(1) computing the index sets $D_1(\boldsymbol{c})$ and $D_0(\boldsymbol{c})$;

(2) computing the correlation discrepancy $L(\boldsymbol{c},\boldsymbol{r})$, the optimality test threshold $G(\boldsymbol{c},w_1)$, and the nearest neighbor test threshold $G(\boldsymbol{c},w_2)$; and

(3) comparing $L(\boldsymbol{c},\boldsymbol{r})$ with $G(\boldsymbol{c},w_1)$, $G(\boldsymbol{c},w_2)$ and $L(\boldsymbol{c}_{\text{best}},\boldsymbol{r})$.

Let $A(T_{\min})$ denote the fixed number of real operations required to search through the minimum-weight trellis $T_{\min}(\boldsymbol{c})$. Then

$$C_{\max} = A_0 + |E_{\text{rep}}| \cdot A_{\text{loop}} + A(T_{\min}) \leq A_0 + 2^{\lfloor d_H/2 \rfloor} \cdot A_{\text{loop}} + A(T_{\min}). \quad (12.26)$$

Let $\tilde{C}_{\max}$ denote the upper bound on $C_{\max}$ given by (12.26). Note that $\tilde{C}_{\max}$ is independent of SNR. Table 12.1 shows the decoding complexities for some well known codes. Also given in the table are the complexities of the minimum-weight subtrellises and full trellises, $A(T_{\min})$ and $A(T)$, of the codes in terms of real operations to be performed. The Viterbi algorithm is used for searching through the subtrellises and full trellises of the codes. We see that $A(T_{\min})$ is much smaller than $A(T)$, especially for codes of length 64 or longer. Consider the $(64,45)$ extended BCH code. Viterbi decoding based on the full code trellis requires $4,301,823$ real operations to decode a received sequence of 64 symbols. However, the worst-case maximum number of real operations required by the iterative decoding based on the minimum-weight trellis is upper bounded by $57,182$ while $A(T_{\min}) = 48,830$. We see that there is a tremendous reduction in decoding complexity. Table 12.1 also lists the average number $|E_{\text{rep}}|$ of test error pattern representatives for each code at bit-error-rate (BER) of 10^{-5}. We see that $|E_{\text{rep}}|$ is much smaller than $|E|$, the size of the basic set of test error patterns.

Table 12.1. Complexity of the minimum-weight trellis iterative decoding algorithm.

Code	$A(T)$	$A(T_{\min})$	L	$\tilde{C}_{\max}$	C_{ave} @BER $= 10^{-5}$	N_{ave} @BER $= 10^{-5}$	$\lvert E_{\text{rep}}\rvert$ @BER $= 10^{-5}$
Golay$(23,12,7)$	2,559	1,415	3	2,767	< 50	< 1	2
RM$(32,16,8)$	4,016	1,543	4	5,319	10	0.7	2
ex-BCH$(32,21,6)$	30,156	5,966	16	6,350	25	0.35	2
RM$(32,26,4)$	1,295	1,031	8	1,951	< 10	0.1	1
RM$(64,22,16)$	131,071	11,039	4	146,719	5,000	5	29
ex-BCH$(64,24,16)$	524,287	11,039	4	146,719	4,000	1.8	20
RM$(64,42,8)$	544,640	8,111	8	16,495	310	1.3	2
ex-BCH$(64,45,8)$	4,301,823	48,830	8	57,182	275	0.72	1
ex-BCH$(64,51,6)$	448,520	50,750	8	52,760	200	0.45	1
RM$(64,57,4)$	6,951	3,079	8	5,151	< 1	0.09	2

L: Number of section in trellis.

N_{ave}: Average number of iterations.

$|E_{\text{rep}}|$: Average size of the set of representatives of equivalence classes in E.

The average computational complexity of the iterative decoding algorithm can also be analyzed based on the decoding flowchart shown in Figure 12.1. Define the following event:

(1) For $1 \le i \le 2^{|E_{\text{rep}}|}$, let $\mathcal{B}_i$ denote the event that the condition $L(\boldsymbol{c},\boldsymbol{r}) \le G(\boldsymbol{c}; w_1)$ holds for the first time during the i-th iteration;

(2) For $1 \le i \le 2^{|E_{\text{rep}}|}$, let $\mathcal{C}_i$ denote the event that the condition $G(c, w_1) < L(\boldsymbol{c},\boldsymbol{r}) \le G(\boldsymbol{c}; w_2)$ is satisfied during the i-th decoding iteration;

(3) Let $\mathcal{B} \triangleq \cup_{\forall i}\mathcal{B}_i$, $\mathcal{C} \triangleq \cup_{\forall i}\mathcal{C}_i$ and $\mathcal{E} \triangleq \mathcal{B} \cup \mathcal{C} \setminus (\mathcal{B}_1 \cup \mathcal{C}_1)$; and

(4) Let $\mathcal{D}$ denote the event that neither $\mathcal{B}$ nor $\mathcal{C}$ occurs during the $|E_{\text{rep}}|$ runs through the decoding loop and the decoding process is terminated.

Let J be the average number of iterations preceding the event $\mathcal{E}$. Let $Pr(X)$ denote the probability of event X. Then the average number of real operations, denoted C_{ave}, required to decode a received word using the above iterative decoding algorithm is a function of the average size of the test error pattern set E_{rep}, denoted M, and is given by

$$\begin{aligned} C_{\text{ave}} &= A_0 + (Pr(\mathcal{B}_1) + Pr(\mathcal{C}_1))A_{\text{loop}} + (Pr(\mathcal{C}) + Pr(\mathcal{D}))A(T_{\min}) \\ &+ Pr(\mathcal{D})MA_{\text{loop}} + Pr(\mathcal{E})(J+1)A_{\text{loop}}. \end{aligned} \tag{12.27}$$

The probabilities of the above events can be estimated using the Monte-Carlo simulation technique [86].

The average computational complexity of the iterative decoding algorithm is very small compared to the worst-case upper bound $\tilde{C}_{\max}$ and the computational complexity $A(T)$ by using the full code trellis with the Viterbi algorithm, as shown in Table 12.1. For example, consider the $(64, 45, 8)$ extended BCH code. At SNR = 2 dB, the average number of real operations required to decode a received sequence of 64 symbols is 40,000 compared to $\tilde{C}_{\max} = 57,182$ and $A(T) = 4,301,823$. At SNR = 5 dB, $C_{\text{ave}} = 750$.

12.6 ERROR PERFORMANCE

Since the iterative decoding algorithm uses the Chase algorithm-II to generate candidate codewords for optimality test and minimum-weight trellis search, it may be regarded as an improved Chase algorithm-II and hence achieves asymptotically optimal error performance with a faster rate. An upper bound on the block error probability can be found in [101]. Simulation results show that the iterative decoding algorithm achieves near-optimum error performance and significantly outperforms the Chase algorithm-II.

The iterative decoding algorithm has been simulated for all the codes given in Table 12.1. For RM codes, majority-logic decoding is used for generating candidate codewords and otherwise a bounded distance-t decoding algorithm is used. The bit error rates of some codes listed in Table 12.1 are shown in Figures 12.2(a)-12.6(a) and their average computational complexities and average numbers of decoding iterations are shown in Figures 12.2(b)-12.6(b).

Consider the $(32, 16, 8)$ extended primitive BCH code (also an RM code). The iterative decoding algorithm achieves practically the same error performance as the optimum MLD as shown in Figure 12.3(a). To achieve a bit error rate of 10^{-5}, it requires a SNR of 5.6 dB. We also see that the iterative decoding algorithm achieves a 0.55 dB coding gain over the Chase decoding algorithm-II at the bit error rate 10^{-5}. While the Chase decoding algorithm-II requires 16

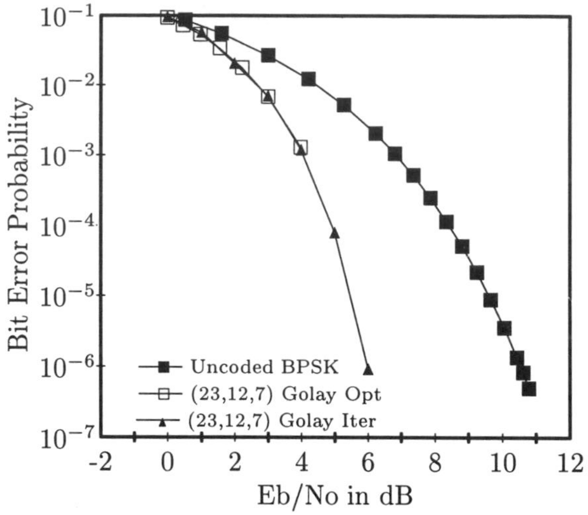

(a) Bit error probability

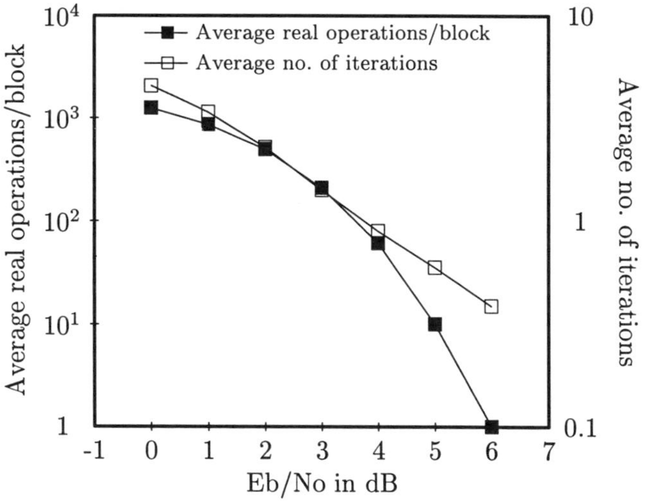

(b) Average computational complexity and average numbers of decoding iterations

Figure 12.2. Error performance and computational complexity of the iterative decoding algorithm for Golay $(23, 12, 7)$ code.

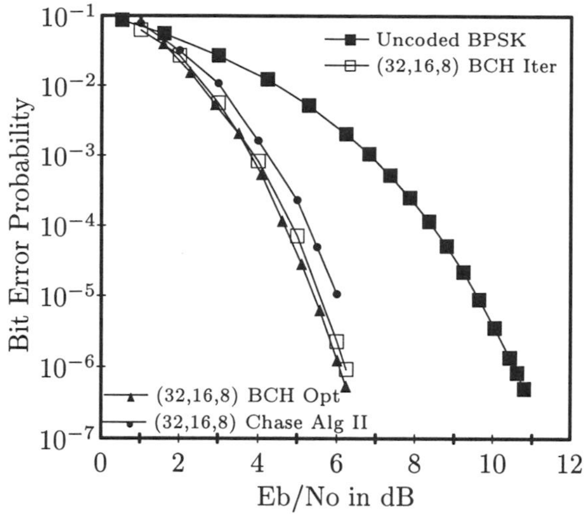

(a) Bit error probability

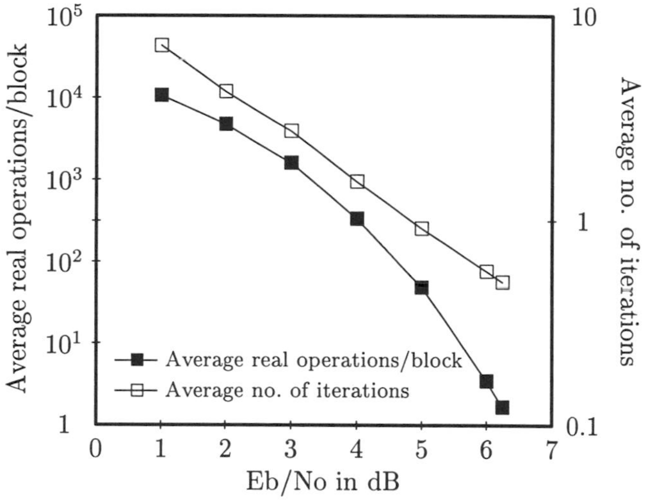

(b) Average computational complexity and average numbers of decoding iterations

Figure 12.3. Error performance and computational complexity of the iterative decoding algorithm for $(32, 16, 8)$ extended primitive BCH code.

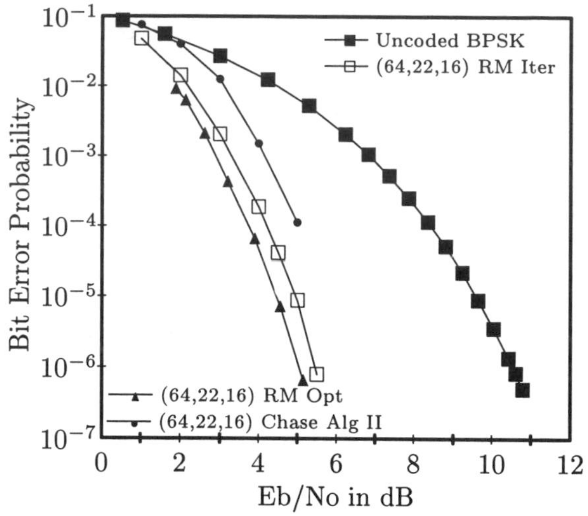

(a) Bit error probability

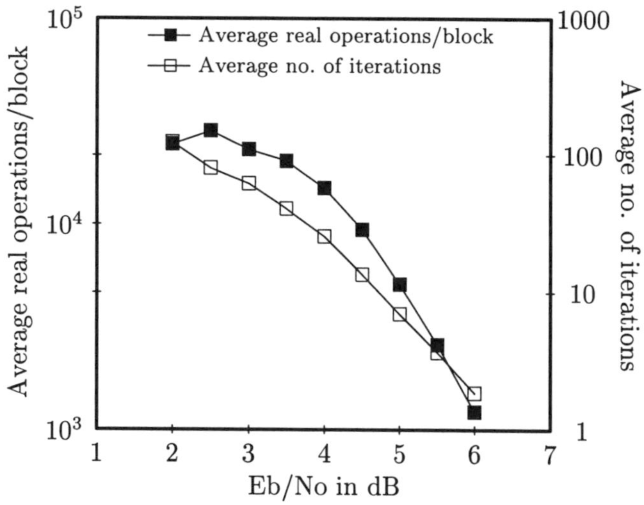

(b) Average computational complexity and average numbers of decoding iterations

Figure 12.4. Error performance and computational complexity of the iterative decoding algorithm for the $(64, 22, 16)$ RM code.

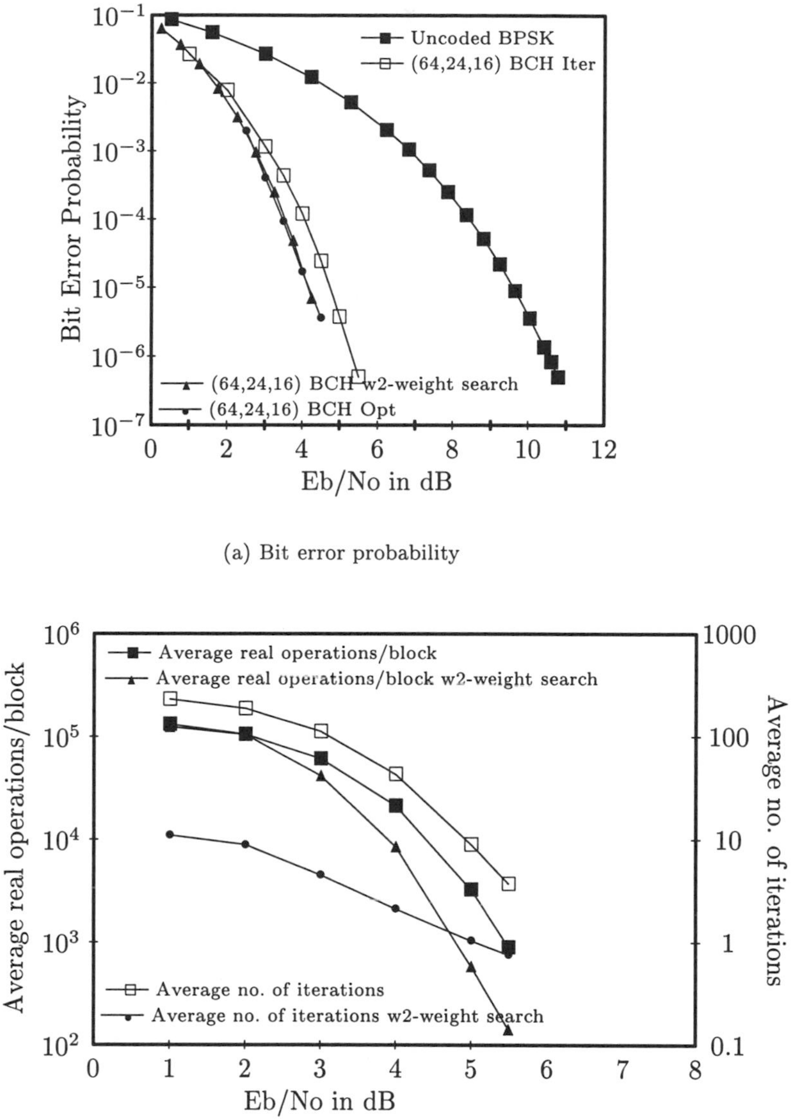

(a) Bit error probability

(b) Average computational complexity and average numbers of decoding iterations

Figure 12.5. Error performance and computational complexity of the iterative decoding algorithm for the $(64, 24, 16)$ extended BCH code.

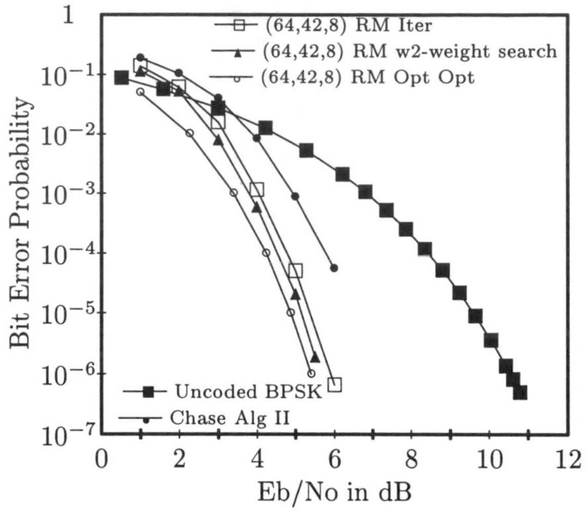

(a) Bit error probability

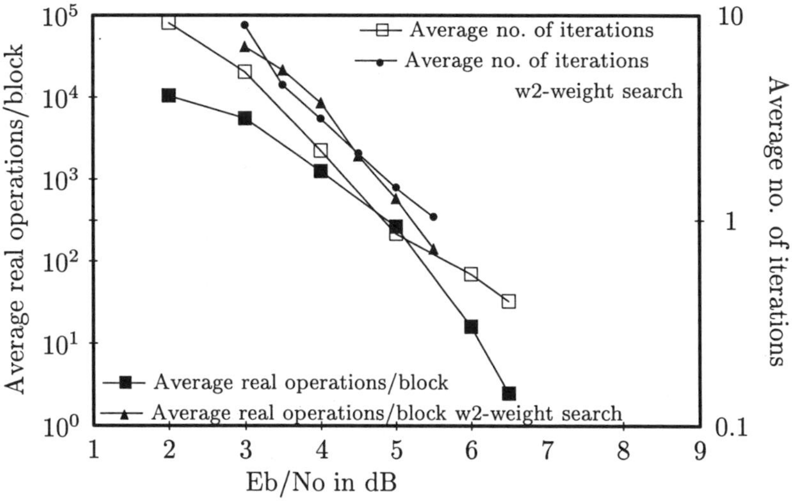

(b) Average computational complexity and average numbers of decoding iterations

Figure 12.6. Error performance and computational complexity of the iterative decoding algorithm for the $(64, 42, 8)$ RM code.

decoding operations, the iterative decoding algorithm takes only one decoding operation (or iteration) on average at the BER of $= 10^{-5}$.

For all simulated codes of length 64 except the extended Hamming code, the iterative decoding algorithm loses less than 0.5 dB with respect to optimal MLD decoding at the BER of 10^{-6}. However, the iterative decoding algorithm yields a drastic reduction in computational complexity compared with the optimal Viterbi decoding based on the full trellis of the code. As shown in Figures 12.6(a) and 12.6(b), for the $(64, 42, 8)$ RM code, the performance degradation of the iterative decoding algorithm compared with the optimum Viterbi decoding based on the full trellis of the code is 0.5 dB at the BER of 10^{-6}. The SNR required to achieve a BER of 10^{-6} by the iterative decoding algorithm is 5.9 dB. At this SNR, the average number of real operations required to decode a received word is about 25, however, the optimal Viterbi decoding based on the full trellis diagram of the code requires 544,640 real operations. This is a tremendous reduction in computational complexity with a small degradation in error performance. The average number of iterations required to complete the decoding at this SNR of 5.9 dB is 0.6. At the BER of 10^{-4}, the iterative decoding algorithm achieves a gain of 1.1 dB over the Chase algorithm-II with an average of 400 real operations and an average of 1.2 decoding iterations.

The error performance of the iterative decoding algorithm can be improved by using a larger search region for the optimal MLD solution $\boldsymbol{c}_{\text{opt}}$. For example, we may use the region $R(\boldsymbol{c}, w_3)$ which consists of all the codewords at the distances w_1 (minimum distance) and w_2 (next to minimum distance) from the current tested codeword for searching $\boldsymbol{c}_{\text{opt}}$. In the case that $G(\boldsymbol{c}, w_2) < L(\boldsymbol{c}, \boldsymbol{r}) \leq G(\boldsymbol{c}, w_3)$, the decoder searches the purged trellis diagram $T_{w_2}(\boldsymbol{c})$ centered around $\boldsymbol{c}$ for finding $\boldsymbol{c}_{\text{opt}}$, where $T_{w_2}(\boldsymbol{c})$ consists of $\boldsymbol{c}$ and all the paths of the overall trellis diagram of the code which are at distances w_1 and w_2 from $\boldsymbol{c}$. We call this search the w_2-weight trellis search. Of course, the improvement is achieved with some additional computational complexity.

Consider the $(64, 24, 16)$ extended BCH code. The w_2-weight trellis search achieves practically optimum performance of MLD as shown in Figure 12.5(a). It recovers the 0.6 dB loss in performance based on the minimum-weight trellis search. The SNR required to achieve the BER of 10^{-5} is 4.2 dB. At this SNR, the average number of real operations and the average number of iterations for decoding a received sequence are about 6,000 and 2, respectively as shown in Figure 12.5(b). The maximum number of real operations required for the improved decoding algorithm is 175,232. At the same BER of 10^{-5}, the minimum-weight trellis search algorithm requires an average of 5,000 real operations and 15 iterations. We see that 0.6 dB gain in performance over the minimum-weight trellis search algorithm is achieved at the expense of a modest additional complexity. If the modified algorithm with a larger search region is applied to the (64, 42, 8) RM code, a 0.3 dB coding gain over the minimum-weight trellis search algorithm at the BER of 10^{-5} is achieved with a very modest additional computational complexity on average as shown in Figure 12.6(a) and 12.6(b).

12.7 SHORTCOMINGS

The above iterative decoding algorithm is very simple and provides good error performance with large reduction in decoding complexity. However, it has several major shortcomings and for long codes, there is a significant performance degradation compared to MLD for low to medium SNRs. First, the algebraic decoder used in the algorithm may fail to decode the modified hard-decision received sequence to generate a candidate codeword for testing. Therefore, the decoding algorithm may have a decoding failure. Second, some test error patterns may result in the same candidate codeword and hence useless decoding iterations unless some preprocessing is done to rule out or reduce the repetitions. Third, there is no guarantee that the candidate codewords are generated in the order of increasing improvement, i.e., the next generated candidate codeword has larger correlation metric than the previous ones. This may result in unnecessary decoding iterations and prevent a fast decoding convergence. Fourth, the sufficient conditions for optimality and nearest neighbor tests are based on only the current candidate codeword, and information from previously tested candidate codewords is not being used. This information may help to narrow down the search of the ML codeword and reduce the possibility that it slips through the test without being detected. Finally, the performance degradation is large for codes whose minimum weight codewords do not span the codes. For example, the minimum weight codewords of the extended (64, 24, 16) BCH code do not span the code and their is a 0.6 dB coding gain loss compared to MLD at the BER of 10^{-5}.

13 THE MAP AND RELATED DECODING ALGORITHMS

In a coded communication system with equiprobable signaling, MLD minimizes the word error probability and delivers the most likely codeword associated with the corresponding received sequence. This decoding has two drawbacks. First, minimization of the word error probability is not equivalent to minimization of the bit error probability. Therefore, MLD becomes suboptimum with respect to the bit error probability. Second, MLD delivers a hard-decision estimate of the received sequence, so that information is lost between the input and output of the ML decoder. This information is important in coded schemes where the decoded sequence is further processed, such as concatenated coding schemes, multi-stage and iterative decoding schemes.

In this chapter, we first present a decoding algorithm which both minimizes bit error probability, and provides the corresponding soft information at the output of the decoder. This algorithm is referred to as the MAP (maximum a-posteriori probability) decoding algorithm [1]. Unfortunately, the trellis-based implementation of the MAP algorithm is much more complex than that of the trellis-based MLD algorithms presented in the previous chapters. Consequently, suboptimum versions of the MAP algorithm with reduced decoding complexity must be considered for many practical applications. In Section 13.2, we present a near-optimum modification of the MAP algorithm, referred to as the Max-Log-MAP (or SOVA) decoding algorithm [34, 39, 40]. This near-optimum algorithm performs within only few tenths of a dB of the optimum MAP decoding algorithm while reduces decoding complexity drastically. Finally, the minimization of the bit error probability in trellis-based MLD is discussed in the last section.

13.1 THE MAP DECODING ALGORITHM

Consider a binary (N, K) linear block code C. Let $\boldsymbol{u} = (u_1, u_2, \ldots, u_N)$ be a codeword in C. Define $P(A|B)$ as the conditional probability of the event A given the occurrence of the event B. The MAP decoding algorithm evaluates the most likely bit value u_i at a given bit position i based on the received sequence $\boldsymbol{r} = (r_1, r_2, \ldots, r_N)$. It first computes the log-likelihood ratio

$$L_i \triangleq \log \frac{P(u_i = 1|\boldsymbol{r})}{P(u_i = 0|\boldsymbol{r})} \tag{13.1}$$

for $1 \leq i \leq N$, and then compares this value to a zero-threshold to decode u_i as

$$u_i = \begin{cases} 1 & \text{for } L_i > 0, \\ 0 & \text{for } L_i \leq 0. \end{cases} \tag{13.2}$$

The value L_i represents the soft information associated with the decision on u_i. It can be used for further processing of the sequence $\boldsymbol{u}$ delivered by the MAP decoder.

In the N-section trellis diagram for the code, let $B_i(C)$ denote the set of all branches (σ_{i-1}, σ_i) that connect the states in the state space $\Sigma_{i-1}(C)$ at time-$(i-1)$ and the states in the state space $\Sigma_i(C)$ at time-i for $1 \leq i \leq N$. Let $B_i^0(C)$ and $B_i^1(C)$ denote the two disjoint subsets of $B_i(C)$ that correspond to the output code bits $u_i = 0$ and $u_i = 1$, respectively, given by (3.3). Clearly

$$B_i(C) = B_i^0(C) \cup B_i^1(C) \tag{13.3}$$

for $1 \leq i \leq N$. For $(\sigma', \sigma) \in B_i(C)$, we define the joint probability

$$\lambda_i(\sigma', \sigma) \triangleq P(\sigma_{i-1} = \sigma'; \sigma_i = \sigma; \boldsymbol{r}) \tag{13.4}$$

for $1 \leq i \leq N$. Then

$$P(u_i = 0; \boldsymbol{r}) = \sum_{(\sigma',\sigma)\in B_i^0(C)} \lambda_i(\sigma',\sigma), \tag{13.5}$$

$$P(u_i = 1; \boldsymbol{r}) = \sum_{(\sigma',\sigma)\in B_i^1(C)} \lambda_i(\sigma',\sigma). \tag{13.6}$$

The MAP decoding algorithm computes the probabilities $\lambda_i(\sigma',\sigma)$ which are then used to evaluate $P(u_i = 0|\boldsymbol{r})$ and $P(u_i = 1|\boldsymbol{r})$ in (13.1) from (13.5) and (13.6).

For $1 \leq i \leq N$, $1 \leq l \leq m \leq N$, and $\boldsymbol{r}_l^m \triangleq (r_l, r_{l+1}, \ldots, r_m)$, we define the probabilities

$$\alpha_i(\sigma) \triangleq P(\sigma_i = \sigma; \boldsymbol{r}_1^i), \tag{13.7}$$

$$\beta_i(\sigma) \triangleq P(\boldsymbol{r}_{i+1}^N|\sigma_i = \sigma), \tag{13.8}$$

$$\begin{aligned} \gamma_i(\sigma',\sigma) &\triangleq P(\sigma_i = \sigma; r_i|\sigma_{i-1} = \sigma') \\ &= P(r_i|(\sigma_{i-1},\sigma_i) = (\sigma',\sigma))\, P(\sigma_i = \sigma|\sigma_{i-1} = \sigma'). \end{aligned} \tag{13.9}$$

Then, for a memoryless channel,

$$\lambda_i(\sigma',\sigma) = \alpha_{i-1}(\sigma')\, \gamma_i(\sigma',\sigma)\, \beta_i(\sigma) \tag{13.10}$$

for $1 \leq i \leq N$, which shows that the values $\lambda_i(\sigma',\sigma)$ can be evaluated by computing all values $\alpha_i(\sigma)$, $\beta_i(\sigma)$ and $\gamma_i(\sigma',\sigma)$. Based on the total probability theorem, we can express $\alpha_i(\sigma)$, for $1 \leq i \leq N$, as follows:

$$\begin{aligned} \alpha_i(\sigma) &= \sum_{\sigma'\in\Sigma_{i-1}(C)} P(\sigma_{i-1} = \sigma'; \sigma_i = \sigma; \boldsymbol{r}_1^{i-1}; r_i) \\ &= \sum_{\sigma'\in\Sigma_{i-1}(C)} \alpha_{i-1}(\sigma')\, \gamma_i(\sigma',\sigma). \end{aligned} \tag{13.11}$$

Similarly, for $1 \leq i < N$,

$$\begin{aligned} \beta_i(\sigma) &= \sum_{\sigma'\in\Sigma_{i+1}(C)} P(r_{i+1}; \boldsymbol{r}_{i+2}^N; \sigma_{i+1} = \sigma'|\sigma_i = \sigma) \\ &= \sum_{\sigma'\in\Sigma_{i+1}(C)} \beta_{i+1}(\sigma')\, \gamma_{i+1}(\sigma,\sigma'). \end{aligned} \tag{13.12}$$

From (13.11), we see that the probabilities $\alpha_i(\sigma)$ with $1 \leq i \leq N$ can be computed recursively from the initial state σ_0 to the final state σ_f of the N-section trellis for code C, once $\gamma_i(\sigma',\sigma)$'s are computed. This is called the **forward recursion**. From (13.12), we see that the probabilities $\beta_i(\sigma)$ with $1 \leq i \leq N$ can be computed recursively in backward direction from the final state σ_f to the initial state σ_0 of the N-section trellis of C. This is called the **backward recursion**.

For the AWGN channel with BPSK transmission, we have

$$P(r_i|\sigma_{i-1} = \sigma'; \sigma_i = \sigma) = (\pi N_0)^{-1/2} \exp(-(r_i - c_i)^2/N_0)\, \delta_i(\sigma',\sigma), \quad (13.13)$$

where

$$\delta_i(\sigma',\sigma) = \begin{cases} 1 & \text{if } (\sigma',\sigma) \in B_i(C), \\ 0 & \text{if } (\sigma',\sigma) \notin B_i(C). \end{cases} \quad (13.14)$$

For $\delta_i(\sigma',\sigma) = 1$, $c_i = 2u_i - 1$ is the transmitted signal corresponding to the label u_i of the branch (σ',σ). Based on (13.13), $\gamma_i(\sigma',\sigma)$ is proportional to the value

$$w_i(\sigma',\sigma) = \exp(-(r_i - c_i)^2/N_0)\, \delta_i(\sigma',\sigma)\, P(\sigma_i = \sigma|\sigma_{i-1} = \sigma'). \quad (13.15)$$

Note that in many applications such as MAP trellis-based decoding of linear codes, the a-priori probability of each information bit is the same, so that all states $\sigma \in \Sigma_i(C)$ are equiprobable. Consequently, $P(\sigma_i = \sigma|\sigma_{i-1} = \sigma')$ becomes a constant that can be discarded in the definition of $w_i(\sigma',\sigma)$. However, this is not true in general. For example, in iterative or multi-stage decoding schemes $P(\sigma_i = \sigma|\sigma_{i-1} = \sigma')$ has to be evaluated after the first iteration, or after the first decoding stage. Since in (13.1), we are interested only in the ratio between $P(u_i = 1|\boldsymbol{r})$ and $P(u_i = 0|\boldsymbol{r})$, $\lambda_i(\sigma',\sigma)$ can be scaled by any value without modifying the decision on u_j. Based on these definitions, $\alpha_i(\sigma)$ can be computed recursively based on (13.11) using the trellis diagram from the initial state σ_0 to the final state σ_f as follows:

(1) Assume that $\alpha_{i-1}(\sigma')$ has been computed for all states $\sigma' \in \Sigma_{i-1}(C)$.

(2) In the i-th section of the trellis diagram, associate the weight $w_i(\sigma',\sigma)$ with each branch $(\sigma',\sigma) \in B_i(C)$.

(3) For each state $\sigma \in \Sigma_i(C)$, evaluate and store the weighted sum

$$\alpha_i(\sigma) = \sum_{\sigma' \in \Sigma_{i-1}(C):\delta_i(\sigma',\sigma)=1} w_i(\sigma', \sigma)\, \alpha_{i-1}(\sigma'). \tag{13.16}$$

The initial conditions for this recursion are $\alpha_0(\sigma_0) = 1$ and $\alpha_0(\sigma) = 0$ for $\sigma \neq \sigma_0$.

Similarly, $\beta_i(\sigma)$ can be computed recursively based on (13.12) using the trellis diagram from the final state σ_f to the initial state σ_0 as follows:

(1) Assume that $\beta_{i+1}(\sigma')$ has been computed for all states $\sigma' \in \Sigma_{i+1}(C)$.

(2) In the $(i+1)$-th section of the trellis diagram, associate the weight $w_{i+1}(\sigma, \sigma')$ with each branch $(\sigma, \sigma') \in B_{i+1}(C)$.

(3) For each state $\sigma \in \Sigma_i(C)$, evaluate and store the weighted sum

$$\beta_i(\sigma) = \sum_{\sigma' \in \Sigma_{i+1}(C):\delta_{i+1}(\sigma,\sigma')=1} w_{i+1}(\sigma, \sigma')\, \beta_{i+1}(\sigma'). \tag{13.17}$$

The corresponding initial conditions are $\beta_N(\sigma_f) = 1$ and $\beta_N(\sigma) = 0$ for $\sigma \neq \sigma_f$.

The MAP decoding algorithm requires one forward recursion from σ_0 to σ_f, and one backward recursion from σ_f to σ_0 to evaluate all values $\alpha_i(\sigma)$ and $\beta_i(\sigma)$ associated with all states $\sigma_i \in \Sigma_i(C)$, for $1 \leq i \leq N$. These two recursions are independent of each other. Therefore, the forward and backward recursions can be executed simultaneously in both directions along the trellis of the code C. This bidirectional decoding reduces the decoding delay. Once all values of $\alpha_i(\sigma)$ and $\beta_i(\sigma)$ for $1 \leq i \leq N$ have been determined, the values L_i in (13.1) can be computed from (13.5), (13.6), (13.9) and (13.10).

13.2 THE SOVA DECODING ALGORITHM

The MAP decoding algorithm presented in the previous section requires a large number of computations and a large storage to compute and store the probabilities $\alpha_i(\sigma)$, $\beta_i(\sigma)$ and $\gamma_i(\sigma', \sigma)$ for all the states σ and state pairs (σ', σ) in the trellis for the code to be decoded. For a long code with large trellis, the implementation of the MAP decoder is practically impossible. Also, the MAP decoding algorithm computes probability values, which require much more complicated real value operations than the real additions performed by trellis-based MLD algorithms, such as the Viterbi decoding and RMLD algorithms.

In this section, we present an algorithm for which the optimum bit error performance associated with the MAP algorithm is traded with a significant reduction in decoding complexity. This algorithm is known as the **Max-Log-MAP** algorithm or **soft-output Viterbi algorithm** (**SOVA**), as it performs the same operations as the Viterbi algorithm, with additional real additions and storages.

For BPSK transmission, (13.1) can be rewritten as

$$L_i = \log\left[\left(\sum_{\boldsymbol{c}:c_i=1} P(\boldsymbol{c}|\boldsymbol{r})\right) / \left(\sum_{\boldsymbol{c}:c_i=-1} P(\boldsymbol{c}|\boldsymbol{r})\right)\right], \tag{13.18}$$

for $1 \leq i \leq N$ and $c_i = 2u_i - 1$. Based on the approximation

$$\log(\sum_{j=1}^{N} \delta_j) \approx \log(\max_{j\in\{1,2,\ldots,N\}} \{\delta_j\}), \tag{13.19}$$

we obtain from (13.18)

$$L_i \approx \log(\max_{\boldsymbol{c}:c_i=1} P(\boldsymbol{c}|\boldsymbol{r})) - \log(\max_{\boldsymbol{c}:c_i=-1} P(\boldsymbol{c}|\boldsymbol{r})). \tag{13.20}$$

For each code bit u_i, the Max-Log-MAP algorithm [40] approximates the corresponding log-likelihood ratio L_i based on (13.20).

For the AWGN channel with BPSK transmission, we have

$$\begin{aligned} P(\boldsymbol{c}|\boldsymbol{r}) &= P(\boldsymbol{r}|\boldsymbol{c})P(\boldsymbol{c})/P(\boldsymbol{r}) \\ &= (\pi N_0)^{-N/2} e^{-\sum_{j=1}^{N}(r_j-c_j)^2/N_0} P(\boldsymbol{c})/P(\boldsymbol{r}). \end{aligned} \tag{13.21}$$

If $\boldsymbol{c}^1 = (c_1^1, c_2^1, \ldots, c_{2j-1}^1, c_{2j}^1, \ldots)$ and $\boldsymbol{c}^0 = (c_1^0, c_2^0, \ldots, c_{2j-1}^0, c_{2j}^0, \ldots)$ represent the codewords corresponding to the first term and the second term of (13.20), respectively, it follows from (13.21) that for equiprobable signaling, the approximation of L_i given in (13.20) is proportional to the value

$$r_i + \sum_{j:j\neq i, c_j^0 \neq c_j^1} c_j^1 r_j. \tag{13.22}$$

We observe that one of the terms in (13.20) corresponds to the MLD solution, while the other term corresponds to the most likely codeword which differs from the MLD solution in u_i. Consequently, in Max-Log-MAP or SOVA decoding, the hard-decision codeword corresponding to (13.20) is the MLD codeword and (13.22) is proportional to the difference of squared Euclidean distances (SED) $||\boldsymbol{r} - \boldsymbol{c}^1||^2 - ||\boldsymbol{r} - \boldsymbol{c}^0||^2$. For any two codewords $\boldsymbol{c}$ and $\boldsymbol{c}'$, we define $\big|||\boldsymbol{r} - \boldsymbol{c}||^2 - ||\boldsymbol{r} - \boldsymbol{c}'||^2\big|$ as the **reliability difference** between $\boldsymbol{c}$ and $\boldsymbol{c}'$.

For simplicity, we consider the trellis diagram of a rate-1/2 antipodal convolutional code C. Hence, the two branches that merge into each state have different branch labels, as described in Figure 10.1. Also, we assume that the trellis diagram for the code C is terminated so that N encoded bits are transmitted. Generalization of the derived results to other trellis diagrams is straightforward, after proper modification of the notations. At each state σ_{i-1} of the state space $\Sigma_{i-1}(C)$ at time-$(i-1)$, the SOVA stores the cumulative correlation metric value $M(\sigma_{i-1})$ and the corresponding decoded sequence

$$\hat{\boldsymbol{c}}(\sigma_{i-1}) = \left(\hat{c}_1(\sigma_{i-1}), \hat{c}_2(\sigma_{i-1}), \ldots, \hat{c}_{2(i-1)-1}(\sigma_{i-1}), \hat{c}_{2(i-1)}(\sigma_{i-1})\right), \tag{13.23}$$

as for the Viterbi algorithm. In addition, it also stores the reliability measures

$$\hat{\boldsymbol{L}}(\sigma_{i-1}) = \left(\hat{L}_1(\sigma_{i-1}), \hat{L}_2(\sigma_{i-1}), \ldots, \hat{L}_{2(i-1)-1}(\sigma_{i-1}), \hat{L}_{2(i-1)}(\sigma_{i-1})\right), \tag{13.24}$$

associated with the corresponding decision $\hat{\boldsymbol{c}}(\sigma_{i-1})$.

At the decoding time-i, for each state σ_i in the state space $\Sigma_i(C)$, the SOVA first evaluates the two cumulative correlation metric candidates $M(\sigma_{i-1}^1, \sigma_i)$ and $M(\sigma_{i-1}^2, \sigma_i)$ corresponding to the two paths terminating in state σ_i with transitions from states σ_{i-1}^1 and σ_{i-1}^2, respectively. As for the Viterbi algorithm, the SOVA selects the cumulative correlation metric

$$M(\sigma_i) = \max_{l \in \{1,2\}} \{M(\sigma_{i-1}^l, \sigma_i)\}, \tag{13.25}$$

and updates the corresponding pair $(\hat{c}_{2i-1}(\sigma_i), \hat{c}_{2i}(\sigma_i))$ in the surviving path $\hat{\boldsymbol{c}}(\sigma_i)$ at state σ_i. Next, $\hat{\boldsymbol{L}}(\sigma_i)$ has to be updated. To this end, we define

$$\Delta_i \triangleq \max_{l \in \{1,2\}} \{M(\sigma_{i-1}^l, \sigma_i)\} - \min_{l \in \{1,2\}} \{M(\sigma_{i-1}^l, \sigma_i)\}. \tag{13.26}$$

Based on (13.22) and the fact that the code considered is antipodal, we set

$$\hat{L}_{2i-1}(\sigma_i) = \hat{L}_{2i}(\sigma_i) = \Delta_i, \tag{13.27}$$

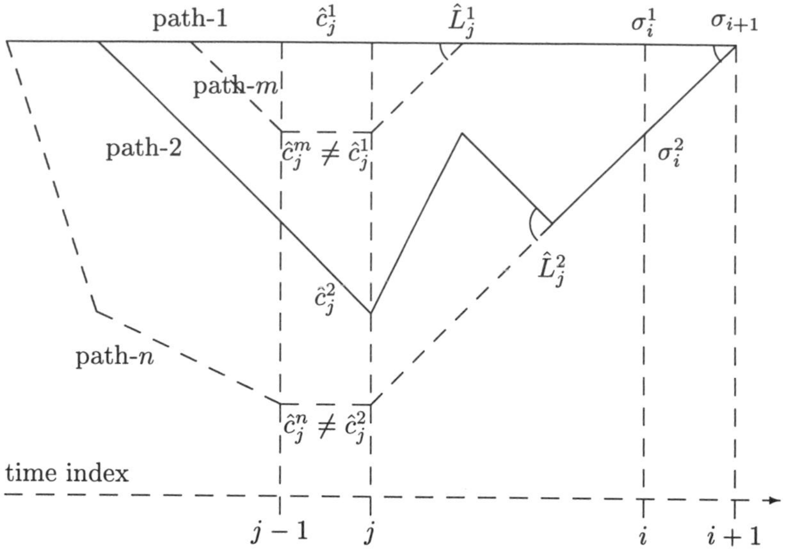

Figure 13.1. Trellis description with respect to reliability values associated with bit-j at state σ_{i+1}.

since Δ_i represents the reliability difference between the two most likely code-sequences terminating at state σ_i with different values for both $\hat{c}_{2i-1}$ and $\hat{c}_{2i}$. The remaining values $\hat{L}_j(\sigma_i)$ for $j = 1, \ldots, 2(i-1)$ of the surviving $\hat{\boldsymbol{L}}(\sigma_i)$ at state σ_i have to be updated.

In the following, we simplify the above notations and define

$$\hat{\boldsymbol{L}}(\sigma_{i-1}^l) = (\hat{L}_1^l, \hat{L}_2^l, \ldots, \hat{L}_{2(i-1)-1}^l, \hat{L}_{2(i-1)}^l) \tag{13.28}$$

for $l = 1, 2$, as the two sets of reliability measures corresponding to the two candidate paths merging into state σ_i with transitions from states σ_{i-1}^1 and σ_{i-1}^2, respectively. We refer to these two paths as path-1 and path-2, and without loss of generality assume that path-1 is the surviving path. Similarly, for $l = 1, 2$,

$$\hat{\boldsymbol{u}}(\sigma_{i-1}^l) = (\hat{c}_1^l, \hat{c}_2^l, \ldots, \hat{c}_{2(i-1)-1}^l, \hat{c}_{2(i-1)}^l) \tag{13.29}$$

represent the two sets of decisions corresponding to path-1 and path-2, respectively.

First, we consider the case $\hat{c}_j^1 \neq \hat{c}_j^2$, for some $j \in \{1, \ldots, 2(i-1)\}$, and recall that path-1 and path-2 have a reliability difference equal to Δ_i. Also,

$\hat{L}_j^1$ represents the reliability difference between path-1 and a code-sequence represented by a path-m merging with path-1 between the decoding steps j and $(i-1)$, with $\hat{c}_j^m \neq \hat{c}_j^1$, as shown in Figure 13.1. On the other hand, $\hat{L}_j^2$ represents the reliability difference between path-2 and a code-sequence represented by a path-n merging with path-2 between the decoding steps j and $(i-1)$, with $\hat{c}_j^n = \hat{c}_j^1$. Hence, $\hat{L}_j^2$ does not need to be considered to update $\hat{L}_j(\sigma_i)$ in $\hat{\boldsymbol{L}}(\sigma_i)$. Since no additional reliability information is available at state σ_i, we update

$$\hat{L}_j(\sigma_i) = \min\{\Delta_i, \hat{L}_j^1\}. \tag{13.30}$$

Next, we consider the case $\hat{c}_j^1 = \hat{c}_j^2$, for some $j \in \{1, \ldots, 2(i-1)\}$, so that path-2 is no longer considered to update $\hat{L}_j(\sigma_i)$ in $\hat{\boldsymbol{L}}(\sigma_i)$. However, path-$n$ previously defined now satisfies $\hat{c}_j^n \neq \hat{c}_j^1$. Since the reliability difference between path-1 and path-n is $\Delta_i + \hat{L}_j^2$ (i.e. the reliability difference between path-1 and path-2 plus the reliability difference between path-2 and path-n), we obtain

$$\hat{L}_j(\sigma_i) = \min\{\Delta_i + \hat{L}_j^2, \hat{L}_j^1\}. \tag{13.31}$$

The first version of SOVA that is equivalent to the above development was introduced by Battail in 1987 [3]. This algorithm was later reformulated in conjunction with the MAP algorithm in [9, 61] and formally shown to be equivalent to the Max-Log-MAP decoding algorithm in [34]. Consequently, the Max-Log-MAP or SOVA decoding algorithm can be summarized as follows.
For each state σ_i of the state space $\Sigma_i(C)$:

Step 1: Perform the Viterbi decoding algorithm to determine the survivor metric $M(\sigma_i)$ and the corresponding code-sequence $\hat{\boldsymbol{c}}(\sigma_i)$.

Step 2: For $j \in \{1, \ldots, 2(i-1)\}$, set $\hat{L}_j(\sigma_i)$ in $\hat{\boldsymbol{L}}(\sigma_i)$ either to the value $\min\{\Delta_i, \hat{L}_j^1\}$ if $\hat{c}_j^1 \neq \hat{c}_j^2$, or to the value $\min\{\Delta_i + \hat{L}_j^2, \hat{L}_j^1\}$ if $\hat{c}_j^1 = \hat{c}_j^2$.

Step 3: Set $\hat{L}_{2i-1}(\sigma_i)$ and $\hat{L}_{2i}(\sigma_i)$ in $\hat{\boldsymbol{L}}(\sigma_i)$ to the value Δ_i.

In [39], a simplified version of SOVA is presented. It is proposed to update $\hat{L}_j(\sigma_i)$, for $j = 1, 2, \ldots, 2(i-1)$, only when $c_j^1 \neq c_j^2$. Hence (13.30) remains unchange while (13.31) simply becomes:

$$\hat{L}_j(\sigma_i) = \hat{L}_j^1. \tag{13.32}$$

Consequently, the values $\hat{L}_j^2$ are no longer needed in the updating rule, so that this simplified version of SOVA is easier to implement. At the BER 10^{-4}, its performance degradation with respect to the MAP decoding algorithm is about $0.6 - 0.7$ dB coding gain loss, against $0.3 - 0.4$ dB loss for the Max-Log-MAP decoding algorithm.

13.3 BIT ERROR PROBABILITY OF MLD

In many practical applications, soft output information is not needed and only the binary decoded codeword is delivered by the decoder. However, it is still desirable to minimize the bit error probability rather than the word error probability. In such cases, the SOVA has no advantage over MLD since both algorithms deliver the same binary decoded sequence. Although the MAP decoding algorithm minimizes the decoding bit error probability, no significant improvement is observed over the bit error probability associated with MLD if properly implemented. Consequently, MLD remains to be the practical solution due to its much lower computational cost and implementation flexibility. However, when a word is in error, different mappings between the information sequences and the code sequences may result in a different number of bits in error, and hence a different average bit error probability.

As described in Chapter 3, the trellis diagram for an (N, K) linear block code is constructed from its TOGM. A trellis-based ML decoder simply finds the most likely path and its corresponding codeword among the 2^K possible paths that represent all the codewords generated by the TOGM. Therefore, a trellis-based decoder can be viewed as a device which searches for the most likely codeword out of the set of the 2^K codewords generated by the TOGM, independent of the mapping between information sequences and codewords. It follows that the mapping between information sequences and the 2^K codewords generated by the TOGM, or equivalently the encoder, can be modified without modifying the trellis-based determination of the most likely codeword. The corresponding information sequence is then retrieved from the knowledge of the mapping used by the encoder. Consequently, a trellis-based ML decoder can be viewed as the cascade of two elements: (1) a trellis search device which delivers the most likely codeword of the code generated by the TOGM; and (2) an inverse mapper which retrieves the information sequence from the delivered codeword. Although all mappings produce the same word error probability, different mappings may produce different bit error probabilities.

Different mappings can be obtained from the TOGM by applying elementary row additions and row permutations to this matrix. Let G_t denote the TOGM of the code considered, and let G_m denote the new matrix. If G_m is used for encoding, then the inverse mapper is represented by the right inverse of G_m. Since this mapping is bijective (or equivalently, G_m has full-rank K) and thus invertible, the right inverse of G_m is guaranteed to exist. In [33], it is shown that for many good codes, the best strategy is to have the K columns of the $K \times K$ identity matrix I_K in G_m (in reduced echelon form). Based on the particular structure of the TOGM G_t, this is readily realized in K steps of Gaussian elimination as follows: For $1 \le i \le K$, assume that $G_m(i-1)$ is the matrix obtained at step-$(i-1)$, with $G_m(0) = G_t$ and $G_m(K) = G_m$. Let $\boldsymbol{c}^i = (c_1^i, c_2^i, \ldots, c_K^i)^T$ denote the column of $G_m(i-1)$ that contains the leading '1' of the i-th row of $G_m(i-1)$. Then $c_i^i = 1$ and $c_{i+1}^i = c_{i+2}^i = \cdots = c_K^i = 0$. For $1 \le j \le i-1$, add row-i to row-j in $G_m(i-1)$ if $c_j^i = 1$. This results in matrix $G_m(i)$. For $i = K$, $G_m(K) = G_m$ contains the K columns of I_K with the same order of appearance. This matrix is said to be in reduced echelon form (REF), and is referred to as the **REF matrix**.

Now we perform the encoding based on G_m in REF instead of the TOGM G_t. Since both matrices generate the same 2^K codewords, any trellis-based decoder using the trellis diagram constructed from the TOGM G_t can still be used to deliver the most likely codeword. From the knowledge of this codeword and the fact that G_m in REF was used for encoding, the corresponding information sequence is easily recovered by taking only the positions which correspond to the columns of I_K. Note that this strategy is intuitively correct since whenever a code sequence delivered by the decoder is in error, the best strategy to recover the information bits is simply to determine them independently. Otherwise, errors propagate.

Example 13.1 In Example 3.1, the TOGM of the (8,4) RM code is given by

$$G_t = \begin{bmatrix} 1 & 1 & 1 & 1 & 0 & 0 & 0 & 0 \\ 0 & 1 & 0 & 1 & 1 & 0 & 1 & 0 \\ 0 & 0 & 1 & 1 & 1 & 1 & 0 & 0 \\ 0 & 0 & 0 & 0 & 1 & 1 & 1 & 1 \end{bmatrix}.$$

After adding rows 2 and 3 to row-1, we obtain

$$G_m = \begin{bmatrix} 1 & 0 & 0 & 1 & 0 & 1 & 1 & 0 \\ 0 & 1 & 0 & 1 & 1 & 0 & 1 & 0 \\ 0 & 0 & 1 & 1 & 1 & 1 & 0 & 0 \\ 0 & 0 & 0 & 0 & 1 & 1 & 1 & 1 \end{bmatrix}.$$

We can readily verify that G_t and G_m generate the same 16 codewords, so that there is a one-to-one correspondence between the codewords generated by G_m and the paths in the trellis diagram constructed from G_t. Once the trellis-based decoder delivers the most likely codeword $\hat{\boldsymbol{u}} = (\hat{u}_1, \hat{u}_2, \ldots, \hat{u}_8)$, the corresponding information sequence $\hat{\boldsymbol{a}} = (\hat{a}_1, \hat{a}_2, \hat{a}_3, \hat{a}_4)$ is retrieved by identifying

$$\begin{aligned} \hat{a}_1 &= \hat{u}_1, \\ \hat{a}_2 &= \hat{u}_2, \\ \hat{a}_3 &= \hat{u}_3, \\ \hat{a}_4 &= \hat{u}_8. \end{aligned}$$

Figure 13.2 depicts the bit error probabilities for the $(32, 26)$ RM code with encoding based on the TOGM and the REF matrix, respectively. The corresponding union bounds obtained in [33] are also shown in this figure. We see that there is a gap in error performance of about 1.0 dB and 0.2 dB at the BERs $10^{-1.5}$ and 10^{-6}, respectively. Similar results have been observed for other good block codes [33, 71]. The situation is different for good convolutional codes of short to medium constraint lengths, for which the feedforward non-systematic realizations outperform their equivalent feedback systematic realizations [80]. This can be explained by the fact that for short to medium constraint length convolutional codes in feedforward form, the bit error probability is dominated by error events of the same structures. Due to the small number of such structures, an efficient mapping that minimizes the bit error probability can be

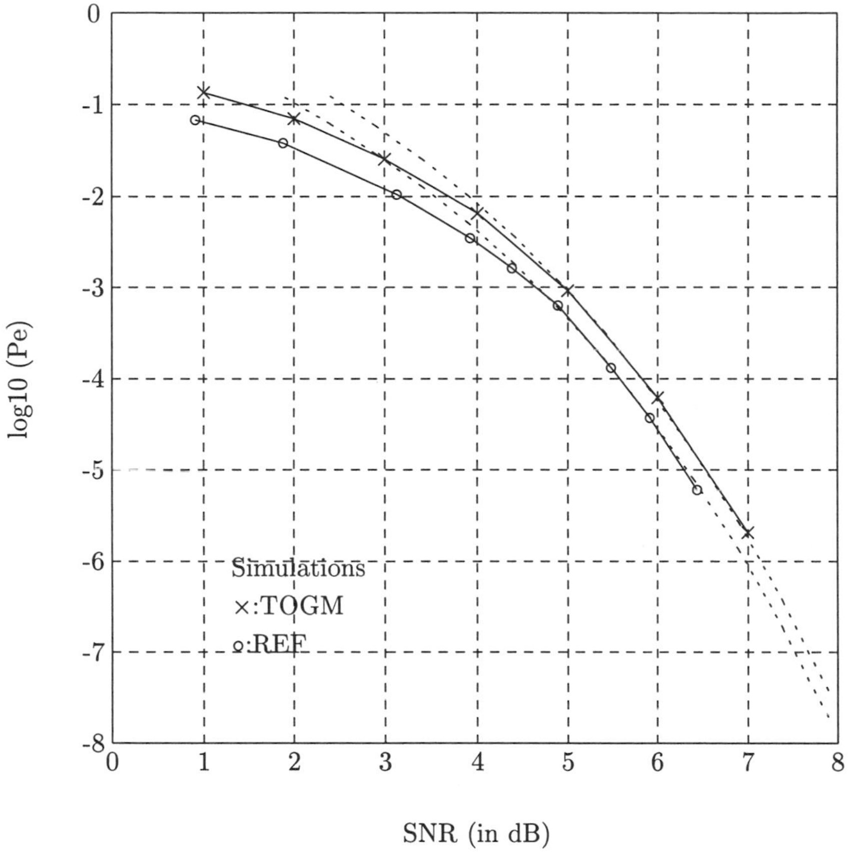

Figure 13.2. The bit error probabilities for the $(32, 26)$ RM code.

devised. This is no longer possible when the error performance is dominated by numerous unstructured error events, such as for long constraint length good convolutional codes or good block codes.

Based on these results, we may conclude: (1) although modest, the difference in bit error performance between encoding with the TOGM and the REF is of the same order as the difference in bit error performances between MLD and some sub-optimum low-complexity decoding methods; (2) the overall error performance of a conventional concatenated scheme with a RS outer code performing algebraic decoding is subject to these differences; and most importantly (3) **the gain is free**, since only the encoding circuit and the retrieving of the information sequence have to be modified. Furthermore this approach can be used for trellis-based MAP or SOVA decodings if a likelihood measure associated with each bit of the decoded information sequence, rather than each bit of the decoded codeword is needed, as in [34, 39, 40].

This approach can be generalized to any soft decision decoding method. In general, a particular decoding algorithm is based on a particular structure of the code considered. For example, majority-logic-decoding of RM codes is based on the generator matrices of these codes in their original form (presented in Section 2.5), or trellis-based decoding of linear codes is based on the TOGM of the code considered. Two cases are possible depending on whether the decoder delivers a codeword as in trellis-based decoding or directly an information sequence as in majority-logic-decoding of RM codes. In the first case, the procedure previously described is generalized in a straightforward way, while in the second case, the row additions performed to obtain G_m from the generator matrix corresponding to the decoding method considered are applied to the delivered information sequence by the inverse mapper [33].

APPENDIX A
A Trellis Construction Procedure

To decode a linear block code with a trellis-based decoding algorithm, the code trellis must be constructed to be used effectively in the decoding process. Therefore, the construction must meet a number of basic requirements. In the implementation of a trellis-based decoder, every state in the code trellis is labeled. The label of a state is used as the index to the memory where the state metric and the survivor into the state are stored. For efficient indexing, the sequence required to label a state must be as short as possible. Furthermore, the labels of two states at two boundary locations of a trellis section must provide complete information regarding the adjacency of the two states and the label of the composite branch connecting the two states, if they are adjacent, in a simple way. In general, a composite branch label appears many times in a section (see (6.13)). In order to compute the branch metrics efficiently, all the distinct composite branch labels in a trellis section must be generated systematically without duplication and stored in a block of memory. Then, for each pair of adjacent states, the index to the memory storing the composite branch label (or composite branch metric) between the two adjacent states must be derived readily from the labels of the two states. To achieve this, we must derive a condition that two composite branches have the same label, and partition the parallel components in a trellis section into blocks such that the parallel components in a block have the same set of composite branch labels (see Section 6.4). The composite branch label sets for two different blocks are disjoint. This localizes the composite branch metric computation. We can compute the composite branch metrics for each representative parallel component in a block independently. In this appendix, we present an efficient procedure for constructing a sectionalized trellis for a linear block code that meets all the above requirements. The construction makes use of the parallel structure of a trellis section presented in Section 6.4.

A.1 A BRIEF REVIEW OF THE TRELLIS ORIENTED GENERATOR MATRIX FOR A BINARY LINEAR BLOCK CODE

We first give a brief review of the trellis oriented generator matrix (TOGM) for a binary linear block code introduced in Section 3.4. Let G be a binary $K \times N$ generator matrix of a binary (N, K) linear code C. For $1 \leq i \leq K$, let $\mathrm{ld}(i)$ and $\mathrm{tr}(i)$ denote the column numbers of the leading '1' and the trailing '1' of the i-th row of G, respectively. G is called a TOGM if and only if for $1 \leq i < i' \leq K$,

$$\mathrm{ld}(i) \quad < \quad \mathrm{ld}(i'), \tag{A.1}$$

$$\mathrm{tr}(i) \quad \neq \quad \mathrm{tr}(i'). \tag{A.2}$$

Let M be a matrix with r rows. Hereafter, for a submatrix M' of M consisting of a subset of the rows in M, the order of rows in M' is assumed to be the same as in M. For a submatrix M' of M consisting of the i_1-th row, the i_2-th row, ..., the i_p-th row of M, let us call the set $\{i_1, i_2, \ldots, i_p\}$ as the **row number set** of M' (as a submatrix of M). M is said to be partitioned into the submatrices $M_1, M_2, \ldots, M_\mu$ if each row of M is contained in exactly one submatrix M_i with $1 \leq i \leq \mu$.

In Section 3.4, a TOGM G of C is partitioned into three submatrices, G_h^p, G_h^f, and G_h^s (also shown in Figure 6.3), for $0 \leq h \leq N$. The row number sets of G_h^p, G_h^f, and G_h^s are $\{i : \mathrm{tr}(i) \leq h\}$, $\{i : h < \mathrm{ld}(i)\}$, $\{i : \mathrm{ld}(i) \leq h < \mathrm{tr}(i)\}$, respectively. G_h^p and G_h^f generate the past and future codes at time-h, $C_{0,h}$ and $C_{h,N}$, respectively (see Section 3.7). That is,

$$C_{0,h} = \Gamma(G_h^p), \tag{A.3}$$

$$C_{h,N} = \Gamma(G_h^f), \tag{A.4}$$

where for a matrix M, $\Gamma(M)$ denotes the linear space generated by M.

Since from (A.1) $\mathrm{ld}(i) < \mathrm{ld}(i')$ for $1 \leq i < i' \leq K$, the order of information bits corresponds to the order of rows in G, and therefore, the row number sets of G_h^p, G_h^s and G_h^f as submatrices of G correspond to A_h^p, A_h^s and A_h^f (refer to Section 3.4), respectively. In this appendix, we put the TOGM in **reverse direction** such that for $1 \leq i < i' \leq K$,

$$\mathrm{ld}(i) \neq \mathrm{ld}(i'), \tag{A.5}$$

$$\mathrm{tr}(i) > \mathrm{tr}(i'). \tag{A.6}$$

Using a TOGM in reverse order, we can store the states (or state metrics) at the left end of a parallel component in consecutive memory cells using a simple addressing method. This method is very useful for designing IC decoder. It also reduces the actual computation time for a software decoder, since (the metrics of) the states at the left ends are accessed consecutively and computers have cache memories.

For a binary m-tuple $\boldsymbol{u} = (u_1, u_2, \ldots, u_m)$ and a set $I = \{i_1, i_2, \ldots, i_p\}$ of p positive integers with $1 \leq i_1 < i_2 < \cdots < i_p \leq m$, define

$$p_I(\boldsymbol{u}) \triangleq (u_{i_1}, u_{i_2}, \ldots, u_{i_p}). \tag{A.7}$$

For convenience, for $I = \emptyset$, $p_I(\boldsymbol{u}) \triangleq \epsilon$ (the null sequence). For a set U of m-tuples, define

$$p_I(U) \triangleq \{p_I(\boldsymbol{u}) : \boldsymbol{u} \in U\}. \tag{A.8}$$

For $I = \{h+1, h+2, \ldots, h'\}$, p_I is denoted by $p_{h,h'}$.

Then, for a partition $\{M_1, M_2, \ldots, M_\mu\}$ of M with r rows and $\boldsymbol{v} \in \{0,1\}^r$, it follows from the definition of a row number set and (A.7) that

$$\boldsymbol{v}M = p_{I_1}(\boldsymbol{v})M_1 + p_{I_2}(\boldsymbol{v})M_2 + \cdots + p_{I_\mu}(\boldsymbol{v})M_\mu, \tag{A.9}$$

where I_i denotes the row number set of M_i for $1 \leq i \leq \mu$. If M_i is further partitioned into submatrices $M_{i_1}, M_{i_2}, \ldots$, then

$$p_{I_i}(\boldsymbol{v})M_i = p_{I_{i_1}}(p_{I_i}(\boldsymbol{v}))M_{i_1} + p_{I_{i_2}}(p_{I_i}(\boldsymbol{v}))M_{i_2} + \cdots, \tag{A.10}$$

where $I_{i_1}, I_{i_2}, \ldots$, denote the row number sets of $M_{i_1}, M_{i_2}, \ldots$ as submatrices of M_i.

In the construction of the trellis section from time-h to time-h', the submatrix G_h^s of G takes a key role. Let I_h denote the row number set of G_h^s as a submatrix G. Then, $\rho_h = |I_h|$, and the following property of state labeling by the state defining information set (Section 4.1) holds.

The Key Property of State Labeling: Let $L(\sigma_0, \sigma_h, \sigma_f) \triangleq L(\sigma_0, \sigma_h) \circ L(\sigma_h, \sigma_f)$ be the set of paths in a code trellis for C that connects the initial state σ_0 to the final state σ_f through the state σ_h at time-h. There is a **one-to-one mapping** l from $\Sigma_h(C)$ to $\{0,1\}^{\rho_h}$ with $0 < h < N$ such that for $\sigma_h \in \Sigma_h(C)$ and $\boldsymbol{a} \in \{0,1\}^K$,

$$\boldsymbol{a}G \in L(\sigma_0, \sigma_h, \sigma_f), \tag{A.11}$$

if and only if

$$p_{I_h}(\boldsymbol{a}) = l(\sigma_h). \tag{A.12}$$

$\triangle\triangle$

Here, $l(\sigma_h) \in \{0,1\}^{\rho_h}$ is called the label of state σ_h. This state labeling is simply the state labeling by the state defining information set given in Section 4.1. We can readily see that

$$L(\sigma_0, \sigma_h, \sigma_f) = l(\sigma_h)G_h^s \oplus C_{0,h} \oplus C_{h,N}. \tag{A.13}$$

Note that if $\rho_h = 0$, G_h^s is the empty matrix and $l(\sigma_h) = \epsilon$. For convenience, we define the product of ϵ and the empty matrix as the zero vector.

A.2 STATE LABELING BY THE STATE DEFINING INFORMATION SET AND COMPOSITE BRANCH LABEL

For $\sigma_h \in \Sigma_h(C)$ and $\sigma_{h'} \in \Sigma_{h'}(C)$ with $L(\sigma_h, \sigma_{h'}) \neq \emptyset$, the composite branch label $L(\sigma_h, \sigma_{h'})$ can be expressed in terms of the labels of σ_h and $\sigma_{h'}$. Since

$$\begin{aligned} L(\sigma_0, \sigma_h, \sigma_{h'}, \sigma_f) &\triangleq L(\sigma_0, \sigma_h) \circ L(\sigma_h, \sigma_{h'}) \circ L(\sigma_{h'}, \sigma_f) \\ &= L(\sigma_0, \sigma_h, \sigma_f) \cap L(\sigma_0, \sigma_{h'}, \sigma_f), \end{aligned}$$

it follows from the key property of state labeling that for $\boldsymbol{a} \in \{0,1\}^K$,

$$\boldsymbol{a}G \in L(\sigma_0, \sigma_h, \sigma_{h'}, \sigma_f), \tag{A.14}$$

if and only if

$$p_{I_h}(\boldsymbol{a}) = l(\sigma_h), \tag{A.15}$$
$$p_{I_{h'}}(\boldsymbol{a}) = l(\sigma_{h'}). \tag{A.16}$$

Define $I_{h,h'}$ and $\rho_{h,h'}$ as follows:

$$I_{h,h'} = I_h \cap I_{h'} \tag{A.17}$$
$$\rho_{h,h'} = |I_{h,h'}|. \tag{A.18}$$

Then $I_{h,h'} = \{i : \mathrm{ld}(i) \leq h < h' < \mathrm{tr}(i)\}$ is the row number set of the submatrix of G which consists of those rows in both G_h^s and $G_{h'}^s$. Let $G_{h,h'}^{s,s}$ denote this submatrix. Let $G_{h,h'}^{s,p}$, $G_{h,h'}^{f,s}$ and $G_{h,h'}^{f,p}$ denote the submatrices of G whose row number sets as submatrices of G are $I_h \backslash I_{h,h'} = \{i : \mathrm{ld}(i) \leq h < \mathrm{tr}(i) \leq h'\}$, $I_{h'} \backslash I_{h,h'} = \{i : h < \mathrm{ld}(i) \leq h' < \mathrm{tr}(i)\}$ and $\{i : h < \mathrm{ld}(i) \leq \mathrm{tr}(i) \leq h'\}$, respectively. Then,

$$C_{h,h'} = \Gamma(G_{h,h'}^{f,p}), \tag{A.19}$$

and G is partitioned into G_h^p, $G_{h,h'}^{s,s}$, $G_{h,h'}^{s,p}$, $G_{h,h'}^{f,s}$, $G_{h,h'}^{f,p}$ and $G_{h'}^f$ (see Figure 6.3). From (A.3), (A.4), (A.9) and (A.19), we have that for $\boldsymbol{a} \in \{0,1\}^K$, $\boldsymbol{a}G \in C$ can be expressed as

$$\boldsymbol{a}G = p_{I_{h,h'}}(\boldsymbol{a})G_{h,h'}^{s,s} + p_{I_h \backslash I_{h,h'}}(\boldsymbol{a})G_{h,h'}^{s,p} + p_{I_{h'} \backslash I_{h,h'}}(\boldsymbol{a})G_{h,h'}^{f,s} + \boldsymbol{u}, \tag{A.20}$$

where $\boldsymbol{u} \in C_{0,h} \oplus C_{h,h'} \oplus C_{h',N}$.

Since h and h' are fixed hereafter, we abbreviate $p_{h,h'}(G_{h,h'}^{x,y})$ as $G^{x,y}$ where $x \in \{s, f\}$ and $y \in \{s, p\}$. If $\boldsymbol{a}G \in L(\sigma_0, \sigma_h, \sigma_{h'}, \sigma_f)$, then $p_{h,h'}(\boldsymbol{a}G) = \boldsymbol{a}p_{h,h'}(G) \in L(\sigma_h, \sigma_{h'})$. Since $L(\sigma_h, \sigma_{h'})$ is a coset in $p_{h,h'}(C)/C_{h,h'}^{\mathrm{tr}}$ (see (3.18)), it follows from (A.20) that

$$L(\sigma_h, \sigma_{h'}) = p_{I_{h,h'}}(\boldsymbol{a})G^{s,s} + p_{I_h \backslash I_{h,h'}}(\boldsymbol{a})G^{s,p} + p_{I_{h'} \backslash I_{h,h'}}(\boldsymbol{a})G^{f,s} + C_{h,h'}^{\mathrm{tr}}. \tag{A.21}$$

In the following, we will derive a relation between $L(\sigma_h, \sigma_{h'})$ and the labels of states σ_h and $\sigma_{h'}$, $l(\sigma_h)$ and $l(\sigma_{h'})$. Lemma A.1 gives a simple relation between $I_{h,h'}$ and I_h.

Lemma A.1 $I_{h,h'}$ consists of the smallest $\rho_{h,h'}$ integers in I_h.

Proof: For $i \in I_{h,h'}$, we have that $h' < \text{tr}(i)$. For $i' \in I_h \backslash I_{h,h'}$, we have that $\text{tr}(i') \leq h'$. Hence, $\text{tr}(i') < \text{tr}(i)$. From (A.6), we have $i < i'$.

△△

In general, there is no such a simple relation between $I_{h,h'}$ and $I_{h'}$.

Let $I_{h'} = \{i_1, i_2, \ldots, i_{\rho_{h'}}\}$ with $1 \leq i_1 < i_2 < \cdots < i_{\rho_{h'}}$ be the row number set of $G^s_{h'}$, and let $I_{h,h'} = \{i_{j_1}, i_{j_2}, \ldots, i_{j_{\rho_{h,h'}}}\}$ with $1 \leq j_1 < j_2 \cdots < j_{\rho_{h,h'}} \leq \rho_{h'}$ be the row number set of $G^{s,s}_{h,h'}$. By definition, the ρ-th row of $G^s_{h'}$ is the i_ρ-th row of G for $1 \leq \rho \leq \rho_{h'}$ and the μ-th row of $G^{s,s}_{h,h'}$ is the i_{j_μ}-th row of G for $1 \leq \mu \leq \rho_{h,h'}$. Hence, the μ-th row of $G^{s,s}_{h,h'}$ is the j_μ-th row of $G^s_{h'}$. That is, the row number set of $G^{s,s}_{h,h'}$ as a submatrix of $G^s_{h'}$, denoted $J_{h'}$, is $J_{h'} = \{j_1, j_2, \ldots, j_{\rho_{h,h'}}\}$, and $\bar{J}_{h'} \triangleq \{1, 2, \ldots, \rho_{h'}\} \setminus J_{h'}$ is the row number set of $G^{f,s}_{h,h'}$.

Suppose (A.14) holds. Then, from (A.15), (A.16) and Lemma A.1,

$$p_{I_{h,h'}}(\boldsymbol{a}) = p_{0,\rho_{h,h'}}(l(\sigma_h)) = p_{J_{h'}}(l(\sigma_{h'})), \tag{A.22}$$

$$p_{I_h \backslash I_{h,h'}}(\boldsymbol{a}) = p_{\rho_{h,h'},\rho_h}(l(\sigma_h)), \tag{A.23}$$

$$p_{I_{h'} \backslash I_{h,h'}}(\boldsymbol{a}) = p_{\bar{J}_{h'}}(l(\sigma_{h'})). \tag{A.24}$$

For simplicity, define

$$l^{(s)}(\sigma_h) \triangleq p_{0,\rho_{h,h'}}(l(\sigma_h)), \tag{A.25}$$

$$l^{(p)}(\sigma_h) \triangleq p_{\rho_{h,h'},\rho_h}(l(\sigma_h)), \tag{A.26}$$

$$l^{(s)}(\sigma_{h'}) \triangleq p_{J_{h'}}(l(\sigma_{h'})), \tag{A.27}$$

$$l^{(f)}(\sigma_{h'}) \triangleq p_{\bar{J}_{h'}}(l(\sigma_{h'})). \tag{A.28}$$

From (A.25) and (A.26), the label $l(\sigma_h)$ of the state σ_h with $l^{(s)}(\sigma_h) = \boldsymbol{\alpha}$ and $l^{(p)}(\sigma_h) = \boldsymbol{\beta}$ is given by

$$l(\sigma_h) = \boldsymbol{\alpha} \circ \boldsymbol{\beta}. \tag{A.29}$$

The label $l(\sigma_{h'})$ of the state $\sigma_{h'}$ with $l^{(s)}(\sigma_h) = \boldsymbol{\alpha}$ and $l^{(f)}(\sigma_{h'}) = \boldsymbol{\gamma}$ can be easily obtained from $\boldsymbol{\alpha}$ and $\boldsymbol{\gamma}$ using (A.27) and (A.28).

By summarizing (A.14) to (A.16), (A.22) to (A.26) and (A.28), a condition for the adjacency between two states at time-h and time-h', and the composite branch label between them (shown in Section 6.3) are given in Theorem A.1.

Theorem A.1 For $\sigma_h \in \Sigma_h(C)$ and $\sigma_{h'} \in \Sigma_{h'}(C)$ with $0 \leq h < h' \leq N$, $L(\sigma_h, \sigma_{h'}) \neq \emptyset$ if and only if

$$l^{(s)}(\sigma_h) = l^{(s)}(\sigma_{h'}).$$

Also if $L(\sigma_h, \sigma_{h'}) \neq \emptyset$, then $L(\sigma_h, \sigma_{h'})$ is given by

$$L(\sigma_h, \sigma_{h'}) = l^{(s)}(\sigma_h)G^{s,s} + l^{(p)}(\sigma_h)G^{s,p} + l^{(f)}(\sigma_{h'})G^{f,s} + C^{\mathrm{tr}}_{h,h'}. \qquad \text{(A.30)}$$

△△

We call $l^{(s)}(\sigma_h)$ (or $l^{(s)}(\sigma_{h'})$) the first label part of σ_h (or $\sigma_{h'}$) and $l^{(p)}(\sigma_h)$ (or $l^{(f)}(\sigma_{h'})$) the second label part of σ_h (or $\sigma_{h'}$). Now we partition $\Sigma_h(C)$ and $\Sigma_{h'}(C)$ into $2^{\rho_{h,h'}}$ blocks of the same size, respectively, in such a way that two states are in the same block if and only if they have the same first label part. For $\boldsymbol{\alpha} \in \{0,1\}^{\rho_{h,h'}}$, a block, denoted $\Sigma_h^{\boldsymbol{\alpha}}$, (or $\Sigma_{h'}^{\boldsymbol{\alpha}}$) in the above partition of $\Sigma_h(C)$ (or $\Sigma_{h'}(C)$) is defined as

$$\Sigma_h^{\boldsymbol{\alpha}} \triangleq \{\sigma_h \in \Sigma_h(C) : l^{(s)}(\sigma_h) = \boldsymbol{\alpha}\}, \qquad \text{(A.31)}$$

$$\Sigma_{h'}^{\boldsymbol{\alpha}} \triangleq \{\sigma_{h'} \in \Sigma_{h'}(C) : l^{(s)}(\sigma_{h'}) = \boldsymbol{\alpha}\}. \qquad \text{(A.32)}$$

The blocks $\Sigma_h^{\boldsymbol{\alpha}}$ and $\Sigma_{h'}^{\boldsymbol{\alpha}}$ correspond to $S_L(\boldsymbol{a}_h)$ and $S_R(\boldsymbol{a}_h)$ in Section 6.4, respectively. That is, for any $\boldsymbol{\alpha} \in \{0,1\}^{\rho_{h,h'}}$, a subgraph which consists of the set of state at time-h, $\Sigma_h^{\boldsymbol{\alpha}}$, the set of states at time-h', $\Sigma_{h'}^{\boldsymbol{\alpha}}$, and all the composite branches between them form a parallel component. This parallel component is denoted $\Lambda_{\boldsymbol{\alpha}}$. The total number of parallel components in a trellis section from time-h to time-h' is given by $2^{\rho_{h,h'}}$.

Since all states in $\Sigma_h^{\boldsymbol{\alpha}} \cup \Sigma_{h'}^{\boldsymbol{\alpha}}$ have the same first label part $\boldsymbol{\alpha}$, there is a one-to-one mapping, denoted $s_{h,\boldsymbol{\alpha}}$, from the set of the second label parts of the states in $\Sigma_h^{\boldsymbol{\alpha}}$, denoted L_h, to $\Sigma_h^{\boldsymbol{\alpha}}$, and there is a one-to-one mapping, denoted $s_{h',\boldsymbol{\alpha}}$, from the set of the second label parts of the states in $\Sigma_{h'}^{\boldsymbol{\alpha}}$, denoted $L_{h'}$, to $\Sigma_{h'}^{\boldsymbol{\alpha}}$. Then, $L_h = \{0,1\}^{\rho_h - \rho_{h,h'}}$, $L_{h'} = \{0,1\}^{\rho_{h'} - \rho_{h,h'}}$ and $s_{h,\boldsymbol{\alpha}}$ and $s_{h',\boldsymbol{\alpha}}$ are the inverse mappings of $l^{(p)}$ and $l^{(f)}$, respectively.

A.3 TRELLIS CONSTRUCTION

Now consider how to construct a trellis section. When we use a trellis diagram for decoding, we have to store the state metric for each state. Therefore, we must assign memory for each state. For a state σ, we use the label $l(\sigma)$ as the index of the allocated memory, because: (1) $l(\sigma)$ is the shortest possible label for σ and is easy to obtain, and (2) as shown in Theorem A.1, the complete information on $L(\sigma_h, \sigma_{h'})$ is provided in a very simple way by $l^{(s)}(\sigma_h)$, $l^{(p)}(\sigma_h)$, $l^{(s)}(\sigma_{h'})$ and $l^{(f)}(\sigma_{h'})$ which can be readily derived from $l(\sigma_h)$ and $l(\sigma_{h'})$ by (A.25) to (A.28).

If instead we use the state labeling by parity-check matrix for the states at time-h, the label length becomes $N-K$. Since $\rho_h \leq \min\{N, N-K\}$ (see (5.4)), if $\rho_h < N-K$, a linear mapping from $\{0,1\}^{N-K}$ to $\{0,1\}^{\rho_h}$ which depends on h in general is necessary to obtain the index of the memory.

The next problem is how to generate branch metrics of adjacent state pairs and give access to them. The set of composite branch labels of the section from time-h to time-h', denoted $L_{h,h'}$, is the set of the following cosets (see (3.18)):

$$L_{h,h'} = p_{h,h'}(C)/C^{\text{tr}}_{h,h'}. \tag{A.33}$$

It follows from (A.30) that

$$\begin{aligned} L_{h,h'} &= \{\boldsymbol{\alpha} G^{s,s} + \boldsymbol{\beta} G^{s,p} + \boldsymbol{\gamma} G^{f,s} + C^{\text{tr}}_{h,h'} : \boldsymbol{\alpha} \in \{0,1\}^{\rho_{h,h'}}, \\ &\qquad \boldsymbol{\beta} \in \{0,1\}^{\rho_h - \rho_{h,h'}}, \boldsymbol{\gamma} \in \{0,1\}^{\rho_{h'} - \rho_{h,h'}}\}. \end{aligned} \tag{A.34}$$

Each composite branch label appears

$$2^{K-k(C_{0,h})-k(C_{h',N})-k(p_{h,h'}(C))}$$

times in the trellis section (see (6.13)).

Next we consider how to generate all the composite branch labels in $L_{h,h'}$ given by (A.34) without duplication. Suppose we choose submatrices $G_1^{s,s}$ of $G^{s,s}$, $G_1^{s,p}$ of $G^{s,p}$ and $G_1^{f,s}$ of $G^{f,s}$ such that the rows in $G_1^{s,s}$, $G_1^{s,p}$, $G_1^{f,s}$ and $G^{f,p}$ are linearly independent and

$$\begin{aligned} L_{h,h'} &= \{\boldsymbol{\alpha}_1 G^{s,s} + \boldsymbol{\beta}_1 G^{s,p} + \boldsymbol{\gamma}_1 G^{f,s} + C^{\text{tr}}_{h,h'} : \boldsymbol{\alpha}_1 \in \{0,1\}^{r_{s,s}}, \\ &\qquad \boldsymbol{\beta}_1 \in \{0,1\}^{r_{s,p}}, \boldsymbol{\gamma}_1 \in \{0,1\}^{r_{f,s}}\}, \end{aligned} \tag{A.35}$$

where $r_{s,s}$, $r_{s,p}$ and $r_{f,s}$ denote the numbers of rows of $G_1^{s,s}$, $G_1^{s,p}$ and $G_1^{f,s}$, respectively. If we generate composite branch labels by using the right-hand side of (A.35) and store $\boldsymbol{\alpha}_1 G^{s,s} + \boldsymbol{\beta}_1 G^{s,p} + \boldsymbol{\gamma}_1 G^{f,s} + C^{\text{tr}}_{h,h'}$ into a memory indexed with $(\boldsymbol{\alpha}_1, \boldsymbol{\beta}_1, \boldsymbol{\gamma}_1)$, then the duplication can be avoided. The next requirement is that the index $(\boldsymbol{\alpha}_1, \boldsymbol{\beta}_1, \boldsymbol{\gamma}_1)$ can be readily obtained from $\boldsymbol{\alpha}, \boldsymbol{\beta}$ and $\boldsymbol{\gamma}$. To provide a good solution of the above problem, we first analyze the linear dependency of rows in $p_{h,h'}(G)$.

For a TOGM G, $p_{h,h'}(G)$ consists of the disjoint submatrices $G^{s,s}$, $G^{s,p}$, $G^{f,s}$, $G^{f,p}$ and the all zero submatrices $p_{h,h'}(G_h^p)$ and $p_{h,h'}(G_{h'}^f)$.

Lemma A.2 In $p_{h,h'}(G)$: (1) the rows in submatrices $G^{f,s}$ and $G^{f,p}$ are linearly independent; and (2) the rows in submatrices $G^{s,p}$ and $G^{f,p}$ are linearly independent.

Proof: If the i-th row of G is in $G^{f,s}$ or $G^{f,p}$, then

$$h < \mathrm{ld}(i) \leq h'. \tag{A.36}$$

Similarly, if the i-th row of G is in $G^{s,p}$ or $G^{f,p}$, then

$$h < \mathrm{tr}(i) \leq h'. \tag{A.37}$$

Hence, (1) and (2) of the lemma follow from (A.1) and (A.2), respectively.

△△

For two binary $r \times m$ matrices M and M' and a binary linear block code C_0 of length m, we write

$$M \equiv M' \qquad (\mathrm{mod} C_0)$$

if and only if every row in the matrix $M - M'$ is in C_0, where "$-$" denotes the component-wise subtraction.

Partition $G^{f,s}$ into two submatrices $G_0^{f,s}$ and $G_1^{f,s}$ (see Figure A.1) by partitioning the rows of $G^{f,s}$ in such a way that

(1) the rows in $G_1^{f,s}$, $G^{s,p}$ and $G^{f,p}$ are linearly independent, and

(2) each row of $G_0^{f,s}$ is a linear combination of rows in $G_1^{f,s}$, $G^{s,p}$ and $G^{f,p}$.

Let ν_1 denote the number of rows of $G_1^{f,s}$, and define $\nu_0 \triangleq \rho_{h'} - \rho_{h,h'} - \nu_1$, which is the number of rows of $G_0^{f,s}$.

Let $\boldsymbol{e}_1, \boldsymbol{e}_1, \ldots, \boldsymbol{e}_{\nu_0}$ be the first to the last rows of $G_0^{f,s}$. From conditions (1) and (2) of the partitioning of $G^{f,s}$, there are unique $\boldsymbol{v}_i^{(1)} \in \{0,1\}^{\nu_1}$, $\boldsymbol{v}_i^{(2)} \in \{0,1\}^{\rho_h - \rho_{h,h'}}$ and $\boldsymbol{u}_i \in C_{h,h'}^{\mathrm{tr}}$ such that

$$\boldsymbol{e}_i = \boldsymbol{v}_i^{(1)} G_1^{f,s} + \boldsymbol{v}_i^{(2)} G^{s,p} + \boldsymbol{u}_i, \quad \text{for } 1 \leq i \leq \nu_0.$$

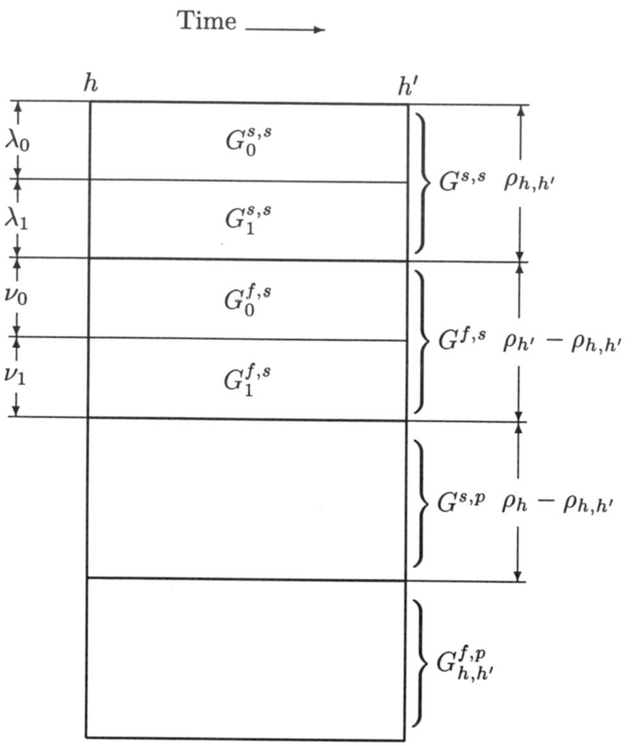

Figure A.1. Further partitioning of the TOGM G.

Let $E^{(1)}$ denote the $\nu_0 \times \nu_1$ matrix whose i-th row is $\boldsymbol{v}_i^{(1)}$ and $E^{(2)}$ denote the $\nu_0 \times (\rho_h - \rho_{h,h'})$ matrix whose i-th row is $\boldsymbol{v}_i^{(2)}$. Then, we have

$$G_0^{f,s} \equiv E^{(1)} G_1^{f,s} + E^{(2)} G^{s,p} \quad (\text{mod} C_{h,h'}^{\text{tr}}). \tag{A.38}$$

$G_0^{f,s}$, $G_1^{f,s}$, $E^{(1)}$ and $E^{(2)}$ can be efficiently derived by using standard row operations.

Next, partition $G^{s,s}$ into two submatrices $G_0^{s,s}$ and $G_1^{s,s}$ (see Figure A.1) by partitioning the rows of $G^{s,s}$ in such a way that:

(1) the rows in $G_1^{s,s}$, $G_1^{f,s}$, $G^{s,p}$ and $G^{f,p}$ are linearly independent; and

(2) each row of $G_0^{s,s}$ is a linear combination of rows in $G_1^{s,s}$, $G_1^{f,s}$, $G^{s,p}$ and $G^{f,p}$.

Let λ_1 denote the number of rows of $G_1^{s,s}$, and define $\lambda_0 \triangleq \rho_{h,h'} - \lambda_1$, which is the number of rows of $G_0^{s,s}$. In a similar way as the derivation of (A.38), we can find a unique $\lambda_0 \times \lambda_1$ matrix $F^{(1)}$, a unique $\lambda_0 \times \nu_1$ matrix $F^{(2)}$ and a unique $\lambda_0 \times (\rho_h - \rho_{h,h'})$ matrix $F^{(3)}$ such that

$$G_0^{s,s} \equiv F^{(1)} G_1^{s,s} + F^{(2)} G_1^{f,s} + F^{(3)} G^{s,p} \pmod{C_{h,h'}^{\text{tr}}}. \tag{A.39}$$

Let R and Q denote the row number sets of $G_1^{f,s}$ and $G_1^{s,s}$, respectively. Define $\bar{R} \triangleq \{1, 2, \ldots, \rho_{h'} - \rho_{h,h'}\} \backslash R$ and $\bar{Q} \triangleq \{1, 2, \ldots, \rho_{h,h'}\} \backslash Q$. Then, from (A.9)

$$\begin{aligned} \boldsymbol{\alpha} G^{s,s} &= p_{\bar{Q}}(\boldsymbol{\alpha}) G_0^{s,s} + p_Q(\boldsymbol{\alpha}) G_1^{s,s}, \quad \text{for } \boldsymbol{\alpha} \in \{0,1\}^{\rho_{h,h'}}, & \text{(A.40)} \\ \boldsymbol{\gamma} G^{f,s} &= p_{\bar{R}}(\boldsymbol{\gamma}) G_0^{f,s} + p_R(\boldsymbol{\gamma}) G_1^{f,s}, \quad \text{for } \boldsymbol{\gamma} \in \{0,1\}^{\rho_{h'} - \rho_{h,h'}}. & \text{(A.41)} \end{aligned}$$

It follows from (A.38) to (A.41) that for $\boldsymbol{\alpha} \in \{0,1\}^{\rho_{h,h'}}$, $\boldsymbol{\beta} \in \{0,1\}^{\rho_h - \rho_{h,h'}}$ and $\boldsymbol{\gamma} \in \{0,1\}^{\rho_{h'} - \rho_{h,h'}}$,

$$\begin{aligned} & \boldsymbol{\alpha} G^{s,s} + \boldsymbol{\beta} G^{s,p} + \boldsymbol{\gamma} G^{f,s} + C_{h,h'}^{\text{tr}} \\ &= (p_Q(\boldsymbol{\alpha}) + p_{\bar{Q}}(\boldsymbol{\alpha}) F^{(1)}) G_1^{s,s} \\ &\quad + (p_{\bar{Q}}(\boldsymbol{\alpha}) F^{(3)} + p_{\bar{R}}(\boldsymbol{\gamma}) E^{(2)} + \boldsymbol{\beta}) G^{s,p} \\ &\quad + (p_{\bar{Q}}(\boldsymbol{\alpha}) F^{(2)} + p_R(\boldsymbol{\gamma}) + p_{\bar{R}}(\boldsymbol{\gamma}) E^{(1)}) G_1^{f,s} + C_{h,h'}^{\text{tr}}. \end{aligned} \tag{A.42}$$

Define

$$\begin{aligned} f^{(1)}(\boldsymbol{\alpha}) &\triangleq p_Q(\boldsymbol{\alpha}) + p_{\bar{Q}}(\boldsymbol{\alpha}) F^{(1)}, & \text{(A.43)} \\ f^{(2)}(\boldsymbol{\alpha}, \boldsymbol{\beta}, \boldsymbol{\gamma}) &\triangleq p_{\bar{Q}}(\boldsymbol{\alpha}) F^{(3)} + p_{\bar{R}}(\boldsymbol{\gamma}) E^{(2)} + \boldsymbol{\beta}, & \text{(A.44)} \\ f^{(3)}(\boldsymbol{\alpha}, \boldsymbol{\gamma}) &\triangleq p_{\bar{Q}}(\boldsymbol{\alpha}) F^{(2)} + p_R(\boldsymbol{\gamma}) + p_{\bar{R}}(\boldsymbol{\gamma}) E^{(1)}. & \text{(A.45)} \end{aligned}$$

When $\boldsymbol{\alpha}$, $\boldsymbol{\beta}$ and $\boldsymbol{\gamma}$ run over $\{0,1\}^{\rho_{h,h'}}$, $\{0,1\}^{\rho_h - \rho_{h,h'}}$ and $\{0,1\}^{\rho_{h'} - \rho_{h,h'}}$, respectively, $f^{(1)}(\boldsymbol{\alpha})$ $f^{(2)}(\boldsymbol{\alpha}, \boldsymbol{\beta}, \boldsymbol{\gamma})$ and $f^{(3)}(\boldsymbol{\alpha}, \boldsymbol{\gamma})$ run over $\{0,1\}^{\lambda_1}$, $\{0,1\}^{\rho_h - \rho_{h,h'}}$ and $\{0,1\}^{\nu_1}$, respectively. Hence, the set $L_{h,h'}$ of all composite branch labels is given by

$$\begin{aligned} L_{h,h'} = \{ & \boldsymbol{\alpha}_1 G_1^{s,s} + \boldsymbol{\beta}_1 G^{s,p} + \boldsymbol{\gamma}_1 G_1^{f,s} + C_{h,h'}^{\text{tr}} : \\ & \boldsymbol{\alpha}_1 \in \{0,1\}^{\lambda_1}, \boldsymbol{\beta}_1 \in \{0,1\}^{\rho_h - \rho_{h,h'}}, \boldsymbol{\gamma}_1 \in \{0,1\}^{\nu_1} \}. \end{aligned} \tag{A.46}$$

Since the rows in $G_1^{s,s}$, $G^{s,p}$, $G_1^{f,s}$ and $G^{f,p}$ are linearly independent, for $(\boldsymbol{\alpha}_1, \boldsymbol{\beta}_1, \boldsymbol{\gamma}_1) \neq (\boldsymbol{\alpha}_1', \boldsymbol{\beta}_1', \boldsymbol{\gamma}_1')$, the cosets $\boldsymbol{\alpha}_1 G_1^{s,s} + \boldsymbol{\beta}_1 G^{s,p} + \boldsymbol{\gamma}_1 G_1^{f,s} + C_{h,h'}^{\text{tr}}$ and

$\boldsymbol{\alpha}_1' G_1^{s,s} + \boldsymbol{\beta}_1' G^{s,p} + \boldsymbol{\gamma}_1' G_1^{f,s} + C_{h,h'}^{\rm tr}$ are disjoint. Consequently, the set of composite branch labels of a parallel component $\Lambda_{\boldsymbol{\alpha}}$ with $\boldsymbol{\alpha} \in \{0,1\}^{\rho_{h,h'}}$, denoted $L_{\boldsymbol{\alpha}}$, is given by

$$\begin{aligned} L_{\boldsymbol{\alpha}} \triangleq \; & \{\boldsymbol{\alpha}_1 G_1^{s,s} + \boldsymbol{\beta}_1 G^{s,p} + \boldsymbol{\gamma}_1 G_1^{f,s} + C_{h,h'}^{\rm tr} : \\ & \boldsymbol{\beta}_1 \in \{0,1\}^{\rho_h - \rho_{h,h'}}, \boldsymbol{\gamma}_1 \in \{0,1\}^{\nu_1}\}, \end{aligned} \quad \text{(A.47)}$$

where $\boldsymbol{\alpha}_1 = f^{(1)}(\boldsymbol{\alpha}) = p_Q(\boldsymbol{\alpha}) + p_{\bar{Q}}(\boldsymbol{\alpha})F^{(1)}$.

Theorem A.2 Let Q denote the row number sets of $G_1^{s,s}$, and define $\bar{Q} \triangleq \{1,2,\ldots,\rho_{h,h'}\} \setminus Q$. Two parallel components $\Lambda_{\boldsymbol{\alpha}}$ and $\Lambda_{\boldsymbol{\alpha}'}$ with $\boldsymbol{\alpha}$ and $\boldsymbol{\alpha}'$ in $\{0,1\}^{\rho_{h,h'}}$ are isomorphic up to composite branch labels, if and only if

$$p_Q(\boldsymbol{\alpha} + \boldsymbol{\alpha}') = (p_{\bar{Q}}(\boldsymbol{\alpha} + \boldsymbol{\alpha}'))F^{(1)}, \quad \text{(A.48)}$$

where $F^{(1)}$ is defined in (A.39). If (A.48) does not hold, $L_{\boldsymbol{\alpha}}$ and $L_{\boldsymbol{\alpha}'}$ are disjoint.

Proof: (1) The only-if part: If $\boldsymbol{\alpha}_1 \triangleq p_Q(\boldsymbol{\alpha}) + p_{\bar{Q}}(\boldsymbol{\alpha})F^{(1)} \neq \boldsymbol{\alpha}_1' \triangleq p_Q(\boldsymbol{\alpha}') + p_{\bar{Q}}(\boldsymbol{\alpha}')F^{(1)}$, then $L_{\boldsymbol{\alpha}}$ and $L_{\boldsymbol{\alpha}'}$ are mutually disjoint from (A.47).
(2) The if part: Suppose that (A.48) holds, that is, $\boldsymbol{\alpha}_1 = \boldsymbol{\alpha}_1'$. Let $l_{\boldsymbol{\alpha}+\boldsymbol{\alpha}'}$ denote the binary $(\rho_{h'} - \rho_{h,h'})$-tuple such that

$$p_R(l_{\boldsymbol{\alpha}+\boldsymbol{\alpha}'}) = p_{\bar{Q}}(\boldsymbol{\alpha} + \boldsymbol{\alpha}')F^{(2)}, \quad \text{(A.49)}$$

$$p_{\bar{R}}(l_{\boldsymbol{\alpha}+\boldsymbol{\alpha}'}) = \mathbf{0}. \quad \text{(A.50)}$$

For any given state pair $(\sigma_h, \sigma_{h'}) \in \Sigma_h^{\boldsymbol{\alpha}} \times \Sigma_{h'}^{\boldsymbol{\alpha}}$, define $(\sigma_h', \sigma_{h'}') \in \Sigma_h^{\boldsymbol{\alpha}'} \times \Sigma_{h'}^{\boldsymbol{\alpha}'}$ as

$$\begin{aligned} l^{(p)}(\sigma_h') &= l^{(p)}(\sigma_h) + p_{\bar{Q}}(\boldsymbol{\alpha} + \boldsymbol{\alpha}')F^{(3)}, \\ & \qquad \text{i.e.}, \sigma_h' = s_{h,\boldsymbol{\alpha}'}(l^{(p)}(\sigma_h) + p_{\bar{Q}}(\boldsymbol{\alpha} + \boldsymbol{\alpha}')F^{(3)}), \quad \text{(A.51)} \\ l^{(f)}(\sigma_{h'}') &= l^{(f)}(\sigma_{h'}) + l_{\boldsymbol{\alpha}+\boldsymbol{\alpha}'}, \\ & \qquad \text{i.e}, \sigma_{h'}' = s_{h',\boldsymbol{\alpha}'}(l^{(f)}(\sigma_{h'}) + l_{\boldsymbol{\alpha}+\boldsymbol{\alpha}'}). \quad \text{(A.52)} \end{aligned}$$

Then, $f^{(1)}(\boldsymbol{\alpha}) = f^{(1)}(\boldsymbol{\alpha}')$, $f^{(2)}(\boldsymbol{\alpha}, l^{(p)}(\sigma_h), l^{(f)}(\sigma_{h'})) = f^{(2)}(\boldsymbol{\alpha}', l^{(p)}(\sigma_h'), l^{(f)}(\sigma_{h'}'))$ and $f^{(3)}(\boldsymbol{\alpha}, l^{(f)}(\sigma_{h'})) = f^{(3)}(\boldsymbol{\alpha}', l^{(f)}(\sigma_{h'}'))$. Hence, it follows from (A.30), and (A.42) to (A.45) that

$$L(\sigma_h, \sigma_{h'}) = L(\sigma_h', \sigma_{h'}').$$

Note that when σ_h runs over $\Sigma_h^{\boldsymbol{\alpha}}$, $l^{(p)}(\sigma_h)$ runs over $L_h = \{0,1\}^{\rho_h - \rho_{h,h'}}$ and therefore, $s_{h,\boldsymbol{\alpha}'}(l^{(p)}(\sigma_h) + p_{\bar{Q}}(\boldsymbol{\alpha}+\boldsymbol{\alpha}')F^{(3)})$ defines a permutation of $\Sigma_h^{\boldsymbol{\alpha}'}$. Similarly, $s_{h',\boldsymbol{\alpha}'}(l^{(f)}(\sigma_{h'}) + l_{\boldsymbol{\alpha}+\boldsymbol{\alpha}'})$ defines a permutation of $\Sigma_{h'}^{\boldsymbol{\alpha}'}$.

△△

Corollary A.1 The block of the isomorphic parallel components containing $\Lambda_{\boldsymbol{\alpha}}$ is given by

$$\{\Lambda_{\boldsymbol{\alpha}+\boldsymbol{\alpha}'} : p_Q(\boldsymbol{\alpha}') = p_{\bar{Q}}(\boldsymbol{\alpha}')F^{(1)}, p_{\bar{Q}}(\boldsymbol{\alpha}') \in \{0,1\}^{\lambda_0}\}. \tag{A.53}$$

Each block of the partition consists of 2^{λ_0} identical parallel components, where λ_0 is the number of rows of $G_0^{s,s}$.

△△

It is shown in [44] that λ_0 is equal to $\lambda_{h,h'}(C)$ defined by (6.36).

Example A.1 Consider the $\mathrm{RM}_{3,6}$ code which is a $(64, 42)$ code. The second section of the 8-section minimal trellis diagram $T(\{0, 8, 16, \ldots, 64\})$ for this code consists of 16 parallel components, and they are partitioned into two blocks. Each block of the partition consists of 2^3 identical parallel components, Each parallel components has 8 states at time 8 and 64 states at time 16. Hence, there are $2^{13} = 16 \times 8 \times 64$ composite branches in this trellis section. However, there are only $2^{10} = 2 \times 8 \times 64$ different composite branch labels.

△△

A.4 AN EFFICIENT TRELLIS CONSTRUCTION PROCEDURE

In this section, an efficient procedure for constructing the trellis section from time-h to time-h' is presented. First, we present a subprocedure, denoted **GenerateCBL**($\boldsymbol{\alpha}$), that generates the set of composite branch labels for a representative parallel component $\Lambda_{\boldsymbol{\alpha}}$ in a block. From Corollary A.1, we can choose the parallel component $\Lambda_{\boldsymbol{\alpha}}$ with $p_{\bar{Q}}(\boldsymbol{\alpha}) = \mathbf{0}$ as the representative (in (A.53), for any $\boldsymbol{\alpha}$, $\Lambda_{\boldsymbol{\alpha}+\boldsymbol{\alpha}'}$ with $p_{\bar{Q}}(\boldsymbol{\alpha}') = p_{\bar{Q}}(\boldsymbol{\alpha})$ is such one). Let $\boldsymbol{\alpha}_1 \triangleq p_Q(\boldsymbol{\alpha})$. Then, the subprocedure, GenerateCBL($\boldsymbol{\alpha}_1$), generates the set of the composite branch labels of the parallel component $\Lambda_{\boldsymbol{\alpha}}$, denoted $L_{\boldsymbol{\alpha}_1}$:

$$\begin{aligned} L_{\boldsymbol{\alpha}_1} &= \{\boldsymbol{\alpha}_1 G_1^{s,s} + \boldsymbol{\gamma}_1 G_1^{f,s} + \boldsymbol{\beta} G^{s,p} + C_{h,h'}^{\mathrm{tr}} : \\ &\qquad \boldsymbol{\gamma}_1 \in \{0,1\}^{\nu_1}, \boldsymbol{\beta} \in \{0,1\}^{\rho_h - \rho_{h,h'}}\}, \end{aligned}$$

and stores each composite branch label $\alpha_1 G_1^{s,s} + \gamma_1 G_1^{f,s} + \beta G^{s,p} + C^{\text{tr}}_{h,h'}$, in the memory at the index $\alpha_1 \circ \gamma_1 \circ \beta$. This makes it possible to store the composite branch labels in the parallel component at consecutive memory locations.

It follows from (A.42) that for $\sigma_h \in \Sigma_h^{\alpha}$ and $\sigma_{h'} \in \Sigma_{h'}^{\alpha}$ with $l^{(s)}(\sigma_h) = \alpha$, $l^{(p)}(\sigma_h) = \beta$ and $l^{(f)}(\sigma_{h'}) = \gamma$, the composite branch label, $L(\sigma_h, \sigma_{h'}) = \alpha G^{s,s} + \gamma G^{f,s} + \beta G^{s,p} + C^{\text{tr}}_{h,h'}$ (or the maximum metric of $L(\sigma_h, \sigma_{h'})$) is stored in the memory with the following index:

$$\begin{aligned} \text{indx}(\alpha, \beta, \gamma) \;\triangleq\; & (p_Q(\alpha) + p_{\bar{Q}}(\alpha) F^{(1)}) \circ (p_{\bar{Q}}(\alpha) F^{(2)} + p_R(\gamma) + p_{\bar{R}}(\gamma) E^{(1)}) \\ & \circ (p_{\bar{Q}}(\alpha) F^{(3)} + p_{\bar{R}}(\gamma) E^{(2)} + \beta). \end{aligned} \tag{A.54}$$

[Trellis Construction Procedure Using Isomorphic Parallel Components]

For every $\alpha_1 \in \{0,1\}^{\lambda_1}$ {
    Construct L_{α_1}, by executing GenerateCBL(α_1).
    (* Construct isomorphic parallel components. *)
    For every $\alpha_0 \in \{0,1\}^{\lambda_0}$ {
        Let α be an element in $\{0,1\}^{\rho_{h,h'}}$ such that

$$p_Q(\alpha) = \alpha_1 + \alpha_0 F^{(1)}, \qquad \text{and} \qquad p_{\bar{Q}}(\alpha) = \alpha_0.$$

        Construct Λ_α by executing Construct$\Lambda(\alpha)$ subprocedure stated below.
    }
}

△△

The following subprocedure **Construct$\Lambda(\alpha)$** to construct $\Lambda(\alpha)$ is one to list

$$(l(\sigma_h), l(\sigma_{h'}), \text{indx}(l^{(s)}(\sigma_h), l^{(p)}(\sigma_h), l^{(f)}(\sigma_{h'})))$$

for every state pair $(\sigma_h, \sigma_{h'}) \in \Sigma_h^{\alpha} \times \Sigma_{h'}^{\alpha}$.

Subprocedure Construct$\Lambda(\alpha)$:

(* Construct a parallel component Λ_α. *)
For every $\gamma \in \{0,1\}^{\rho_{h'} - \rho_{h,h'}}$ {
    For every $\beta \in \{0,1\}^{\rho_h - \rho_{h,h'}}$ {

Output $(l(s_{h,\boldsymbol{\alpha}}(\boldsymbol{\beta})), l(s_{h',\boldsymbol{\alpha}}(\boldsymbol{\gamma})), \mathrm{indx}(\boldsymbol{\alpha}, \boldsymbol{\beta}, \boldsymbol{\gamma}))$.

    }

}

$\triangle\triangle$

Note that the labels, $l(s_{h,\boldsymbol{\alpha}}(\boldsymbol{\beta}))$, $l(s_{h',\boldsymbol{\alpha}}(\boldsymbol{\gamma}))$ and $\mathrm{indx}(\boldsymbol{\alpha}, \boldsymbol{\beta}, \boldsymbol{\gamma})$ are given by (A.29), (A.27) and (A.28), and (A.54), respectively.

REFERENCES

[1] L.R. Bahl, J. Cocke, F. Jelinek, and J. Raviv, "Optimal decoding of linear codes for minimizing symbol error rate," *IEEE Trans. Inform. Theory*, vol.IT-20, pp.284–287, 1974.

[2] A.H. Banihashemi and I.F. Blake, "Trellis complexity and minimal trellis diagrams of lattices," *IEEE Trans. Inform. Theory*, submitted for publication, 1996.

[3] G. Battail, "Pondération des symboles décodés par l'algorithm de Viterbi," *Ann. Télécommun.*, vol. 42, pp.31–38, 1987.

[4] E. Berlekamp, *Algebraic Coding Theory*, Agean Park Press, Laguna Hills, Revised edition, 1984.

[5] Y. Berger and Y. Be'ery, "Bounds on the trellis size of linear block codes," *IEEE Trans. Inform. Theory*, vol.39, pp.203–209, 1993.

[6] Y. Berger and Y. Be'ery, "Soft trellis-based decoder for linear block codes," *IEEE Trans. Inform. Theory*, vol.40, pp.764–773, 1994.

[7] Y. Berger and Y. Be'ery, "Trellis-oriented decomposition and trellis-complexity of composite length cyclic codes," *IEEE Trans. Inform. Theory*, vol.41, pp.1185–1191, 1995.

[8] Y. Berger and Y. Be'ery, "The twisted squaring construction, trellis complexity and generalized weights of BCH and QR codes," *IEEE Trans. Inform. Theory*, vol.42, pp.1817–1827, 1996.

[9] C.Berrou, P.Adde, E.Angui and S.Faudeil, "A low complexity soft-output Viterbi decoder architecture," *in Proc. of ICC*, pp.737–740, 1993.

[10] P.J. Black and T.H. Meng, "A 140-Mb/s, 32-state, radix-4, Viterbi decoder," *IEEE J. Solid-State Circuits*, vol.27, December 1992.

[11] R.E. Blahut, *Theory and Practice of Error Correcting Codes*, Addition-Wesley, Reading MA, 1984.

[12] I.F. Blake and V. Tarokh, "On the trellis complexity of the densest lattice packings in $\boldsymbol{R}^n$," SIAM J. Discrete Math., vol.9, pp.597–601, 1996.

[13] A.R. Calderbank, "Multilevel codes and multi-stage decoding," *IEEE Trans. Commun.*, vol.37, no.3, pp.222–229, March 1989.

[14] D. Chase, "A class of algorithms for decoding block codes with channel measurement information," *IEEE Trans. Inform. Theory*, vol.18, no.1, pp.170–181, January 1972.

[15] J.F. Cheng and R.J. McEliece, "Near capacity codes for the Gaussian channel based on low-density generator matrices," in *Proc. 34-th Allerton Conf. on Comm., Control, and Computing*, Monticello, IL., pp.494–503, October 1996.

[16] G.C. Clark and J.B. Cain, *Error Correcting Codes for Digital Communications*, Plenum Press, New York, 1981.

[17] Y. Desaki, T. Fujiwara and T. Kasami, "A Method for Computing the Weight Distribution of a Block Code by Using Its Trellis Diagram," *IEICE Trans. Fundamentals*, vol.E77-A, no.8, pp.1230–1237, August 1994.

[18] S. Dolinar, L. Ekroot, A.B. Kiely, R.J. McEliece, and W. Lin, "The permutation trellis complexity of linear block codes," in *Proc. 32-nd Allerton Conf. on Comm., Control, and Computing*, Monticello, IL., pp.60–74, September 1994.

[19] A. Engelhart, M. Bossert, and J. Maucher, "Heuristic algorithms for ordering a linear block code to reduce the number of nodes of the minimal trellis," *IEEE Trans. Inform. Theory*, submitted for publication, 1996.

[20] J. Feigenbaum, G.D. Forney Jr., B.H. Marcus, R.J. McEliece, and A. Vardy, Special issue on "Codes and Complexity," *IEEE Trans. Inform. Theory*, vol.42, November 1996.

[21] G. Fettweis and H. Meyr, "Parallel Viterbi algorithm implementation: breaking the ACS-bottleneck," *IEEE Trans. Commun.*, vol.37, no.8, pp.785–789, August 1989.

[22] G.D. Forney Jr., *Concatenated Codes*, MIT Press, Cambridge, MA, 1966.

[23] G.D. Forney Jr., "The Viterbi algorithm," *Proc. IEEE*, vol.61, pp.268–278, 1973.

[24] G.D. Forney Jr., "Coset codes II: Binary lattices and related codes," *IEEE Trans. Inform. Theory*, vol.34, pp.1152–1187, 1988.

[25] G.D. Forney Jr., "Dimension/length profiles and trellis complexity of linear block codes," *IEEE Trans. Inform. Theory*, vol.40, pp.1741–1752, 1994.

[26] G.D. Forney Jr., "Dimension/length profiles and trellis complexity of lattices," *IEEE Trans. Inform. Theory*, vol.40, pp.1753–1772, 1994.

[27] G.D. Forney Jr., "The forward-backward algorithm,"in *Proc. 34-th Allerton Conf. on Comm., Control, and Computing, Control, and Computing*, Monticello, IL., pp.432–446, October 1996.

[28] G.D. Forney, Jr. and M.D. Trott, "The dynamics of group codes: state spaces, trellis diagrams and canonical encoders," *IEEE Trans. Inform. Theory*, vol.39, pp.1491–1513, 1993.

[29] G.D. Forney, Jr. and M.D. Trott, "The dynamics of group codes: dual group codes and systems," preprint.

[30] M.P.C. Fossorier and S. Lin, "Coset codes viewed as terminated convolutional codes," *IEEE Trans. Comm.*, vol.44, pp.1096–1106, September 1996.

[31] M.P.C. Fossorier and S. Lin, "Some decomposable codes: the $|a+x|b+x|a+b+x|$ construction," *IEEE Trans. Inform. Theory*, vol.IT-43, pp.1663–1667, September 1997.

[32] M.P.C. Fossorier and S. Lin, "Differential trellis decoding of conventional codes," *IEEE Trans. Inform. Theory*, submitted for publication.

[33] M.P.C. Fossorier, S. Lin, and D. Rhee, "Bit error probability for maximum likelihood decoding of linear block codes," in *Proc. International Symposium on Inform. Theory and its Applications*, Victoria, Canada, pp.606–609, September 1996, and submitted to *IEEE Trans. Inform. Theory*, June 1996.

[34] M.P.C. Fossorier, F. Burkert, S. Lin, and J. Hagenauer, "On the equivalence between SOVA and Max-Log-MAP decoding," *IEEE Comm. Letters*, submitted for publication, August 1997.

[35] B.J. Frey and F.R. Kschischang, "Probability propagation and iterative decoding," in *Proc. 34-th Allerton Conf. on Comm., Control, and Computing*, Monticello, IL., pp.482–493, October 1996.

[36] T. Fujiwara, T. Kasami, R.H. Morelos-Zaragoza, and S. Lin, "The state complexity of trellis diagrams for a class of generalized concatenated codes," in *Proc. 1994 IEEE International Symposium on Inform. Theory*, Trondheim, Norway, p.471, June/July, 1994.

[37] T. Fujiwara, H. Yamamoto, T. Kasami, and S. Lin, "A trellis-based recursive maximum likelihood decoding algorithm for linear codes," *IEEE Trans. Inform. Theory*, vol.44, to appear, 1998.

[38] P.G. Gulak and T. Kailath, "Locally connected VLSI architectures for the Viterbi algorithms," *IEEE J. Select Areas Commun.*, vol.6, pp.526–637, April 1988.

[39] J. Hagenauer and P. Hoeher, "A Viterbi algorithm with soft-decision outputs and its applications," in *Proc. IEEE Globecom Conference*, Dallas, TX, pp.1680–1686, November 1989.

[40] J. Hagenauer, E. Offer, and L. Papke, "Iterative decoding of binary block and convolutional codes," *IEEE Trans. Inform. Theory*, vol.42, pp.429–445, March 1996.

[41] B. Honary and G. Markarian, *Trellis decoding of block codes: a practical approach*, Kluwer Academic Publishers, Boston, MA, 1997.

[42] G.B. Horn and F.R. Kschischang, "On the intractability of permuting a block code to minimize trellis complexity," *IEEE Trans. Inform. Theory*, vol.42, pp.2042–2048, 1996.

[43] H. Imai and S. Hirakawa, "A new multilevel coding method using error correcting codes," *IEEE Trans. Inform. Theory*, vol.23, no.3, May 1977.

[44] T. Kasami, T. Fujiwara, Y. Desaki, and S. Lin, "On branch labels of parallel components of the L-section minimal trellis diagrams for binary linear block codes," *IEICE Trans. Fundamentals*, vol.E77-A, no.6, pp.1058–1068, June 1994.

[45] T. Kasami, T. Takata, T. Fujiwara, and S. Lin, "On the optimum bit orders with respect to the state complexity of trellis diagrams for binary linear codes," *IEEE Trans. Inform Theory*, vol.39, no.1, pp.242–243, January 1993.

[46] T. Kasami, T. Takata, T. Fujiwara, and S. Lin, "On complexity of trellis structure of linear block codes," *IEEE Trans. Inform. Theory*, vol.39, no.3, pp.1057–1064, May 1993.

[47] T. Kasami, T. Takata, T. Fujiwara, and S. Lin, "On structural complexity of the L-section minimal trellis diagrams for binary linear block codes," *IEICE Trans. Fundamentals*, vol.E76-A, no.9, pp.1411–1421, September 1993.

[48] T. Kasami, T. Koumoto, T. Fujiwara, H. Yamamoto, Y. Desaki and S. Lin. "Low weight subtrellises for binary linear block codes and their

applications" *IEICE Trans. Fundamentals*, vol.E80-A, no.11, pp.2095–2103, November 1997.

[49] T. Kasami, T. Sugita and T. Fujiwara, "The split weight (w_L, w_R) enumeration of Reed-Muller codes for $w_L + w_R < 2d_{\min}$," in *Lecture Notes in Comput. Sci. (Proc. 12th International Symposium, AAECC-12)*, vol.1255, pp.197–211, Springer-Verlag, June 1997.

[50] T. Kasami, H. Tokushige, T. Fujiwara, H. Yamamoto and S. Lin, "A recursive maximum likelihood decoding algorithm for Reed-Muller codes and related codes," *Technical Report of NAIST*, no.NAIST-IS-97003, January 1997.

[51] A.B. Kiely, S. Dolinar, R.J. McEliece, L. Ekroot, and W. Lin, "Trellis decoding complexity of linear block codes," *IEEE Trans. Inform. Theory*, vol.42, pp.1687–1697, 1996.

[52] T. Komura, M. Oka, T. Fujiwara, T. Onoye, T. Kasami, and S. Lin, "VLSI architecture of a recursive maximum likelihood decoding algorithm for a $(64, 35)$ subcode of the $(64, 42)$ Reed-Muller code," in *Proc. International Symposium on Inform. Theory and Its applications*, Victoria, Canada, pp.709–712, September 1996.

[53] A.D. Kot and C. Leung, "On the construction and dimensionality of linear block code trellises," in *Proc. IEEE Int. Symp. Inform. Theory*, San Antonio, TX., p.291, 1993.

[54] F.R. Kschischang, "The combinatorics of block code trellises," in *Proc. Biennial Symp. on Commun.*, Queen's University, Kingston, ON., Canada, May 1994.

[55] F.R. Kschischang, "The trellis structure of maximal fixed-cost codes," *IEEE Trans. Inform. Theory*, vol.42, pp.1828–1838, 1996.

[56] F.R. Kschischang and G.B. Horn, "A heuristic for ordering a linear block code to minimize trellis state complexity," in *Proc. 32-nd Allerton Conf. on Comm., Control, and Computing*, pp.75–84, Monticello, IL., September 1994.

[57] F.R. Kschischang and V. Sorokine, "On the trellis structure of block codes," *IEEE Trans. Inform. Theory*, vol.41, pp.1924-1937, 1995.

[58] A. Lafourcade and A. Vardy, "Asymptotically good codes have infinite trellis complexity," *IEEE Trans. Inform. Theory*, vol.41, pp.555-559, 1995.

[59] A. Lafourcade and A. Vardy, "Lower bounds on trellis complexity of block codes," *IEEE Trans. Inform. Theory*, vol.41, pp.1938–1954, 1995.

[60] A. Lafourcade and A. Vardy, "Optimal sectionalization of a trellis," *IEEE Trans. Inform. Theory*, vol.42, pp.689–703, 1996.

[61] L. Lin and R.S. Cheng, "Improvements in SOVA-based decoding for turbo codes," *Proc. of ICC*, pp.1473–1478, 1997.

[62] S. Lin and D.J. Costello, Jr., *Error Control Coding: Fundamentals and Applications*, Prentice Hall, Englewood Cliffs, NY, 1983.

[63] S. Lin, G.T. Uehara, E. Nakamura, and W.P. Chu, "Circuit design approaches for implementation of a subtrellis IC for a Reed-Muller subcode," *NASA Technical Report* 96–001, February 1996.

[64] C.C. Lu and S.H. Huang, "On bit-level trellis complexity of Reed-Muller codes," *IEEE Trans. Inform. Theory*, vol.41, pp.2061–2064, 1995.

[65] H.H. Ma and J.K. Wolf, "On tail biting convolutional codes," *IEEE Trans. Commun.*, vol.34, pp.104–111, 1986.

[66] F.J. MacWilliams and N.J.A. Sloane, *The Theory of Error Correcting Codes*, North-Holland, Amsterdam, 1977.

[67] J.L. Massey, "Foundation and methods of channel encoding," in *Proc. Int. Conf. Inform. Theory and Systems*, NTG-Fachberichte, Berlin, 1978.

[68] J.L. Massey and M.K. Sain, "Inverses of linear sequential circuits," *IEEE Trans. Computers*, vol.COMP-17, pp.330–337, April 1968.

[69] R.J. McEliece, "On the BCJR trellis for linear block codes," *IEEE Trans. Inform. Theory*, vol.42, pp.1072–1092, 1996.

[70] R.J. McEliece and W. Lin, "The trellis complexity of convolutional codes," *IEEE Trans. Inform. Theory*, vol.42, pp.1855–1864, 1996.

[71] A.M. Michelson and D.F. Freeman, "Viterbi decoding of the (63, 57) Hamming codes — Implementation and performance results," *IEEE Trans. Comm.*, vol.43, pp.2653–2656, 1995.

[72] H.T. Moorthy and S. Lin, "On the labeling of minimal trellises for linear block codes," in *Proc. International Symposium on Inform. Theory and its Applications*, Sydney, Australia, vol.1, pp.33–38, November 20–24, 1994.

[73] H.T. Moorthy, S. Lin, and G.T. Uehara, "Good trellises for IC implementation of Viterbi decoders for linear block codes," *IEEE Trans. Comm.*, vol.45, pp.52–63, 1997.

[74] H.T. Moorthy and S. Lin, "On the trellis structure of (64, 40, 8) subcode of the (64, 42, 8) third-order Reed-Muller code," *NASA Technical Report*, 1996.

[75] H.T. Moorthy, S. Lin, and T. Kasami, "Soft-decision decoding of binary linear block codes based on an iterative search algorithm," *IEEE Trans. Inform. Theory*, vol.43, no.3, pp.1030–1040, May 1997.

[76] R.H. Morelos-Zaragoza, T. Fujiwara, T. Kasami, and S. Lin, "Constructions of generalized concatenated codes and their trellis-based decoding complexity," *IEEE Trans. Inform. Theory*, submitted for publication, 1996.

[77] D.J. Muder, "Minimal trellises for block codes," *IEEE Trans. Inform. Theory*, vol.34, pp.1049–1053, 1988.

[78] D.E. Muller, "Application of Boolean algebra to switching circuit design and to error detection," *IEEE Trans. Computers*, vol.3, pp.6–12, 1954.

[79] J.K. Omura, "On the Viterbi decoding algorithm," *IEEE Trans. Inform. Theory*, vol.IT-15, pp.177–179, 1969.

[80] L.C. Perez, "Coding and modulation for space and satellite communications," Ph.D Thesis, University of Notre Dame, December 1994.

[81] W.W. Peterson and E.J. Weldon, Jr., *Error Correcting Codes*, 2nd edition, MIT Press, Cambridge, MA, 1972.

[82] J.G. Proakis, *Digital Communications*, 3rd edition, McGraw-Hill, New York, 1995.

[83] I.S. Reed, "A class of multiple-error-correcting codes and the decoding scheme," *IEEE Trans. Inform. Theory*, vol.IT-4, pp.38–49, 1954.

[84] I. Reuven and Y. Be'ery, "Entropy/length profiles and bounds on trellis complexity of nonlinear codes," *IEEE Trans. Inform. Theory*, submitted for publication, 1996.

[85] D. Rhee and S. Lin, "A low complexity and high performance concatenated scheme for high speed satellite communications," *NASA Technical Report*, October 1993.

[86] R.Y. Rubinstein, *Simulated and the Monte Carlo Method, New York, Wiley*, 1981.

[87] G. Schnabl and M. Bossert, "Soft decision decoding of Reed-Muller codes and generalized multiple concatenated codes," *IEEE Trans. Inform. Theory*, vol.41, no.1, pp.304–308, January 1995.

[88] P. Schuurman, "A table of state complexity bounds for binary linear codes," *IEEE Trans. Inform. Theory*, vol.42, pp.2034–2042, 1996.

[89] V.R. Sidorenko, "The minimal trellis has maximum Euler characteristic $|V| - |E|$," *Problemy Peredachi Informatsii*, to appear.

[90] V.R. Sidorenko, G. Markarian, and B. Honary, "Minimal trellis design for linear block codes based on the Shannon product," *IEEE Trans. Inform. Theory*, vol.42, pp.2048–2053, 1996.

[91] V.R. Sidorenko, I. Martin, and B. Honary, "On separability of nonlinear block codes," *IEEE Trans. Inform. Theory*, submitted for publication, 1996.

[92] G. Solomon and H.C.A. van Tilborg, "A connection between block and convolutional codes," *SIAM J. Appl. Math.*, vol.37, pp.358–369, 1979.

[93] V. Sorokine, F.R. Kschischang, and V. Durand, "Trellis-based decoding of binary linear block codes," in *Lecture Notes in Comput. Sci. (Proc. 3rd Canadian Workshop on Inform. Theory)*, vol.793, pp.270–286, Springer-Verlag, 1994.

[94] D.J. Taipale and M.B. Pursley, "An improvement to generalized minimum distance decoding," *IEEE Trans. Inform. Theory*, vol.37, no.1, pp.167–172, 1991.

[95] T. Takata, S. Ujita, T. Kasami, and S. Lin, "Multistage decoding of multilevel block M-PSK modulation codes and it performance analysis," *IEEE Trans. Inform. Theory*, vol.39, no.4, pp.1024–1218, 1993.

[96] T. Takata, Y. Yamashita, T. Fujiwara, T. Kasami, and S. Lin, "Suboptimum decoding of decomposable block codes," *IEEE Trans. Inform. Theory*, vol.40, no.5, pp.1392–1405, 1994.

[97] V. Tarokh and I.F. Blake, "Trellis complexity versus the coding gain of lattices I," *IEEE Trans. Inform. Theory*, vol.42, pp.1796–1807, 1996.

[98] V. Tarokh and I.F. Blake, "Trellis complexity versus the coding gain of lattices II," *IEEE Trans. Inform. Theory*, vol.42, pp.1808–1816, 1996.

[99] V. Tarokh and A. Vardy, "Upper bounds on trellis complexity of lattices," *IEEE Trans. Inform. Theory*, vol.43, pp.1294–1300, 1997.

[100] H. Thapar and J. Cioffi, "A block processing method for designing high-speed Viterbi detectors," in *Proc. ICC*, vol.2, pp.1096–1100, June 1989.

[101] H. Thirumoorthy, "Efficient near-optimum decoding algorithms and trellis structure for linear block codes," Ph.D dissertation, Dept. of Elec. Engrg., University of Hawaii at Manoa, November 1996.

[102] A. Vardy and Y. Be'ery, "Maximum-likelihood soft decision decoding of BCH codes," *IEEE Trans. Inform. Theory*, vol.40, pp.546–554, 1994.

[103] A. Vardy and F.R. Kschischang, "Proof of a conjecture of McEliece regarding the expansion index of the minimal trellis," *IEEE Trans. Inform. Theory*, vol.42, pp.2027–2033, 1996.

[104] V.V. Vazirani, H. Saran, and B. Sunder Rajan, "An efficient algorithm for constructing minimal trellises for codes over finite abelian groups," *IEEE Trans. Inform. Theory*, vol.42, pp.1839–1854, 1996.

[105] A.J. Viterbi, "Error bounds for convolutional codes and asymptotically optimum decoding algorithm," *IEEE Trans. Inform. Theory*, vol.IT-13, pp.260–269, 1967.

[106] Y.-Y. Wang and C.-C. Lu, "The trellis complexity of equivalent binary (17, 9) quadratic residue code is five," in *Proc. IEEE Int. Symp. Inform. Theory*, San Antonio, TX., p.200, 1993.

[107] Y.-Y. Wang and C.-C. Lu, "Theory and algorithm for optimal equivalent codes with absolute trellis size," preprint.

[108] X.A. Wang and S.B. Wicker, "The design and implementation of trellis decoding of some BCH codes," *IEEE Trans. Commun.*, submitted for publication, 1997.

[109] J.K. Wolf, "Efficient maximum-likelihood decoding of linear block codes using a trellis," *IEEE Trans. Inform. Theory*, vol.24, pp.76–80, 1978.

[110] J. Wu, S. Lin, T. Kasami, T. Fujiwara, and T. Takata, "An upper bound on effective error coefficient of two-stage decoding and good two level decompositions of some Reed-Muller codes," *IEEE Trans. Comm.*, vol.42, pp.813–818, 1994.

[111] H. Yamamoto, "Evaluating the block error probability of trellis codes and designing a recursive maximum likelihood decoding algorithm of linear block codes using their trellis structure," Ph.D dissertation, Osaka University, March 1996.

[112] H. Yamamoto, H. Nagano, T. Fujiwara, T. Kasami, and S. Lin, "Recursive MLD algorithm using the detail trellis structure for linear block codes and its average complexity analysis," in *Proc. 1996 International Symposium on Inform. Theory and Its Applications*, Victoria, Canada, pp.704–708, September 1996.

[113] Ø. Ytrehus, "On the trellis complexity of certain binary linear block codes," *IEEE Trans. Inform. Theory*, vol.40, pp.559–560, 1995.

[114] V.A. Zinoviev, "Generalized cascade codes," *Problemy Peredachi Informatsii*, vol.12.1, pp.2–9, 1976.

[115] V.V. Zyablov and V.R. Sidorenko, "Bounds on complexity of trellis decoding of linear block codes," *Problemy Peredachi Informatsii*, vol.29, pp.3–9, 1993 (in Russian).

INDEX